TEUBNER-TEXTE zur Mathematik   Band 133

H.-J. Schmeisser / H. Triebel (Hrsg.)

Function Spaces, Differential Operators and Nonlinear Analysis

TEUBNER-TEXTE zur Mathematik

Die Reihe soll ein Forum für Beiträge zu aktuellen Problemstellungen der Mathematik sein. Besonderes Anliegen ist die Veröffentlichung von Darstellungen unterschiedlicher methodischer Ansätze, die das Wechselspiel zwischen Theorie und Anwendungen sowie zwischen Lehre und Forschung reflektieren. Thematische Schwerpunkte sind Analysis, Geometrie und Algebra.

In den Texten sollen sich sowohl Lebendigkeit und Originalität von Spezialvorlesungen und Seminaren als auch Diskussionsergebnisse aus Arbeitsgruppen widerspiegeln.

TEUBNER-TEXTE erscheinen in deutscher oder englischer Sprache.

Function Spaces, Differential Operators and Nonlinear Analysis

Edited by
Prof. Dr. Hans-Jürgen Schmeisser
Prof. Dr. Hans Triebel
Friedrich-Schiller-University, Jena

Springer Fachmedien Wiesbaden
GmbH 1993

Prof. Dr. Hans-Jürgen Schmeisser

Born in 1950 in Trusetal (Thuringia). Studied mathematics in Jena from 1968 to 1973 and in Moscow from 1979 to 1980. Received Dr. rer. nat. in 1978 and Dr. sc. nat. in 1982 from the Friedrich-Schiller-University of Jena.
Since 1992 Professor of Analysis at the University of Jena.
Fields of interest: Fourier analysis and function spaces.

Prof. Dr. Hans Triebel

Born in 1936 in Dessau. Studied mathematics in Jena from 1954 to 1959 and in Leningrad (St. Petersburg) from 1963 to 1964. Received Dr. rer. nat. in 1962 and Dr. habil in 1966 from the Friedrich-Schiller-University of Jena.
Since 1970 Professor of Analysis at the University of Jena.
Fields of interest: Function spaces of partial differential equations.

Die Deutsche Bibliothek – CIP-Einheitsaufnahme

Function spaces, differential operators and nonlinear analysis /
ed. by Hans-Jürgen Schmeisser ; Hans Triebel. –
(Teubner-Texte zur Mathematik ; 133)
ISBN 978-3-8154-2045-4 ISBN 978-3-663-11336-2 (eBook)
DOI 10.1007978-3-663-11336-2
NE: Schmeisser, Hans-Jürgen [Hrsg.]; GT

Ursprünglich erschienen bei B. G. Teubner Verlagsgesellschaft Leipzig 1993
Umschlaggestaltung: E. Kretschmer, Leipzig

Preface

The present Teubner-Text contains invited surveys and shorter communications connected with the International Conference "Function Spaces, Differential Operators and Nonlinear Analysis", which took place in Friedrichroda (Thuringia, Germany) from September 20–26, 1992. The main subjects are well reflected by the table of contents. 55 mathematicians attended the conference, many of them from eastern countries. We take the opportunity to thank DFG for financial support, which enabled us to invite mathematicians from the former socialist countries, and especially from the former Soviet Union, and which gave us the possibility to maintain and to strengthen our contacts to these centers of the theory of function spaces and its application to PDE's, ΨDE's and approximation theory.
The organization of the conference as well as the final preparation of this text was mostly done by our co-workers in Jena. We wish to thank all of them for the generous support they gave us far beyond their duties (whatever this means in connection with the organization of a conference). The final preparation of this text was mainly done by Dr. M. Malarski. Furthermore Dr. M. Geisler, Ms. D. Haroske and Dr. W. Sickel converted some manuscript in readable papers on TEX-standard level. We wish to thank them for doing this time-consuming work.

Jena, May 13, 1993

H.-J. Schmeisser
H. Triebel

Contents

Nonhomogeneous Linear and Quasilinear Elliptic and Parabolic Boundary Value Problems

Herbert Amann

It is the purpose of this paper to describe some of the recent developments in the mathematical theory of linear and quasilinear elliptic and parabolic systems with nonhomogeneous boundary conditions. For illustration we use the relatively simple set-up of reaction-diffusion systems which are — on the one hand — typical for the whole class of systems to which the general theory applies and — on the other hand — still simple enough to be easily described without too many technicalities. In addition, quasilinear reaction-diffusion equations are of great importance in applications and of actual mathematical and physical interest, as is witnessed by the examples we include.

In particular, we try to elucidate the rôles which are played in the theory of quasilinear parabolic systems by the modern theory of function spaces, interpolation and extrapolation techniques, and semigroup theory. Many of the proofs will be sketched only and we will be rather brief at times. However, we try to explain the basic underlying ideas and give references to the research literature where proofs can be found. A complete, detailed, and coherent presentation will be given in the forthcoming monograph [Ama94] which will also contain many additional results and extensions of the theory described in this paper.

1 Model Problems

Throughout this text Ω denotes a bounded domain in $\mathbb{R}^n$ with a smooth boundary so that $\overline{\Omega}$ is a compact n-dimensional C^∞-submanifold of $\mathbb{R}^n$ with boundary $\partial\Omega$. We denote by $\vec{\nu} = (\nu^1, \ldots, \nu^n)$ the outer unit normal vector field to $\partial\Omega$.

A great many physical (physico-chemical, biological, etc.) theories can be described mathematically by conservation laws of the form

$$\partial_t u_r + \operatorname{div} \vec{\jmath}_r = f_r \qquad \text{in } \Omega , \tag{1.1$_r$}$$

where $r = 1, \ldots, N$ and $u := (u_1, \ldots, u_N)$ is a vector of densities (concentrations, population densities, etc.), $\vec{\jmath}_r$ is the flux vector, and f_r is the production rate for the rth species. With this interpretation $(1.1)_r$ is simply a mass balance law. Of course, other interpretations of $(1.1)_r$ are possible too. For example, u_r may describe (for some fixed r) the temperature of a system, $\vec{\jmath}_r$ the heat flux, and f_r the heat source density, etc.

Supported in part by Schweizerischer Nationalfonds (20-25263.88) and European Community (Science Plan, Project "Evolutionary Systems")

Generally, the production rate f_r will be a known function of the position $x \in \Omega$ and the vector u, that is,

$$f_r = f_r(x, u) \ . \qquad (1.2)_r$$

(In general, f_r will depend on the time variable t too. However, for simplicity we restrict ourselves here to the autonomous case or simply suppress the variable t.) Note that $(1.2)_r$ represents the 'reaction rate' of the species r in the physico-chemical context.

The flux vector $\vec{\jmath}_r$ of the species r is specified by phenomenological laws. In many applications it is of the form

$$\vec{\jmath}_r = -\sum_s \left(a^{rs}(\cdot, u)\nabla u_s + \vec{\alpha}^{rs}(\cdot, u)u_s \right) \ , \qquad (1.3)_r$$

where a^{rs} and $\vec{\alpha}^{rs}$are known functions of x and u (and t, in general). The first sum in $(1.3)_r$ is the so-called diffusion term, where a^{rs} for $s \neq r$ describe cross-diffusion effects occurring in multicomponent diffusion problems (e.g., [deGM84]). In the simplest case, where there are no cross-diffusion terms and $\vec{\alpha}^{rs} = 0$ for all s, the flux vector reduces to

$$\vec{\jmath}_r = -a^{rr}(\cdot, u)\nabla u \ ,$$

which is nothing else than Fick's law (Fourier's law, Darcy's law, etc.) provided

$$a^{rr} > 0 \ . \qquad (1.4)_r$$

In the general situation we shall impose a condition upon the matrix $[a^{rs}]_{1 \le r,s \le N}$ generalizing $(1.4)_r$ suitably and guaranteeing the parabolicity of the resulting system.

The terms $\vec{\alpha}^{rs}u_s$ in $(1.3)_r$ describe drifts in the direction of the vector fields $\vec{\alpha}^{rs}$ proportional to the concentration of the species s. These drift vectors, which are often gradients of 'outer potentials', are sometimes specified by additional 'stationary equations'. In those cases the vector fields $\vec{\alpha}_{rs}$ are nonlocal functions (i.e., nonlinear operators) of u.

By inserting $(1.2)_r$ and $(1.3)_r$ in $(1.1)_r$ we obtain a quasilinear system of partial differential equations for u in Ω, a quasilinear reaction-diffusion system. In order to describe the time evolution of the vector u completely we have to specify the behavior of u on the boundary, and initial conditions. In practice there occur two types of boundary conditions. Most prominently, we find 'prescribed (in-) flux conditions' for the species r of the form

$$-(\vec{\jmath}_r | \vec{\nu}) = g_r^1(\cdot, u) \qquad \text{on } \partial_{1,r}\Omega \ , \qquad (1.5)_r$$

where g_r^1 is a known function of x and u (and t, in general), $(\cdot|\cdot)$ denotes the Euclidean inner product, and $\partial_{1,r}\Omega$ is a union of components of connectedness of $\partial\Omega$. Note that $(1.5)_r$ reduces to a 'no flux' boundary condition on $\partial_{1,r}\Omega$ if $g_r^1 = 0$. The second class of boundary conditions which are often considered, comprises Dirichlet boundary conditions where the value of u_r is prescribed on $\partial_{0,r}\Omega := \partial\Omega \backslash \partial_{1,r}\Omega$, that is,

$$u_r = g_r^0 \qquad \text{on } \partial_{0,r}\Omega \ , \qquad (1.6)_r$$

where g_r^0 is a given function of x (and t, in general).

Note that the species r satisfies a prescribed flux boundary condition $(1.5)_r$ on one part of the boundary, namely on $\partial_{1,r}\Omega$, and a Dirichlet condition $(1.6)_r$ on the remaining part $\partial_{0,r}\Omega$.

In order to have a unified simple representation we introduce functions δ_r assuming the two values 0 and 1 only, such that $\delta_r|\partial_{1,r}\Omega = 1$ and $\delta_r|\partial_{0,r}\Omega = 0$. We also put

$$\mathcal{B}_r(u)u := -\delta_r(\vec{\jmath}_r|\vec{\nu}) + (1-\delta_r)u_r$$

and

$$g_r(\cdot,u) := \delta_r g_r^1(\cdot,u) + (1-\delta_r)g_r^0 \ .$$

Then the boundary conditions $(1.5)_r$ and $(1.6)_r$ can be rewritten in the concise form

$$\mathcal{B}_r(u)u = g_r(\cdot,u) \qquad \text{on } \partial\Omega \ . \tag{1.7$_r$}$$

By introducing Euclidean coordinates we rewrite now the complete quasilinear reaction-diffusion system $(1.1)_r$–$(1.6)_r$, $r = 1,\ldots,N$, in the form

$$\begin{aligned} \partial_t u + \mathcal{A}(u)u &= f(u) \qquad \text{in } \Omega \ , \\ \mathcal{B}(u)u &= g(u) \qquad \text{on } \partial\Omega \ , \end{aligned} \tag{1.8}$$

where

$$\mathcal{A}(u)v := -\partial_j\big(a(\cdot,u)\partial_j v + a_j(\cdot,u)v\big) \tag{1.9}$$

and

$$\mathcal{B}(u)v := \delta\big[a(\cdot,u)\partial_\nu v + \nu^j a_j(\cdot,u)v\big] + (1-\delta)v \ . \tag{1.10}$$

Here and in the following we use the summation convention with j (and k) running from 1 to n. Moreover,

$$a(\cdot,u) := \big[a^{rs}(\cdot,u)\big]_{1\le r,s\le N}$$

is the 'diffusion matrix',

$$a_j(\cdot,u) := \big[\alpha_j^{rs}(\cdot,u)\big]_{1\le r,s\le N} \ , \qquad j = 1,\ldots,n \ ,$$

where $\vec{\alpha}^{rs} = (\alpha_1^{rs},\ldots,\alpha_n^{rs})$, and $\partial_\nu u := \nu^j\partial_j u$ is the normal derivative, and $\delta := \operatorname{diag}[\delta_r]_{1\le r\le N}$ is the diagonal matrix with diagonal elements δ_r. Of course, f and g have the obvious meanings.

We include now a number of examples describing standard and not so standard reaction-diffusion systems, which are used to illustrate the scope and the generality of our approach. We will come back to these examples in Section 17 below and show how the general theory developed in this paper applies in these concrete instances.

1.1 Population Dynamics In the theoretical study of spacial segregation of interacting species Shigesada, Kawasaki, and Teramoto [ShKT79] introduced a two population model of the form

$$\begin{aligned} \partial_t u_r + \operatorname{div}\vec{\jmath}_r(u) &= u_r g_r(u) \qquad \text{in } \Omega \ , \\ (\vec{\jmath}_r(u)|\vec{\nu}) &= 0 \qquad \text{on } \partial\Omega \ , \end{aligned} \tag{1.11}$$

$r = 1,2$, where $u = (u_1,u_2)$ is the vector of population densities (i.e., $N = 2$) and Ω is two-dimensional (i.e., $n = 2$). Moreover, the 'flux vectors' $\vec{\jmath}_r(u)$ are of the form

$$\vec{\jmath}_r(u) := -\nabla\big[(\alpha_r + \beta_{r1}u_1 + \beta_{r2}u_2)u_r\big] - \gamma_r u_r \nabla\varphi \ ,$$

where $\alpha_r > 0$, $\beta_{rs} \geq 0$, $\gamma_r \geq 0$ are constants and φ is a known outer potential. Thus the terms $(\beta_{r1}u_1 + \beta_{r2}u_2)u_r$ model 'social friction and competition'. It is easy to rewrite $\vec{\jmath}_r$ in the form $(1.3)_r$. Then, letting

$$a(\cdot, u) := \begin{bmatrix} \alpha_1 + 2\beta_{11}u_1 + \beta_{12}u_2 & \beta_{12}u_1 \\ \beta_{21}u_2 & \alpha_2 + \beta_{21}u_1 + 2\beta_{22}u_2 \end{bmatrix} \tag{1.12}$$

and

$$a_j(\cdot, u) := \begin{bmatrix} \gamma_1 \partial_j \varphi & 0 \\ 0 & \gamma_2 \partial_j \varphi \end{bmatrix} , \tag{1.13}$$

it follows that (1.11) is of the form (1.8)–(1.10) with $\delta = 1$. Since the u_r represent densities, nonnegative solutions are of interest only. □

1.2 Phase Transition Models Penrose and Fife [PeF90] propose thermodynamical consistent models of phase-field type for the study of phase transitions. They involve an 'order parameter' φ and the absolute temperature T, which are supposed to satisfy the differential equations

$$\begin{aligned} \partial_t \varphi &= \nabla \cdot \big(\kappa(T)\nabla\varphi\big) + f_1(\varphi, T) \\ \mu(\varphi)\partial_t T - \lambda(\varphi, T)\partial_t\varphi &= -\nabla \cdot \big(M(\varphi, T)\nabla(1/T)\big) \end{aligned} \qquad \text{in } \Omega \; . \tag{1.14}$$

Here κ, λ, μ, M, and f_1 are given smooth functions such that κ, μ, and M are positive. In fact, various simplifying assumptions are given in [PeF90].

Let $u = (u_1, u_2) := (\varphi, T)$ and solve (1.14) algebraically for $\partial_t u = (\partial_t\varphi, \partial_t T)$. Then (1.14) takes the form $\partial_t u + \mathcal{A}(u)u = f(u)$, where $\mathcal{A}$ is given by (1.9) with

$$a(\cdot, u) := \begin{bmatrix} \kappa & 0 \\ \lambda\kappa/\mu & M/\mu T^2 \end{bmatrix} \tag{1.15}$$

and $a_j = 0$, and where

$$f(u) := \Big(f_1, (\lambda/\mu)f_1 - (\nabla(\lambda/\mu)|\kappa\nabla\varphi) - (\nabla(1/\mu)|(M/T^2)\nabla T)\Big) \; . \tag{1.16}$$

Note that

$$(\nabla(\lambda/\mu)|\kappa\nabla\varphi) = \partial_1(\lambda/\mu)\kappa\,|\nabla\varphi|^2 + \partial_2(\lambda/\mu)\kappa(\nabla T|\nabla\varphi) \tag{1.17}$$

and

$$(\nabla(1/\mu)|(M/T^2)\nabla T) = -(M/\mu T^2)\partial\mu(\nabla\varphi|\nabla T) \tag{1.18}$$

so that f is 'quadratic in the gradient ∇u'.

In [PeF90] there is no discussion of boundary conditions. However, Sprekels and Zheng [SpZ92] study the particular case of (1.14) with

$$\kappa := \text{const} \; , \quad M := \text{const} \; , \quad \mu := 1 \; , \quad \lambda(\varphi, T) := a\varphi \tag{1.19}$$

for some positive constant a, and with

$$f_1(\varphi, T) := s'(\varphi) + a\varphi/T \; , \tag{1.20}$$

where s is a given smooth function (such that $-s$ is a 'double-well potential'). These authors consider the Neumann boundary conditions

$$\partial_\nu \varphi = 0 \ , \quad \partial_\nu T + \alpha(T - T_\partial) = 0 \qquad \text{on } \partial\Omega \ , \tag{1.21}$$

with α a positive constant and T_∂ a positive function. Observe that these boundary conditions are of the form $\mathcal{B}(u)u = g(u)$, where $\mathcal{B}$ is given by (1.10) with $\delta = 1$, and

$$g(u) := \big(0, -(\alpha M/T^2\mu)(T - T_\partial)\big) \ . \tag{1.22}$$

In the general case (1.14) we consider prescribed flux conditions too, that is, we assume that

$$\left.\begin{aligned} \kappa(T)\partial_\nu\varphi &= g_1(\varphi, T) \\ (M(\varphi,T)/T^2)\partial_\nu T + \lambda(\varphi,T)\kappa(T)\partial_\nu\varphi &= \mu(\varphi)g_2(\varphi,T) \end{aligned}\right\} \quad \text{on } \partial\Omega \ , \tag{1.23}$$

where $g := (g_1, g_2)$ is a given smooth function. Then the system (1.14), (1.23) is equivalent to a system of the form (1.8)–(1.10) where, in general, f depends quadratically on ∂u. Of course, solutions satisfying $T > 0$ are of interest only. □

1.3 Flows in Porous Media In [CDL92] the authors consider the problem of the mixing of fresh and salt groundwater, due to mechanical dispersion and molecular diffusion. They are led to a single mass balance equation

$$\partial_t u + \operatorname{div} \vec{\jmath} = 0 \qquad \text{in } \Omega \ , \tag{1.24}$$

together with the no flux boundary condition

$$(\vec{\jmath}\,|\,\vec{\nu}) = 0 \qquad \text{on } \partial\Omega \ , \tag{1.25}$$

where

$$\vec{\jmath} = -M(\vec{q})\nabla u + \vec{q}u \tag{1.26}$$

and $n = 2$. Here $M(\vec{q})$ is the hydrodynamical dispersion matrix, given (in dimensionless form) by

$$M(\vec{q})_{jk} := \begin{cases} (a\,|\vec{q}| + m)\delta_{jk} + bq_j q_k/|\vec{q}| & \text{if } \vec{q} \neq 0 \ , \\ m\delta_{jk} & \text{if } \vec{q} = 0 \ , \end{cases} \tag{1.27}$$

where m is a positive and a and b are nonnegative constants. Here u is the (suitably normalized) specific weight of the fluid (the mixture of fresh and salt water) which is supposed to be incompressible. Lastly, $\vec{q}$ is the vector of specific discharge of the fluid. The functions u and $\vec{q}$ are related, via a (suitably normalized) stream function v, by

$$\vec{q} = \operatorname{curl} v \quad \big(= (\partial_2 v, -\partial_1 v)\big) \ . \tag{1.28}$$

Darcy's law and the particular geometry of the problem imply finally that u and v are connected by the elliptic boundary value problem

$$\begin{aligned} -\Delta v &= \partial_1 u \qquad && \text{in } \Omega \ , \\ v &= 0 && \text{on } \partial\Omega \ . \end{aligned} \tag{1.29}$$

Thus the model under consideration consists of the system (1.24)–(1.29) forming a parabolic-elliptic boundary value problem.

By solving (1.29) and substituting the solution via (1.27), (1.28) into (1.26) we arrive (at least formally) at a single quasilinear parabolic 'reaction-diffusion' system which is slightly more general than the form (1.8)–(1.10). More precisely, let

$$\vec{Q} := \mathrm{curl} \circ (-\Delta_D)^{-1} \circ \partial_1 \ , \qquad a := M \circ \vec{Q} \ , \qquad a_j := -Q_j \ , \quad j = 1,2 \ , \tag{1.30}$$

where Δ_D denotes the Laplacian under zero Dirichlet boundary conditions, it follows that (1.24)–(1.29) is (formally) equivalent to the problem

$$\begin{aligned} \partial_t u - \partial_j \big(a_{jk}(u)\partial_k u + a_j(u)u\big) &= 0 \qquad \text{in } \Omega \ , \\ \nu^j \big(a_{jk}(u)\partial_k u + a_j(u)u\big) &= 0 \qquad \text{on } \partial\Omega \ . \end{aligned} \tag{1.31}$$

Note that in (1.31) the coefficients are nonlocal functions of u and that the dispersion matrix M is Lipschitz continuous but not continuously differentiable. Thus the coefficients a_{jk} have little regularity only. □

1.4 Moving Boundaries In [CoHS89] and [CoHS90] there is investigated a solid-liquid system of a small spherical grain exchanging matter with a liquid solution by a dissolution-growth process. Its mathematical formulation leads to a 'moving boundary problem'. To be more precise, denote by $c = c(r,t)$ the concentration in the liquid phase, where r is the distance from the center of the grain, and by $R = R(t)$ the radius of the grain. The authors study the following (rescaled) equations which describe the growth of the spherical grain.

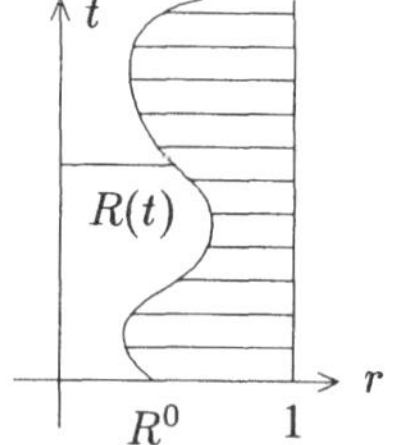

$$\begin{aligned} c_t - r^{-2}(r^2 c_r)_r &= F_1(c) \qquad && \text{in } \big(R(t),1\big) \ , \\ c_r &= (1-c)H(r,c) && \text{for } r = R(t) \ , \\ c &= \varphi && \text{for } r = 1 \ , \\ \dot{R} &= H\big(R, c(R,\cdot)\big) \ , \end{aligned} \tag{1.32}$$

where F_1, H, and φ are given functions.

We transform (1.32) to a 'fixed domain' problem by introducing $x := (1-r)/(1-R)$ as new independent space variable. We also put

$$u := (u_1, u_2) := (c - \varphi, R) \ .$$

Then (1.32) goes over into the following equivalent system:

$$\begin{aligned} \partial_t u_1 + \mathcal{A}_{11}(u)u_1 &= f_1(u) \qquad && \text{in } \Omega \ , \\ \mathcal{B}_{11}(u)u_1 &= g_1(u) && \text{on } \partial\Omega \ , \\ \dot{u}_2 &= f_2(u) \ , \end{aligned} \tag{1.33}$$

where

$$\mathcal{A}_{11}(u)v := -\big(a(u)v_x\big)_x + b(\cdot,u)v_x \tag{1.34}$$

and

$$\mathcal{B}_{11}(u)v := \delta a(u)\partial_\nu v + (1-\delta)v$$

with $\Omega := (0,1)$, $\delta(0) := 0$, $\delta(1) := 1$,

$$a(u) := (u_2 - 1)^{-2} , \tag{1.35}$$

and

$$b(x,u) := (1-u_2)^{-1}\{2[(u_2-1)x+1]^{-1} - xH(u_2,\gamma_1 u_1 + \varphi)\} , \tag{1.36}$$

with γ_1 being the restriction operator to the boundary point $x = 1$, and where

$$f_1(u) := F_1(u_1 + \varphi) - \dot{\varphi} , \quad f_2(u) := H(u_2, \gamma_1 u_1 + \varphi) , \tag{1.37}$$

and

$$g_1(u) := \delta(u_2 - 1)^{-1}(1 - u_1 - \varphi)H(u_2, \gamma_1 u_1 + \varphi) . \tag{1.38}$$

Note that b and f_2 are nonlocal operators and that $\mathcal{A}_{11}$ differs from an operator of the form (1.9) by the additive term $b\partial_x$ only.

We introduce now matrices of operators

$$\mathcal{A}(u) := \begin{bmatrix} \mathcal{A}_{11}(u) & 0 \\ 0 & 0 \end{bmatrix} , \quad \mathcal{B}(u) := \begin{bmatrix} \mathcal{B}_{11}(u) & 0 \\ 0 & 0 \end{bmatrix} \tag{1.39}$$

and put

$$f(u) := \begin{bmatrix} f_1(u) \\ f_2(u) \end{bmatrix} , \quad g(u) := \begin{bmatrix} g_1(u) \\ 0 \end{bmatrix} . \tag{1.40}$$

Then problem (1.33) has *formally* the appearance (1.8). Note that the operator matrices (1.39) are degenerate. □

1.5 Diffusion in Polymers In the study of diffusion and viscoelastic relaxation in polymers D.S. Cohen introduced a class of nonstandard reaction-diffusion models which he analyzed — together with coworkers — by means of formal asymptotic techniques and numerical methods (cf. [ChC89], [ChW89], [ChW91], for example). There are two unknown functions, the density c of a quantity (a fluid, for example) penetrating a polymer entanglement network, and the stress σ induced by the penetrating moleculs. These quantities satisfy the system of differential equations

$$\begin{aligned} c_t + j_x &= 0 \\ \sigma_t - \varphi(c)c_t &= \rho(c)c - \beta(c)\sigma \end{aligned} \qquad \text{in } \Omega , \tag{1.41}$$

where $\Omega \subset \mathbb{R}$ is a bounded open interval (i.e., $n = 1$), subject to the Dirichlet boundary condition

$$c = \psi \qquad \text{on } \partial\Omega . \tag{1.42}$$

The flux 'vector' is given by

$$j := -D(c)c_x - H(c)\sigma_x - M(c,\sigma)c ,$$

where $D > 0$ and H, φ, ρ, and β are nonnegative. Moreover, all these functions, as well as M and ψ, are smooth.

Let Φ be the primitive of φ satisfying $\Phi(0) = 0$ and assume without loss of generality that $\Omega = (0,1)$. Put

$$\Psi(x) := (1-x)\psi(0) + x\psi(1) \ , \qquad 0 \le x \le 1 \ ,$$

and

$$u_1 := c - \Psi \ , \quad u_2 := \sigma - \Phi(c) \ , \quad u := (u_1, u_2) \ .$$

Then (1.41), (1.42) is transformed into the equivalent system

$$\begin{aligned} \partial_t u_1 + \mathcal{A}_{11}(u)u_1 + \mathcal{A}_{12}(u)u_2 &= f_1(u) \\ \partial_t u_2 &= f_2(u) \end{aligned} \quad \text{in } \Omega \ , \qquad u_1 = 0 \quad \text{on } \partial\Omega \ , \tag{1.43}$$

where $\mathcal{A}_{11}$ and $\mathcal{A}_{12}$ are of the form (1.9), the top order coefficient of $\mathcal{A}_{11}$ being

$$(D + \varphi H) \circ c(u_1) \ge D(c(u_1)) \ . \tag{1.44}$$

Also note that

$$f_1(u) = \big(h(u)\big)_x \tag{1.45}$$

with a smooth function h. By introducing the operator matrices

$$\mathcal{A}(u) := \begin{bmatrix} \mathcal{A}_{11}(u) & \mathcal{A}_{12}(u) \\ 0 & 0 \end{bmatrix} , \quad \mathcal{B} := \begin{bmatrix} 1 & 0 \\ 0 & 0 \end{bmatrix}$$

the system (1.43) takes (formally) the form (1.8) with $g = 0$. Of course, this problem is again degenerate. In fact, it is even more degenerate than (1.33) since, in general, the differential expression $\mathcal{A}_{12}$ operating on u_2 is of the same strength — namely a second order operator — as $\mathcal{A}_{11}$ operating on u_1. Of course, only solutions satisfying $c = u_1 + \Psi \ge 0$ are of interest. □

1.6 Reaction and Diffusion in Electrolysis Consider a vessel Ω, whose boundary consists of the outer wall $\partial_W\Omega$ and the surfaces $\partial_E\Omega$ of electrodes, an anode A and a cathode C, such that $\partial\Omega$ is the disjoint union of $\partial_W\Omega$ and $\partial_E\Omega$. In Ω there is contained a solute of N dissolved species, which can be ions or electrically neutral, whose nonnegative concentrations are denoted by $u_1, \ldots, u_N$. Each one of the species r satisfies the mass balance law $(1.1)_r$. The flux $\vec{\jmath}_r$ is due to the concentration gradient of species r, to convection, and to migration in an electric field φ. Thus

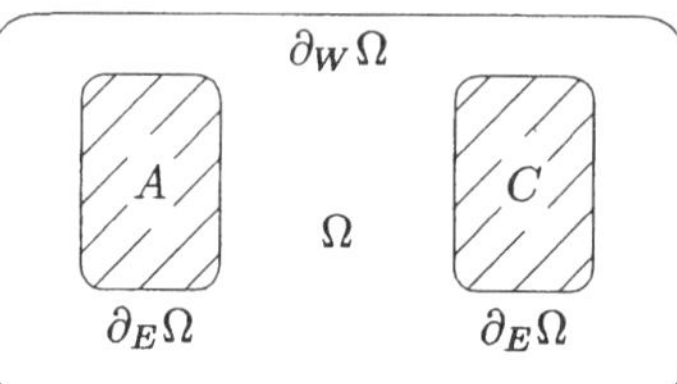

$$\vec{\jmath}_r := -a_r(\cdot, u)\nabla u_r + u_r\vec{\omega} - m_r u_r \nabla\varphi \ ,$$

where the 'diffusion coefficients' a_r are positive, $\vec{\omega}$ is a given velocity vector, and m_r is the 'mobility' of species r. On the outer wall there are the no flux conditions

$$(\vec{\jmath}_r | \vec{\nu}) = 0 \qquad \text{on } \partial_W\Omega \ , \quad r = 1, \ldots, N \ . \tag*{$(1.46)_r$}$$

On the surface of the electrodes electrochemical reactions occur which are modelled by nonlinear boundary conditions of the form

$$-(\vec{\jmath}_r|\vec{\nu}) = g_r(\cdot, u, \varphi) \qquad \text{on } \partial_E\Omega\ , \quad 1 \le r \le N\ , \tag{1.47$_r$}$$

where the g_r are highly nonlinear functions of u and φ (e.g., [RMK86], [New68]). All occurring functions are smooth.

Letting

$$a(\cdot, u) := \operatorname{diag}\big[a_r(\cdot, u)\big]\ , \quad a_j(\cdot, u, \varphi) := \operatorname{diag}[-\omega_j + m_r \partial_j \varphi]\ ,$$

the resulting reaction-diffusion system $(1.1)_r$, $(1.46)_r$, $(1.47)_r$ is of the form (1.8)–(1.10) with $\delta = 1$ (and $g|\partial_W\Omega = 0$, of course). Note that φ, the electric potential in the solute, is an unknown too, so that the system is underdetermined as it stands. However, there is an additional algebraic condition which has to be satisfied. Indeed, let z_r be the charge of species r so that z_r is a given real number which equals zero if species r is electrically neutral. Then $\sum z_r u_r$ is the total charge in the solute, and the 'electroneutrality condition'

$$\textstyle\sum_r z_r u_r = 0 \tag{1.48}$$

requires that the total charge be zero at any place and at any time. It is an important information — as we shall see in Section 15 — that

$$z_r m_r \ge 0\ , \qquad r = 1, \ldots, N\ , \tag{1.49}$$

that is, the mobility of each species, has the same sign as the corresponding charge.

Thus the full system, describing the mathematical model for reactions and diffusion in the electrolytical model envisaged, is given by $(1.1)_r$, $(1.46)_r$, $(1.47)_r$, and (1.48). Note again that only solutions satisying $u_r \ge 0$ are of interest. □

1.7 Stefan Problems Consider a one-dimensional heat conducting medium occupying the interval $[0, 1]$ and suppose that a phase change takes place, where the 'interface' between the two phases (a solid and a liquid one, for example) is at time t at the point $s(t) \in (0, 1)$. Also suppose that in each phase the temperature T satisfies a nonlinear heat equation

$$\partial_t T - \partial_x\big(k_i(\cdot, T)\partial_x T\big) = g_i(\cdot, T) \quad \text{in } \Omega_i(t)\ , \qquad i = 0, 1\ , \tag{1.50}$$

where

$$\Omega_i(t) := \begin{cases} \big(0, s(t)\big)\ , & i = 0\ , \\ \big(s(t), 1\big)\ , & i = 1\ , \end{cases}$$

and the boundary conditions

$$\begin{aligned} -\delta(0)k_0(0, T)\partial_x T + \big(1 - \delta(0)\big)T = 0 \qquad & \text{at } x = 0\ , \\ \delta(1)k_1(1, T)\partial_x T + \big(1 - \delta(1)\big)T = 0 \qquad & \text{at } x = 1\ . \end{aligned} \tag{1.51}$$

Of course, we suppose that the k_j are everywhere positive and — for simplicity — that all occurring functions are smooth.

At the interface the temperature must be equal to the phase change temperature, which we normalize to be zero. Hence

$$T\big(s(t)-\big) = T\big(s(t)+\big) = 0\ . \tag{1.52}$$

Let L be the heat absorbed or released for melting or solidifying (that is, for the mechanism of the phase change) per unit volume of the material. Then $L\dot{s}$ is the local rate of heat absorption or release in the process. The heat balance states that $L\dot{s}$ must equal the difference between the heat flux coming to the interphase from each side, that is, the 'Stefan condition'

$$L\dot{s} = -k_0\partial_x T|_{x=s(t)-} + k_1\partial_x T|_{x=s(t)+} \tag{1.53}$$

has to be satisfied (e.g., [Cra84], [Rub71]). The one-dimensional two phase Stefan problem consists in finding the temperature distribution T and the interface s such that (1.50)–(1.53) are satisfied, provided, of course, initial conditions are specified.

Observe that the coordinate transformations

$$\varphi_i(x,t) := (y,\tau) := \begin{cases} (x/s(t),t) & \text{if } i=0\ , \\ ((1-x)/(1-s(t)),t) & \text{if } i=1\ , \end{cases}$$

transform each

$$D_i := \{\,(x,t)\ ;\ x\in\Omega_i(t)\,\}\ , \qquad i=0,1\ ,$$

onto $\Omega\times\mathbb{R}^+$, where $\Omega := (0,1)$, such that

$$\varphi_0(0,t)=(0,t)\ ,\quad \varphi_i(s(t),t)=(1,t)\ ,\quad \varphi_1(1,t)=(0,t)\ ,$$

(provided, of course, $0 < s(t) < 1$ for $t\in\mathbb{R}^+$).

Let

$$u_i := T\circ\varphi_i^{-1}$$

and note that

$$\partial_x = -(i-s)^{-1}\partial_y\ ,\quad \partial_t = \partial_\tau + \big(y\dot{s}/(i-s)\big)\partial_y\ .$$

Put

$$a_i(\cdot,u_i,s) := (i-s)^{-2}k_i(\cdot,T)\circ\varphi_i^{-1}$$

and

$$f_i(\cdot,u_i) := g_i(\cdot,T)\circ\varphi_i^{-1}\ .$$

Then (1.50) transforms into

$$\partial_\tau u_i - \partial_y\big(a_i(\cdot,u_i,s)\partial_y u_i\big) + \big(y\dot{s}/(i-s)\big)\partial_y u_i = f_i(u_i) \qquad \text{in } \Omega\ . \tag{1.54}$$

The boundary conditions (1.51) and (1.52) are equivalent to

$$\begin{aligned} -\delta_i a_i(\cdot,u_i,s)\partial_y u_i + (1-\delta_i)u_i &= 0 && \text{at } x=0\ , \\ u_i &= 0 && \text{at } x=1\ , \end{aligned} \tag{1.55}$$

where $\delta_0 := \delta(0)$ and $\delta_1 := \delta(1)$. Lastly, the Stefan condition (1.53) transforms into

$$L\dot{s} = -sa_0\big(1,u_0(1,\cdot),s\big)\partial_y u_0(1,\cdot) - (1-s)a_1\big(1,u_1(1,\cdot),s\big)\partial_y u_1(1,\cdot)\ . \tag{1.56}$$

Define a nonlocal map h by

$$h(s,u) := (1/L)\sum_i(-1)^i(i-s)a_i(1,\gamma_1 u_i,s)\gamma_1\partial_y u_i$$

and put $u := (u_0, u_1)$,

$$a(\cdot, u, s) := \operatorname{diag}\big[a_0(\cdot, u_0, s), a_1(\cdot, u_1, s)\big] \tag{1.57}$$

and

$$b(\cdot, u, s) := y \operatorname{diag}\big[-h(s,u)/s, h(s,u)/(1-s)\big] \ . \tag{1.58}$$

Then we can rewrite (1.54)–(1.56) as (writing again t for τ)

$$\begin{aligned} \partial_t u + \mathcal{A}(u,s)u &= f(u) && \text{in } \Omega \ , \\ \mathcal{B}(u,s)u &= 0 && \text{on } \partial\Omega \ , \\ \dot{s} &= h(s,u) \ , \end{aligned} \tag{1.59}$$

where $\mathcal{A}$ and $\mathcal{B}$ are of the form (1.9) (or rather of the form (1.34)) and (1.10), respectively. Thus we have transformed the Stefan problem into an equivalent form which has — superficially — the same appearance as (1.33). □

1.8 Dynamic Boundary Conditions In the mathematical description of chemical kinetics and chemical technology there occur also 'dynamic boundary conditions' for the reaction-diffusion systems $(1.1)_r$, which are of the form

$$\partial_t u_r - (\vec{\jmath}_r | \vec{\nu}) = g_r \qquad \text{on } \partial\Omega \ , \quad 1 \le r \le N \ , \tag{1.60$_r$}$$

(e.g., [Cra56], [SlH72]). Although the reaction-diffusion problem $(1.1)_r$, $(1.60)_r$ does not fit immediately into the framework discussed so far, the more general abstract setting described in this paper encompasses these problems too. □

2 Informal Discussion

We return to problem (1.8)–(1.10) and consider first the particular case where

$$g = 0 \ . \tag{2.1}$$

If u is a sufficiently regular fixed function, $\mathcal{B}(u)$ is a well defined linear operator (on a suitable space of distributions). Thus it makes sense to put

$$A(u) := \mathcal{A}(u) | \ker \mathcal{B}(u) \ . \tag{2.2}$$

With this definition the system (1.8) with $g = 0$ can be written as an abstract quasilinear ordinary differential equation of the form

$$\dot{u} + A(u)u = f(u) \tag{2.3}$$

in a suitable Banach space. To be more specific, we fix $p \in (1, \infty)$ and put

$$E_0 := L_p(\Omega, \mathbb{R}^N) \ .$$

We also put

$$E_1(u) := W^2_{p,\mathcal{B}(u)} := \big\{ v \in W^2_p(\Omega, \mathbb{R}^N) \ ; \ \mathcal{B}(u)v = 0 \big\} = \ker \mathcal{B}(u) \ ,$$

where $\mathcal{B}(u)$ has to be interpreted in the sense of traces, of course. Then $A(u)$, defined by (2.2), is a densely defined linear operator in E_0 with domain $E_1(u)$, and equation (2.3) is meaningful if f is sufficiently nice.

In order to study equation (2.3) we have to impose regularity conditions and structural assumptions upon $A(u)$. In the most elementary case, where $N = 1$ and $\vec{\jmath} := -\nabla u$, where f and g are zero, and where $\delta = 1$ (resp. $\delta = 0$) on $\partial\Omega$, the system (1.8) reduces to the linear heat equation

$$\partial_t u - \Delta u = 0 \qquad \text{in } \Omega \tag{2.4}$$

under homogeneous Neumann (resp. Dirichlet) boundary conditions. In this case it is well known that the operator A, that is, $-\Delta$ with Neumann (resp. Dirichlet) boundary conditions, is the negative infinitesimal generator of a strongly continuous analytic semigroup on E_0 and that the domain of A, equipped with its graph norm, equals (except for equivalent norms) $E_1 = W^2_{p,\mathcal{B}}$. Of course, the general results should contain (2.4) as a special case. For this reason we impose the 'abstract parabolicity condition' that $A(u)$ be, for each u which may occur at any given instant of time as a solution to the system (1.8), the negative infinitesimal generator of an analytic semigroup on E_0 with domain $E_1(u)$; in symbols:

$$A(u) \in \mathcal{H}\big(E_1(u), E_0\big) \ . \tag{2.5}$$

The formulation of the reaction-diffusion system (1.8) in the form (2.3) has the advantage of being very flexible and of comprising a great many different problems like higher order systems, problems with nonlocal nonlinearities, and problems with dynamic boundary conditions, to name a few.

On the other hand, it has serious shortcomings. First, it requires condition (2.1) and there is no easy way to incorporate nonlinearities on the boundary of the form $(1.7)_r$ with $\delta_r \neq 0$. Second, in general, the domains of the operators $A(u)$, namely $E_1(u)$, depend on the sought-for solution u itself. At present, there is no good theory which is sufficiently flexible and well developed to treat abstract evolution equations of the form (2.3) in the case of 'variable domains' of the negative generators $A(u)$.

It is an important feature of the reaction-diffusion system (1.8) that it is of 'divergence form'. This suggests a weak formulation of (1.8). For this we introduce a (generalized) 'Dirichlet form' by

$$\boldsymbol{a}(u)(w,v) := \big\langle \partial_j w, a(\cdot,u)\partial_j v + a_j(\cdot,u)v \big\rangle \ , \qquad w \in W^1_{p'}(\Omega,\mathbb{R}^N) \ , \quad v \in W^1_p(\Omega,\mathbb{R}^N) \ , \tag{2.6}$$

where

$$\langle w, v\rangle := \int_\Omega \big\langle w(x), v(x)\big\rangle_{\mathbb{K}^N}\, dx \tag{2.7}$$

is the standard duality pairing

$$\langle\cdot,\cdot\rangle : L_{p'}(\Omega,\mathbb{K}^N) \times L_p(\Omega,\mathbb{K}^N) \to \mathbb{K} \ ,$$

with $p' := p/(p-1)$ and $\langle \xi,\eta\rangle_{\mathbb{K}^N} := \sum_{r=1}^N \xi_r\eta_r$ for $\xi,\eta \in \mathbb{K}^N$, and with $\mathbb{K} := \mathbb{R}$ or $\mathbb{C}$.

We also put

$$E_{1/2} := W^1_{p,\mathcal{B}} := \big\{\, u \in W^1_p(\Omega,\mathbb{R}^N) \ ; \ (1-\delta)\gamma_\partial u = 0 \,\big\} \ ,$$

where γ_∂ denotes the trace operator. Note that $W^1_{p,\mathcal{B}} = W^1_p(\Omega, \mathbb{R}^N)$ if $\delta = 1 = 1_{\mathbb{R}^N}$, that is, if the Neumann type boundary condition $(1.5)_r$ holds for each r and everywhere on $\partial\Omega$.

Lastly, we assume instead of (2.1) that $g^0 = 0$, that is,

$$(1-\delta)g = 0 . \tag{2.8}$$

In other words, we assume that the homogeneous Dirichlet boundary conditions $u_r = 0$ on $\partial_{0,r}\Omega$ are satisfied if $\partial_{0,r}\Omega \neq \emptyset$. Thus (2.8) is no restriction at all if $\delta = 1$, that is, if the prescribed flux condition $(1.5)_r$ is satisfied for each species r and on all of $\partial\Omega$. Putting

$$\langle w, v\rangle_\partial := \int_{\partial\Omega} \langle w(x), v(x)\rangle_{\mathbb{K}^N}\, d\sigma(x) ,$$

where $d\sigma$ is the standard Riemann-Lebesgue volume measure on $\partial\Omega$, we associate with (1.8) the following 'weak W^1_p-formulation':

$$\begin{aligned}&\text{find } u : \mathbb{R}^+ \to W^1_{p,\mathcal{B}} \text{ such that}\\ &\langle w, \partial_t u\rangle + \boldsymbol{a}(u)(w,u) = \langle w, f(u)\rangle + \langle \gamma_\partial w, g(u)\rangle_\partial\\ &\text{for each } w \in W^1_{p',\mathcal{B}}.\end{aligned} \tag{2.9}$$

Of course, $\partial_t u$ has to exist and be sufficiently regular so that $\langle w, \partial_t u\rangle$ makes sense.

If $a(\cdot,u)$ and $a_j(\cdot,u)$ belong to $L_\infty(\Omega, \mathcal{L}(\mathbb{R}^N))$, the Dirichlet form (2.6) is a continuous bilinear form on $W^1_{p',\mathcal{B}} \times W^1_{p,\mathcal{B}}$. Thus it defines a bounded linear operator

$$A_{-1/2}(u) \in \mathcal{L}(E_{1/2}, E_{-1/2}) , \qquad \text{where} \quad E_{-1/2} := (W^1_{p',\mathcal{B}})' , \tag{2.10}$$

by

$$\langle w, A_{-1/2}(u)v\rangle_{E_{-1/2}} := \boldsymbol{a}(u)(w,v) , \qquad w \in W^1_{p',\mathcal{B}} , \quad v \in W^1_{p,\mathcal{B}} . \tag{2.11}$$

Here and in the following $\langle\cdot,\cdot\rangle_E$ denotes the duality pairing

$$\langle\cdot,\cdot\rangle_E : E' \times E \to \mathbb{K}$$

between the dual E' of a Banach space E and E itself. Defining $F(u) \in E_{-1/2}$ by

$$\langle w, F(u)\rangle_{E_{-1/2}} := \langle w, f(u)\rangle + \langle \gamma_\partial w, g(u)\rangle_\partial , \qquad w \in W^1_{p',\mathcal{B}} , \tag{2.12}$$

we can rewrite (2.9), formally, as an abstract evolution equation in the Banach space $E_{-1/2}$:

$$\dot{u} + A_{-1/2}(u)u = F(u) . \tag{2.13}$$

The weak formulation (2.9), resp. its abstract formulation (2.13), has two important advantages over its 'strong' counterpart (2.3). First, the domains of the operators $A_{-1/2}(u)$ are independent of u since they equal the Banach space $E_{1/2} = W^1_{p,\mathcal{B}}$. Second, the weak formulation handles with relative ease nonlinear boundary conditions satisfying (2.8), that is, nonlinear prescribed flux conditions. Of course, in order to treat (2.13) efficiently one has to show that $A_{-1/2}(u)$ is the negative infinitesimal generator of a strongly continuous analytic semigroup on $E_{-1/2}$, that is,

$$A_{-1/2}(u) \in \mathcal{H}(E_{1/2}, E_{-1/2}) .$$

Traditionally, weak formulations of partial differential equations are W_2^1-formulations, that is, one considers the case $p = p' = 2$. In addition, in this connection it is always assumed that the quadratic form $\boldsymbol{a}(u)(\cdot,\cdot)$ is coercive in the sense that

$$\boldsymbol{a}(u)(v,v) \geq \alpha \|v\|_{1,2}^2 - \beta \|v\|_2^2 , \qquad v \in W_{2,\mathcal{B}}^1 , \tag{2.14}$$

where $\alpha > 0$ and $\beta \geq 0$, and where $\|\cdot\|_{s,q}$ is the norm in the Sobolev space W_q^s with $\|\cdot\|_q := \|\cdot\|_{0,q}$. Condition (2.14) amounts to an ellipticity condition for the differential operator which is too restrictive and unnatural in many applications. Moreover, the W_2^1-setting requires severe growth restrictions for the nonlinearities a, a_j, f, and g in order that the problem be well-posed.

It is the purpose of the following considerations

- to show that the weak W_p^1-formulation (2.9) is meaningful in the sense that the corresponding evolution equation (2.13) is well-posed and generates a local semiflow provided, of course, suitable regularity conditions are satisfied. Since the coerciveness condition (2.14) is not available in the general case, we have to develop a method which is completely different from the standard Galerkin approximation technique usually employed in the weak W_2^1-setting (e.g., [DauL90, Chapter XVIII, § 3]). Our method is based upon semigroup theory, abstract evolution equations, and extrapolation techniques and will be explained in the following.

- to have a unified treatment of weak and strong solutions which provides an 'automatic regularity theory'. In particular, it will be shown that weak solutions are automatically classical, in fact C^∞-smooth, provided all data are in this regularity class. This is by no means obvious since it is well known that weak W_2^1-solutions to *systems* are — in general — not even Hölder continuous. The reason why we succeed in proving smoothness stems from the fact that we can study weak W_p^1-solutions for $p > n$, which are automatically Hölder continuous.

- to be able to treat nonlinear boundary conditions which occur rather naturally in many applications.

- to have a general framework which allows us to handle, with the same ease, general problems which are not parabolic systems in the narrow sense but are 'abstractly parabolic', be it in the sense that they involve nonlocal operators or that they are degenerate in the sense of some of the model problems of Section 1. The general theory presented below is applicable whenever the problem can be cast in the form (2.3), where $-A(u)$ is the infinitesimal generator of a strongly continuous analytic semigroup such that $\mathrm{dom}\big(A(u)\big)$ is independent of u, and if A and F satisfy certain mild continuity hypotheses. Thus the whole discussion will essentially be reduced to the task of finding the correct function space setting and proving generation and continuity theorems.

- to present a 'dynamic theory' of quasilinear parabolic evolution equations which is the basis for further qualitative studies of the semiflow, like questions of stability, global behavior, bifurcation phenomena, periodic orbits, etc. To prevent this rather long paper from being even longer and more demanding, we restrict ourselves here to the basic existence, uniqueness, and continuity questions for maximal solutions. We give only a few references to these qualitative questions. More of this will be found in [Ama94].

Although we are ultimately interested in nonlinear problems, large parts of the theory belong to the 'linear field'. For this reason a considerable amount of this paper is devoted to the study of function spaces and linear differential operators. Even in the case of linear elliptic and parabolic boundary value problems the results about weak and 'very weak' solvability given below are new.

3 Notations and Conventions

In general, we use standard notation. Nevertheless, a few explanations are in order.

Let X and Y be topological spaces. If $X \subset Y$ and the topology of X is stronger than the one of Y, that is, if the natural injection $i: X \subset Y$ is continuous, we denote this by $X \hookrightarrow Y$. We write $X \overset{d}{\hookrightarrow} Y$ if X is also dense in Y. If X and Y are topological vector spaces, $X \hookrightarrow Y$ also means that X is a vector subspace of Y.

Suppose $X \hookrightarrow Y$ and let Z be a nonempty subset of Y. Then $Z \cap X$ is given the topology induced by X, unless explicitly stated otherwise.

Let E and F be Banach spaces over $\mathbb{K}$. Then $\mathcal{L}(E,F)$ is the Banach space of all bounded linear operators from E into F, and $\mathcal{L}(E) := \mathcal{L}(E,E)$. We denote by $\mathcal{L}\text{is}(E,F)$ the open set of isomorphisms in $\mathcal{L}(E,F)$. If G is a third Banach space, $\mathcal{L}(E,F;G)$ is the Banach space of all continuous bilinear maps $E \times F \to G$ with its usual norm.

If $E \hookrightarrow F$ and $F \hookrightarrow E$, we write $E \doteq F$, that is, $E \doteq F$ iff E and F coincide as vector spaces and carry equivalent norms. If $E \hookrightarrow F$ and the natural injection is compact, we denote this by $E \overset{c}{\hookrightarrow} F$.

We write $A: \text{dom}(A) \subset E \to F$ if A is a linear operator with domain, $\text{dom}(A)$, in E and image, $\text{im}(A)$, in F. We denote by $D(A)$ the domain of A endowed with its graph norm. Then $D(A) \hookrightarrow E$ and $A \in \mathcal{L}(D(A),F)$. Recall that $D(A)$ is a Banach space iff A is closed.

If $E \hookrightarrow F$ and $A: \text{dom}(A) \subset F \to F$ is a linear operator in F, the **E-realization** (the maximal restriction of A to E, or the part of A in E) is the linear operator A_E in E defined by

$$\text{dom}(A_E) := \{\, x \in E \cap \text{dom}(A) \ ; \ Ax \in E \,\} , \qquad A_E x := Ax .$$

If no confusion seems likely, we denote A_E again by A.

If $\mathbb{K} = \mathbb{R}$ and in a given formula there occur (implicitly or explicitly) complex numbers, it is always understood that the corresponding formula refers to its complexified version. For example, if A is a linear operator in E then $\sigma(A)$ and $\rho(A)$ denote the spectrum and the resolvent set, respectively, of A if $\mathbb{K} = \mathbb{C}$ and of the complexification of A if $\mathbb{K} = \mathbb{R}$.

If no confusion seems likely, 1 denotes the identity 1_E in $\mathcal{L}(E)$. In general, $\mathcal{L}(\mathbb{K}^n, \mathbb{K}^m)$ is canonically identified with the space $\mathbb{K}^{m \times n}$ of $m \times n$-matrices with entries in $\mathbb{K}$ by means of the canonical bases of $\mathbb{K}^m$ and $\mathbb{K}^n$, respectively.

Let E_0 and E_1 be Banach spaces with $E_1 \overset{d}{\hookrightarrow} E_0$. We denote by

$$\mathcal{H}(E_1, E_0)$$

the set of all $A \in \mathcal{L}(E_1, E_0)$ such that A, considered as a linear operator *in* E_0 with domain E_1, is the negative infinitesimal generator of an analytic semigroup $\{\, e^{-tA} \ ; \ t \geq 0 \,\}$

on E_0, that is, in $\mathcal{L}(E_0)$. (We refer to [CH*87], [Dav80], [Fat83], [Gol85], or [Paz83], for example, for the basic theory of analytic semigroups. However, a deeper understanding of this theory is not required for reading this paper.) It is an important fact that

$$\begin{aligned}&A \in \mathcal{H}(E_1, E_0) \text{ iff there exist constants } \kappa \geq 1 \text{ and } \omega > 0\\ &\text{such that } [\operatorname{Re} z \geq \omega] \subset \rho(-A) \text{ and}\\ &|\lambda|\,\|u\|_0 + \|u\|_1 \leq \kappa\,\|(\lambda + A)u\|_0\ , \qquad u \in E_1\ , \quad \operatorname{Re}\lambda \geq \omega\ ,\\ &\text{where } \|\cdot\|_j := \|\cdot\|_{E_j}.\end{aligned} \tag{3.1}$$

Moreover,

$$\mathcal{H}(E_1, E_0) \text{ is open in } \mathcal{L}(E_1, E_0) \tag{3.2}$$

(cf. [Ama91b, Section 1] for (3.1) and (3.2), and [Ama88a, Lemma 4.1] or [CH*87, Theorem 5.3] for (3.2)).

Let $X := (X, d)$ be a metric space. Then $u : X \to E$ is uniformly ρ-Hölder continuous, where $0 < \rho \leq 1$, if

$$[u]_\rho := [u]_{\rho,X} := \sup_{x \neq y} \frac{\|u(x) - u(y)\|}{\left[d(x,y)\right]^\rho} < \infty\ .$$

It is ρ-Hölder continuous if each point has a neighborhood Y such that $u|Y$ is uniformly ρ-Hölder continuous. We put

$$C^\rho(X, E) := \{\, u \in E^X\ ;\ u \text{ is } \rho\text{-Hölder continuous} \,\}\ , \qquad 0 < \rho < 1\ ,$$

and $C^0(X, E) := C(X, E)$ is the set of all continuous maps from X to E. A 1-Hölder continuous function is Lipschitz continuous, and

$$C^{1\text{-}}(X, E) := \{\, u \in E^X\ ;\ u \text{ is Lipschitz continuous} \,\}\ .$$

In numerical calculations the symbol 1- will be identified with the real number 1.

We denote by $B(X, E)$ the Banach space of all bounded $u \in E^X$, endowed with the supremum norm $\|\cdot\|_\infty$, and $BUC(X, E)$ is the closed subspace of all bounded and uniformly continuous functions. If $\rho \in (0, 1) \cup \{1\text{-}\}$ then $BUC^\rho(X, E)$ is the Banach space of all uniformly ρ-Hölder continuous functions equipped with the norm

$$\|\cdot\|_{C^\rho} := \|\cdot\|_\infty + [\cdot]_\rho\ .$$

Recall that a [ρ-Hölder] continuous function is uniformly [ρ-Hölder] continuous on compact sets. Hence $C^\rho(X, E) = BUC^\rho(X, E)$ for $\rho \in [0, 1) \cup \{1\text{-}\}$, if X is compact.

Lastly, if X is an open subset of some Banach space or a perfect subset of $\mathbb{K}$, the spaces $C^s(X, E)$ and $BUC^s(X, E)$ have the obvious meanings and topologies for $s \in \mathbb{R}^+$ and $s \in \{\, k\text{-}\ ;\ k \in \dot{\mathbb{N}} \,\}$, where $\dot{\mathbb{N}} := \mathbb{N} \backslash \{0\}$ and $(k+1)\text{-} := k + 1\text{-}$ for $k \in \mathbb{N}$.

By an ordered Banach space, $E := (E, \leq) := (E, P)$, we mean a real Banach space E together with a closed, proper, and convex cone P, the positive cone of E, such that $x \leq y$ iff $y - x \in P$. If E and F are ordered Banach spaces, $T \in \mathcal{L}(E, F)$ is said to be positive iff $x \geq 0$ implies $Tx \geq 0$. The subset of all positive elements of $\mathcal{L}(E, F)$ is denoted by $\mathcal{L}^+(E, F)$, and we write $T \geq 0$ iff $T \in \mathcal{L}^+(E, F)$. Note that $\mathcal{L}^+(E, F)$ is a closed convex cone in $\mathcal{L}(E, F)$, which is proper iff P is total, that is, $\overline{P - P} = E$. Of course, $\mathcal{L}^+(E) := \mathcal{L}^+(E, E)$. Moreover,

$$(P')^+ := (P^+)' := \mathcal{L}^+(E, \mathbb{R})$$

is the dual positive cone of E'. Thus $(E',(P')^+)$ is an ordered Banach space if P is total, and the order induced by $(P')^+$ is the canonical order of E'.

Note that, given $x' \in E'$,

$$x' \geq 0 \quad \text{iff} \quad \langle x', x\rangle \geq 0 \ , \qquad x \in P \ . \tag{3.3}$$

Moreover, given $T \in \mathcal{L}(E)$, it follows that $T \geq 0$ iff $T' \geq 0$. (We refer to [Jam70] or [Sch71], for example, for the elementary facts about ordered Banach spaces which are being used in this paper.)

Let $E_0 := (E_0, P_0)$ be an ordered Banach space. If $i : E_1 \hookrightarrow E_0$ then E_1 is given the order whose positive cone equals $i^{-1}(P) = E_1 \cap P_0$. Thus E_1 is an ordered Banach space, whose order is said to be canonically induced by P_0, and $i \in \mathcal{L}^+(E_1, E_0)$. Suppose now that E_1 is dense in E_0 and $A \in \mathcal{H}(E_1, E_0)$. Then A is resolvent positive if there exists $\omega_0 \in \mathbb{R}$ such that $(\lambda + A)^{-1} \geq 0$ for $\lambda > \omega_0$. If $A \in \mathcal{H}(E_1, E_0)$, it is resolvent positive iff

$$e^{-tA} \geq 0 \ , \qquad t \geq 0 \ , \tag{3.4}$$

that is, iff the semigroup generated by $-A$ is positive. In this case

$$\omega_0 = \sup\{\, \mathrm{Re}\,\lambda \ ; \ \lambda \in \sigma(-A)\,\} \tag{3.5}$$

(e.g., [CH*87, Propositions 7.1 and 9.2]).

4 Normally Elliptic Boundary Value Problems

In the following,

$$\mathcal{A}u := -\partial_j(a_{jk}\partial_k u + a_j u) + b_j\partial_j u + a_0 u \tag{4.1}$$

denotes a general second order linear differential operator on Ω 'in divergence form' with $\mathcal{L}(\mathbb{K}^N)$-valued coefficients, that is, operating on $\mathbb{K}^N$-valued functions on Ω. To be precise, we assume that

$$a_{jk}, a_j, b_j \in C(\overline{\Omega}, X) \ , \quad a_0 \in L_\infty(\Omega, X) \ ,$$

where

$$X := \mathcal{L}(\mathbb{K}^N) \ .$$

We fix a diagonal-matrix-valued function

$$\delta := \mathrm{diag}[\delta_r]_{1\leq r\leq N} : \partial\Omega \to X \ , \tag{4.2}$$

a **boundary characterization map**, such that

$$\delta_r \in C(\partial\Omega, \{0,1\}) \ , \qquad 1 \leq r \leq N \ . \tag{4.3}$$

Note that each δ_r is constant on each component of $\partial\Omega$. Then we associate with $\mathcal{A}$ a linear boundary operator of order at most one, defined by

$$\mathcal{B}u := \delta\{\, \nu^j\gamma_\partial(a_{jk}\partial_k u + a_j u) + c\gamma_\partial u\,\} + (1-\delta)\gamma_\partial u \ , \tag{4.4}$$

where γ_∂ denotes the trace operator for $\partial\Omega$ and

$$c \in C(\partial\Omega, X) .$$

Then $(\mathcal{A}, \mathcal{B})$ is said to be a **linear boundary value problem** on Ω of order at most two in divergence form. We topologize the set of these boundary value problems by identifying them with the Banach space

$$\mathbb{E}(\Omega) := \mathbb{E}(\Omega, N) := C(\overline{\Omega}, X)^{n^2+n+n} \times L_\infty(\Omega, X) \times C(\partial\Omega, X)$$

by means of the identification

$$(\mathcal{A}, \mathcal{B}) \longleftrightarrow (a_{jk}, a_j, b_j, a_0, c) . \tag{4.5}$$

With the boundary value problem $(\mathcal{A}, \mathcal{B})$ we associate its **principal symbol**

$$(a_\pi, b_\pi) \in C(\overline{\Omega} \times \mathbb{R}^N, X) \times C(\partial\Omega \times \mathbb{R}^N, X)$$

defined by

$$a_\pi(x, \xi) := a_{jk}(x)\xi^j\xi^k , \qquad (x, \xi) \in \overline{\Omega} \times \mathbb{R}^n ,$$

and

$$b_\pi(y, \xi) := \delta(y)\nu^j(y)a_{jk}(y)\xi^k + \big(1 - \delta(y)\big) , \qquad (y, \xi) \in \partial\Omega \times \mathbb{R}^n .$$

Let S^{n-1} be the unit sphere in $\mathbb{R}^n$. Then $\mathcal{A}$ is said to be **normally elliptic** if

$$\sigma\big(a_\pi(x, \xi)\big) \subset [\operatorname{Re} z > 0] := \{ z \in \mathbb{C} ; \operatorname{Re} z > 0 \} , \qquad (x, \xi) \in \overline{\Omega} \times S^{n-1} . \tag{4.6}$$

The boundary operator $\mathcal{B}$ satisfies the **normal complementing condition** with respect to $\mathcal{A}$ (the parameter-dependent Lopatinskii-Shapiro condition) if zero is, for each $(x, \xi) \in T(\partial\Omega)$ and $\lambda \in [\operatorname{Re} z \geq 0]$ with $(\xi, \lambda) \neq (0, 0)$, the only exponentially decaying solution of the boundary value problem on the half-line:

$$\big[\lambda + a_\pi\big(x, \xi + \nu(x)i\partial_t\big)\big]u = 0 , \quad t > 0 , \qquad b_\pi\big(x, \xi + \nu(x)i\partial_t\big)u(0) = 0 . \tag{4.7}$$

Of course, $T(\partial\Omega)$ is the tangent bundle of $\partial\Omega$. Lastly, the boundary value problem $(\mathcal{A}, \mathcal{B})$ is **normally elliptic** on Ω if $\mathcal{A}$ is normally elliptic and $\mathcal{B}$ satisfies the normal complementing condition with respect to $\mathcal{A}$.

We denote by

$$\mathcal{E}(\Omega) := \mathcal{E}(\Omega, \delta, N)$$

the set of all normally elliptic boundary value problems on Ω. It follows from [Ama90a, Theorem 2] that

$$\mathcal{E}(\Omega) \text{ is open in } \mathbb{E}(\Omega) . \tag{4.8}$$

Besides boundary value problems $(\mathcal{A}, \mathcal{B})$ with 'continuous coefficients', that is, besides $\mathbb{E}(\Omega)$, we consider boundary value problems $(\mathcal{A}, \mathcal{B})$ with 'C^1–coefficients' which we identify by means of (4.5) with

$$\mathbb{E}^1(\Omega) := \big[C^1(\overline{\Omega}, X)\big]^{n^2+n+n} \times L_\infty(\Omega, X) \times C^1(\partial\Omega, X) .$$

Moreover,

$$\mathcal{E}^1(\Omega) := \mathcal{E}(\Omega) \cap \mathbb{E}^1(\Omega) .$$

Since $\mathbb{E}^1(\Omega) \hookrightarrow \mathbb{E}(\Omega)$, it follows from (4.8) that $\mathcal{E}^1(\Omega)$ is open in $\mathbb{E}^1(\Omega)$.

Given $(\mathcal{A}, \mathcal{B}) \in \mathbb{E}^1(\Omega)$,

$$(\mathcal{A}, \mathcal{B}) \in \mathcal{L}(W_p^2, L_p \times L_p(\partial\Omega, \mathbb{K}^N)) , \qquad 1 < p < \infty ,$$

where $W_p^2 := W_p^2(\Omega, \mathbb{K}^N)$ and $L_p := L_p(\Omega, \mathbb{K}^N)$. Thus

$$W_{p,\mathcal{B}}^2 := \ker \mathcal{B} = \{ u \in W_p^2 \ ; \ \mathcal{B}u = 0 \} \tag{4.9}$$

is a closed linear subspace of W_p^2, whence a Banach space. Hence

$$A := A_p := \mathcal{A} | \ker \mathcal{B} \in \mathcal{L}(W_{p,\mathcal{B}}^2, L_p) \tag{4.10}$$

is well defined, the **L_p-realization of the boundary value problem** $(\mathcal{A}, \mathcal{B})$.

The importance of the class $\mathcal{E}^1(\Omega)$ of normally elliptic boundary value problems with C^1–coefficients is seen from the following

4.1 Theorem *If $(\mathcal{A}, \mathcal{B}) \in \mathcal{E}^1(\Omega)$ then $A \in \mathcal{H}(W_{p,\mathcal{B}}^2, L_p)$. Conversely, if $(\mathcal{A}, \mathcal{B}) \in \mathbb{E}^1(\Omega)$ and $A \in \mathcal{H}(W_{p,\mathcal{B}}^2, L_p)$ then $\mathcal{A}$ is normally elliptic.*

For a proof of this theorem we refer to [Ama90a, Section 3] and [Ama94].

4.2 Remarks

(a) $\mathcal{A}$ is normally elliptic iff $\partial_t + \mathcal{A}$ is parabolic in the sense of Petrowskii (cf. [LadSU68, Section VII.8]).

(b) Theorem 4.1 is true for general linear normally parabolic boundary value problems of order m which are not necessarily in divergence form.

More precisely, suppose that $\mathcal{A} := \sum_{|\alpha| \le m} a_\alpha D^\alpha$, where

$$a_\alpha \in C(\overline{\Omega}, X) , \quad |\alpha| = m , \qquad a_\alpha \in L_\infty(\Omega, X) , \quad |\alpha| < m ,$$

and where $D_j := -i\partial_j$, $j = 1, \ldots, n$. Let

$$a_\pi(x, \xi) := \sum_{|\alpha|=m} a_\alpha(x)\xi^\alpha , \qquad (x, \xi) \in \overline{\Omega} \times \mathbb{R}^n ,$$

be the principal symbol of $\mathcal{A}$. Then $\mathcal{A}$ is normally elliptic if condition (4.6) is satisfied. It is easily verified that this implies that m is even.

Let

$$\mathcal{B}_r := \sum_{|\alpha|\le m_r} b_{\alpha,r}\gamma_\partial D^\alpha \ , \qquad r = 1,\ldots,mN/2 \ ,$$

be boundary operators of orders $m_r \in \{0,1,\ldots,m-1\}$, where m_r may vary from component to component of $\partial\Omega$, such that

$$b_{\alpha,r} \in C^{m-m_r}(\partial\Omega, \mathcal{L}(\mathbb{K}^N,\mathbb{K})) \ .$$

Define the principal boundary symbol of $\mathcal{B} := (\mathcal{B}_1,\ldots,\mathcal{B}_{mN/2})$ by

$$b_\pi(x,\xi) := \Bigl(\sum_{|\alpha|=m_r} b_{\alpha,r}(x)\xi^\alpha \ ; \ 1\le r\le mN/2\Bigr) \ , \qquad (x,\xi)\in\partial\Omega\times\mathbb{R}^n \ .$$

Then $(\mathcal{A},\mathcal{B})$ is said to be normally elliptic iff $\mathcal{A}$ is normally elliptic and (4.7) is satisfied.

The analogue to Theorem 4.1 says that $A \in \mathcal{H}(W^m_{p,\mathcal{B}}, L_p)$ if $(\mathcal{A},\mathcal{B})$ is normally elliptic. Conversely, if $A \in \mathcal{H}(W^m_{p,\mathcal{B}}, L_p)$ then $\mathcal{A}$ has to be normally elliptic (cf. [Ama94]).

(c) The fact that the normal ellipticity of $\mathcal{A}$ is necessary for $-A$ to be the generator of an analytic semigroup on L_p relies on the fact that we restrict ourselves to generators with domains $W^2_{p,\mathcal{B}}$. If we omit this condition, it is only necessary that $\sigma\bigl(a(x,\xi)\bigr) \subset [\mathrm{Re}\, z > 0] \cup \{0\}$, as has been proven in [Sad91], where also an example of a degenerate parabolic equation is given which shows that this case can occur (also cf. [ElF87] for the case where $\mathrm{dom}(A) = W^2_2(\mathbb{R}^n,\mathbb{C}^N)$, and [BoS86]). In general, $A \in \mathcal{H}(W^2_{p,\mathcal{B}}, L_p)$ does not imply that the boundary operator has to have the property that Dirichlet and Neumann type conditions occur on distinct components, that is, mixed boundary conditions are possible (cf. [Tai91], [GrR89]). □

In general, it is not easy to verify that a given boundary value problem $(\mathcal{A},\mathcal{B})$ is normally elliptic. For this reason we present in the following some useful examples and sufficient conditions which are easier to check than the definition above. Proofs are found in [Ama90a, Section 4].

4.3 Examples

(a) The differential operator $\mathcal{A}$ is **uniformly strongly elliptic** (satisfies the uniform Legendre-Hadamard condition) if

$$\mathrm{Re}\,\bigl(a_\pi(x,\xi)\eta|\eta\bigr) > 0 \ , \qquad (x,\xi)\in\overline{\Omega}\times S^{n-1} \ , \quad \eta\in\mathbb{K}^N\backslash\{0\} \ ,$$

that is, iff the symmetric part of $a_\pi(x,\xi)\in\mathcal{L}(\mathbb{K}^N)$ is positive definite, uniformly with respect to $(x,\xi)\in\overline{\Omega}\times S^{n-1}$. It is obvious that

$$\mathcal{A} \text{ is uniformly stronlgy elliptic} \quad\Longrightarrow\quad \mathcal{A} \text{ is normally elliptic.}$$

The converse is true iff $N = 1$.

(b) With $\mathcal{A}$ we associate a quadratic form on $\mathbb{K}^{n\times N}$, a **Legendre form** $a_{\mathcal{L}}$, by

$$a_{\mathcal{L}}(x,\zeta) := \sum_{r,s=1}^{N} a_{jk}^{rs}(x)\zeta_r^j\overline{\zeta}_s^k \ , \qquad (x,\zeta)\in\overline{\Omega}\times\mathbb{K}^{n\times N} \ ,$$

where $a_{jk} = [a_{jk}^{rs}]_{1\le r,s\le N}$. Then $\mathcal{A}$ satisfies a **uniform Legendre condition** on Ω if

$$\operatorname{Re} a_{\mathcal{L}}(x,\zeta) > 0 \ , \qquad (x,\zeta)\in\overline{\Omega}\times\left(\mathbb{K}^{n\times N}\backslash\{0\}\right) \ .$$

It is not difficult to see that

$\mathcal{A}$ satisfies a uniform Legendre condition
$\Longrightarrow$ $\mathcal{A}$ is uniformly strongly elliptic.

The converse is true iff $N=1$.

It should be noted that the Legendre form $a_{\mathcal{L}}$ is uniquely determined by the matrix coefficients of a_{jk} if the principal part $\mathcal{A}_\pi$ of $\mathcal{A}$ is given the representation $-a_{jk}\partial_j\partial_k$. However, there may be other representations of $\mathcal{A}_\pi$ which generate different Legendre forms. Thus, unlike the principal symbol a_π of $\mathcal{A}$, which is invariantly associated with $\mathcal{A}_\pi$, the Legendre form $a_{\mathcal{L}}$ of $\mathcal{A}$ is not an invariant of $\mathcal{A}_\pi$. (The author is grateful to P. Quittner for pointing this out.)

(c) If $\mathcal{A}$ is uniformly strongly elliptic and $\delta=0$, that is, if $\mathcal{B}$ is the Dirichlet boundary operator γ_∂ then $(\mathcal{A},\mathcal{B})$ is normally elliptic.

(d) If $\mathcal{A}$ satisfies a uniform Legendre condition, $(\mathcal{A},\mathcal{B})$ is normally elliptic.

(e) Suppose that $(\mathcal{A},\mathcal{B})\in\mathbb{E}(\Omega)$ is of **separated divergence form**, that is, it satisfies

$$a_{jk} = a\alpha_{jk} \ , \qquad 1\le j,k\le n \ ,$$

where

$$a\in C(\overline{\Omega},X) \ , \quad \alpha := [\alpha_{jk}]_{1\le j,k\le n}\in C\left(\overline{\Omega},\mathcal{L}(\mathbb{R}^n)\right)$$

and α is symmetric and uniformly positive definite. Then $(\mathcal{A},\mathcal{B})$ is normally elliptic provided

$$\sigma\big(a(x)\big)\subset[\operatorname{Re} z>0] \ , \qquad x\in\overline{\Omega} \ ,$$

and

$$\big(1-\delta(y)\big)a(y)\delta(y) = 0 \ , \qquad y\in\partial\Omega \ . \tag{4.11}$$

Observe that (4.11) is automatically satisfied for

$$\delta\in C\big(\partial\Omega,\{0,1_{\mathbb{K}^N}\}\big) \ ,$$

that is, if on each component Γ of $\partial\Omega$ the boundary operator $\mathcal{B}|\Gamma$ is a Dirichlet boundary operator, i.e., $\mathcal{B}|\Gamma=\gamma_\partial|\Gamma$, or a Neumann type boundary operator of the form

$$\mathcal{B}|\Gamma = (a\partial_{\nu_\alpha} + \nu^j a_j + c)|\Gamma \ ,$$

where ∂_{ν_α} is the derivative with respect to the conormal $\alpha\vec{\nu}$.

(f) Suppose that $\mathcal{A}$ is normally elliptic and $\delta = 0$, that is, $\mathcal{B} = \gamma_\partial$, the Dirichlet boundary operator. Then $(\mathcal{A}, \mathcal{B})$ is not necessarily normally elliptic (cf. [LadSU68, p. 625]). If $\mathcal{A}$ is uniformly strongly elliptic and $\delta = 1$ (Neumann type boundary conditions), it does not seem to be known if $(\mathcal{A}, \mathcal{B})$ is normally elliptic. □

The last example illustrates the difficulties in verifying that $(\mathcal{A}, \mathcal{B}) \in \mathbb{E}(\Omega)$ belongs to $\mathcal{E}(\Omega)$ in the general case. Note, however, that the reaction-diffusion systems of Section 1 are of separated divergence form with $\alpha_{jk} = \delta_{jk}$ (the Kronecker symbol). Thus, fortunately enough, in this important case these difficulties are drastically reduced. In particular, it follows from Example (e) that a boundary value problem $(\mathcal{A}, \mathcal{B})$ of separated divergence form is normally elliptic iff $\mathcal{A}$ is normally elliptic, that is, $\partial_t + \mathcal{A}$ is Petrowskii parabolic, provided $\delta \in C(\partial\Omega, \{0, 1_{\mathbb{K}^N}\})$.

5 Generalized Sobolev Spaces

In the following we shall make essential use of Sobolev spaces of noninteger order. There are several possible extensions of the classical Sobolev spaces W_p^k, $k \in \mathbb{N}$, to intermediate spaces of noninteger order. Among them the Slobodeckii spaces constitute certainly the best-known class — at least in the theory of partial differential equations. This is mainly due to the fact that they occur naturally as trace spaces of the classical Sobolev spaces and to the fact that they can be characterized by a relatively simple intrinsic norm. Namely, putting

$$[u]_{s,p} := \Big(\int_{\Omega\times\Omega} \frac{|u(x) - u(y)|^p}{|x - y|^{n+sp}}\, d(x,y) \Big)^{1/p}, \qquad 1 \le p < \infty, \quad 0 < s < 1,$$

and denoting by

$$\|\cdot\|_{k,p} := \Big(\sum_{|\alpha| \le k} \|\partial^\alpha u\|_p^p \Big)^{1/p}, \qquad 1 \le p < \infty, \quad k \in \mathbb{N},$$

the usual Sobolev norms, let

$$\|u\|_{s,p} := \Big(\|u\|_{[s],p}^p + \sum_{|\alpha| = [s]} [\partial^\alpha u]_{s-[s],p}^p \Big)^{1/p}, \qquad 1 \le p < \infty, \quad s \in \mathbb{R}^+ \backslash \mathbb{N},$$

where $[s]$ is the greatest integer smaller than or equal to s. Then the **Slobodeckii spaces** are the Banach spaces defined by

$$W_p^s := W_p^s(\Omega, \mathbb{K}^N) := \big(\{ u \in W_p^{[s]} \,;\, [u]_{s,p} < \infty \}, \|\cdot\|_{s,p} \big), \qquad 1 \le p < \infty, \quad s \in \mathbb{R}^+ \backslash \mathbb{N}.$$

The Slobodeckii spaces are very special cases of the scales of Besov and Triebel-Lizorkin spaces. To introduce the latter spaces we first suppose that $\mathbb{K} = \mathbb{C}$. Then we choose a smooth function φ_0 on $\mathbb{R}^n$ vanishing identically for $|x| > 2$ and being equal to 1 for $|x| \le 1$. We also put

$$\varphi_1(x) := \varphi_0(x/2) - \varphi_0(x), \quad \varphi_k(x) := \varphi_1(2^{-k}x), \qquad x \in \mathbb{R}^n, \quad k = 2, 3, \ldots.$$

Then the sequence (φ_j) is a smooth 'dyadic' partition of unity on $\mathbb{R}^n$, that is,

$$\sum_{k=0}^{\infty} \varphi_k(x) = 1 , \qquad x \in \mathbb{R}^n ,$$

and

$$\operatorname{supp}(\varphi_k) \subset \{ x \in \mathbb{R}^n ; 2^{k-1} < |x| < 2^{k+1} \} , \qquad k \in \dot{\mathbb{N}} := \mathbb{N}\backslash\{0\} .$$

We denote by $\mathcal{F}$ the Fourier transform on $\mathcal{S}'(\mathbb{R}^n, \mathbb{C}^N)$, the Schwartz space of tempered $\mathbb{C}^N$-valued distributions, and we put

$$\varphi_k(D) := \mathcal{F}^{-1}\varphi_k\mathcal{F} , \qquad k \in \mathbb{N} .$$

Given a Banach space E, we use the notations

$$\ell_p(E) := \ell_p(\mathbb{N}, E) \quad \text{and} \quad L_p(E) := L_p(\mathbb{R}^n, E) , \qquad 1 \le p \le \infty ,$$

in the following definitions.

We put

$$\|u\|_{B^s_{p,q}} := \left\| \left(2^{sk}\varphi_k(D)u\right)_{k\in\mathbb{N}} \right\|_{\ell_q(L_p)} , \qquad 1 \le p, q \le \infty , \quad s \in \mathbb{R} ,$$

and

$$\|u\|_{F^s_{p,q}} := \left\| \left(2^{sk}\varphi_k(D)u\right)_{k\in\mathbb{N}} \right\|_{L_p(\ell_q)} , \qquad 1 \le p, q < \infty , \quad s \in \mathbb{R} .$$

Then the **Besov spaces**, $B^s_{p,q}(\mathbb{R}^n, \mathbb{C}^N)$, are defined by

$$B^s_{p,q}(\mathbb{R}^n, \mathbb{C}^N) := \left(\{ u \in \mathcal{S}'(\mathbb{R}^n, \mathbb{C}^N) ; \|u\|_{B^s_{p,q}} < \infty \}, \|\cdot\|_{B^s_{p,q}}\right)$$

for $1 \le p, q \le \infty$ and $s \in \mathbb{R}$. Similarly, the **Triebel-Lizorkin spaces**, $F^s_{p,q}(\mathbb{R}^n, \mathbb{C}^N)$, are given by

$$F^s_{p,q}(\mathbb{R}^n, \mathbb{C}^N) := \left(\{ u \in \mathcal{S}'(\mathbb{R}^n, \mathbb{C}^N) ; \|u\|_{F^s_{p,q}} < \infty \}, \|\cdot\|_{F^s_{p,q}}\right)$$

for $1 \le p, q < \infty$ and $s \in \mathbb{R}$. Lastly, we put

$$\|u\|_{H^s_p} := \left\|(1-\Delta)^{s/2}u\right\|_p , \qquad 1 < p < \infty , \quad s \in \mathbb{R} ,$$

where $(1-\Delta)^{s/2} := \mathcal{F}^{-1}(1+|\xi|^2)^{s/2}\mathcal{F}$. Then the **Bessel potential spaces**, $H^s_p(\mathbb{R}^n, \mathbb{C}^N)$, are defined by

$$H^s_p(\mathbb{R}^n, \mathbb{C}^N) := \left(\{ u \in \mathcal{S}'(\mathbb{R}^n, \mathbb{C}^N) ; \|u\|_{H^s_p} < \infty \}, \|\cdot\|_{H^s_p}\right)$$

for $1 < p < \infty$ and $s \in \mathbb{R}$.

Observe that the usual definition of $\mathcal{S}'(\mathbb{R}^n, \mathbb{C}^N)$ applies equally well to the real case to give the Schwartz space, $\mathcal{S}'(\mathbb{R}^n, \mathbb{R}^N)$, of $\mathbb{R}^N$-valued distributions on $\mathbb{R}^n$. Moreover, $\mathcal{S}'(\mathbb{R}^n, \mathbb{R}^N)$ is a *real* closed linear subspace of $\mathcal{S}'(\mathbb{R}^n, \mathbb{C}^N)$. Hence the following definitions of the scales of real Besov, Triebel-Lizorkin, and Bessel potential spaces,

$$S(\mathbb{R}^n, \mathbb{R}^N) := S(\mathbb{R}^n, \mathbb{C}^N) \cap \mathcal{S}'(\mathbb{R}^n, \mathbb{R}^N) ,$$

where $S := B^s_{p,q}$, or $S := F^s_{p,q}$, or $S := H^s_p$, respectively, are meaningful.

Lastly, we denote by $\mathcal{D} := \mathcal{D}(\Omega, \mathbb{K}^N)$ and $\mathcal{D}' := \mathcal{D}'(\Omega, \mathbb{K}^N)$ the space of $\mathbb{K}^N$-valued test functions and distributions on Ω, respectively, and by

$$r_\Omega \in \mathcal{L}(\mathcal{S}'(\mathbb{R}^n, \mathbb{K}^N), \mathcal{D}')$$

the 'restriction operator for Ω', given by

$$\langle r_\Omega u, \varphi \rangle := \langle u, \varphi \rangle \ , \qquad u \in \mathcal{S}'(\mathbb{R}^n, \mathbb{K}^N) \ , \quad \varphi \in \mathcal{D} \ ,$$

where $\langle \cdot, \cdot \rangle$ denote the obvious duality pairings. Then we define the **Besov, Triebel-Lizorkin, and Bessel potential spaces over Ω** by restriction, that is,

$$S := S(\Omega, \mathbb{K}^N) := r_\Omega S(\mathbb{R}^n, \mathbb{K}^N) \ ,$$

where $S := B^s_{p,q}$, or $S := F^s_{p,q}$, or $S := H^s_p$, respectively, endowed with the obvious quotient topologies.

The spaces introduced above are all Banach spaces (except for $\mathcal{S}'$, $\mathcal{D}$, and $\mathcal{D}'$, of course) and are independent of the function φ_0, except for equivalent norms. Many of the classical function spaces are contained as special cases in the above scales. In particular,

$$B^s_{\infty,\infty} \doteq C^s \ , \qquad s \in \mathbb{R}^+ \backslash \mathbb{N} \ , \tag{5.1}$$

where $C^s := C^s(\overline{\Omega}, \mathbb{K}^N)$,

$$H^k_p \doteq W^k_p \ , \qquad 1 < p < \infty \ , \quad k \in \mathbb{N} \ , \tag{5.2}$$

and

$$B^s_{p,p} \doteq F^s_{p,p} \doteq W^s_p \ , \qquad 1 \le p < \infty \ , \quad s \in \mathbb{R}^+ \backslash \mathbb{N} \ . \tag{5.3}$$

Moreover,

$$F^s_{p,2} \doteq H^s_p \ , \qquad 1 < p < \infty \ , \quad s \in \mathbb{R} \ . \tag{5.4}$$

In the following, we collect some important properties of these spaces which will be needed — implicitly or explicitly — in later sections. For the benefit of the readers more familiar with the theory of partial differential equations than with function spaces, we shall later on restrict our considerations to the Sobolev-Slobodeckii spaces. However, for reasons which will become clear below, for the development of the theory we need both, the scales of Besov and of Bessel potential spaces. Hence we review the needed results for these scales only and leave aside the more general scale of Triebel-Lizorkin spaces.

It is known that, letting $C^k := C^k(\overline{\Omega}, \mathbb{K}^N)$ for $k \in \mathbb{N} \cup \{\infty\}$,

$$C^\infty \overset{d}{\hookrightarrow} B^s_{p,1} \overset{d}{\hookrightarrow} B^s_{p,q} \hookrightarrow B^s_{p,\infty} \overset{d}{\hookrightarrow} B^t_{p,1} \overset{d}{\hookrightarrow} \mathcal{D}' \tag{5.5}$$

for $1 \le p, q < \infty$ and $-\infty < t < s < \infty$, and that

$$B^s_{p,1} \overset{d}{\hookrightarrow} H^s_p \hookrightarrow B^s_{p,\infty} \ , \qquad 1 < p < \infty \ , \quad s \in \mathbb{R} \ . \tag{5.6}$$

If $s_0, s_1 \in \mathbb{R}$ and $p_0, p_1 \in [1, \infty]$ satisfy

$$s_1 - n/p_1 \ge s_0 - n/p_0 \ , \tag{5.7}$$

it follows that

$$B^{s_1}_{p_1,q} \hookrightarrow B^{s_0}_{p_0,q} \,, \qquad 1/p_1 \geq 1/p_0 \,, \quad 1 \leq q \leq \infty \,, \tag{5.8}$$

and

$$H^{s_1}_{p_1} \hookrightarrow H^{s_0}_{p_0} \,, \qquad 1 > 1/p_1 \geq 1/p_0 > 0 \,. \tag{5.9}$$

Moreover,

$$\partial^\alpha \in \mathcal{L}(B^s_{p,q}, B^{s-|\alpha|}_{p,q}) \,, \qquad 1 \leq p,q \leq \infty \,, \quad s \in \mathbb{R} \,, \quad \alpha \in \mathbb{N}^n \,, \tag{5.10}$$

and

$$\partial^\alpha \in \mathcal{L}(H^s_p, H^{s-|\alpha|}_p) \,, \qquad 1 < p < \infty \,, \quad s \in \mathbb{R} \,, \quad \alpha \in \mathbb{N}^n \,. \tag{5.11}$$

It is an important fact that

$$\mathcal{D} \overset{d}{\hookrightarrow} B^s_{p,q} \,, \qquad -\infty < s < 1/p \,, \quad 1 \leq p,q < \infty \,, \tag{5.12}$$

and

$$\mathcal{D} \overset{d}{\hookrightarrow} H^s_p \,, \qquad -\infty < s < 1/p \,, \quad 1 < p < \infty \,. \tag{5.13}$$

As for duality properties,

$$(B^s_{p,q})' \doteq B^{-s}_{p',q'} \,, \qquad -1 + 1/p < s < 1/p \,, \quad 1 < p,q < \infty \,, \tag{5.14}$$

and

$$(H^s_p)' \doteq H^{-s}_{p'} \,, \qquad -1 + 1/p < s < 1/p \,, \quad 1 < p < \infty \,, \tag{5.15}$$

with respect to the duality pairing naturally induced by $\langle \cdot, \cdot \rangle$, that is, with respect to the standard duality pairing between $\mathcal{D}$ and $\mathcal{D}'$.

By means of local coordinates it is not difficult to define Besov spaces, $B^s_{p,q}(\partial\Omega, \mathbb{K}^N)$, on $\partial\Omega$. It is clear that the imbeddings (5.5) and (5.8) carry over to these spaces provided, of course, n is replaced by $n-1$. (If $n = 1$, the boundary of Ω is a two-point set and all spaces on $\partial\Omega$ are isomorphic to $\mathbb{K}^N \times \mathbb{K}^N$.) As for duality,

$$[B^s_{p,q}(\partial\Omega, \mathbb{K}^N)]' \doteq B^{-s}_{p',q'}(\partial\Omega, \mathbb{K}^N) \,, \qquad 1 < p,q < \infty \,, \quad s \in \mathbb{R} \,, \tag{5.16}$$

with respect to the duality pairing induced by $\langle \cdot, \cdot \rangle_\partial$.

Let E and F be Banach spaces. Then R is said to be a **retraction** from E onto F if $R \in \mathcal{L}(E,F)$ and there exists a right inverse $R^c \in \mathcal{L}(F,E)$ for R, a **coretraction**.

The trace theorem asserts that the trace operator, γ_∂, is a retraction

$$\gamma_\partial : B^s_{p,q} \to B^{s-1/p}_{p,q}(\partial\Omega, \mathbb{K}^N) \,, \qquad 1 \leq p,q < \infty \,, \quad s > 1/p \,, \tag{5.17}$$

and a retraction

$$\gamma_\partial : H^s_p \to B^{s-1/p}_{p,p}(\partial\Omega, \mathbb{K}^N) \,, \qquad 1 < p < \infty \,, \quad s > 1/p \,. \tag{5.18}$$

For proofs of the facts listed above we refer to [Tri78], [Tri83], and [Tri92], where the reader finds many useful additional facts and historical remarks.

In the following, we denote by $[\cdot,\cdot]_\theta$, $0 < \theta < 1$, the complex, and by $(\cdot,\cdot)_{\theta,q}$, $0 < \theta < 1$, $1 \leq q \leq \infty$, the standard real interpolation functors. We refer to [BeL76] and [Tri78], for example, for the general theory of interpolation spaces and assume that the reader has a working knowledge of that theory.

Let (E_0, E_1) be a real interpolation couple, that is, E_0 and E_1 are real Banach spaces and there exists a locally convex space F such that $E_j \hookrightarrow F$ for $j = 0, 1$. Then we define the 'complex interpolation spaces $[E_0, E_1]_\theta$ in the real setting' by

$$[E_0, E_1]_\theta := \big[(E_0)_{\mathbb{C}}, (E_1)_{\mathbb{C}}\big]_\theta \cap (E_0 + E_1) , \qquad 0 < \theta < 1 ,$$

equipped with the topology induced by $\big[(E_0)_{\mathbb{C}}, (E_1)_{\mathbb{C}}\big]_\theta$. Of course, $(E_j)_{\mathbb{C}}$ denotes the complexification of E_j. Thus $[E_0, E_1]_\theta$ is simply the 'real subspace' of $\big[(E_0)_{\mathbb{C}}, (E_1)_{\mathbb{C}}\big]_\theta$. It is not difficult to verify that $[\cdot, \cdot]_\theta$ is an exact interpolation functor of exponent θ in the category of real interpolation couples. Of course, it inherits the properties of the standard complex interpolation functor, namely the reiteration, commutation, and duality properties. These 'complex interpolation functors' are used throughout if $[\cdot, \cdot]_\theta$ appears in connection with real Banach spaces.

5.1 Remark Suppose that $(F, \|\cdot\|_F)$ is a complex Banach space of $\mathbb{C}^N$-valued distributions on Ω and $(E, \|\cdot\|)$ is the real subspace of $\mathbb{R}^N$-valued distributions in F. Then $E_{\mathbb{C}} \doteq F$, but easy examples show that $\|\cdot\|_{E_{\mathbb{C}}} \neq \|\cdot\|_F$, in general. □

We collect now some important interpolation properties of Besov and Bessel potential spaces. For proofs we refer again to Triebel's books cited above (also cf. [BeL76]).

Suppose that $-\infty < s_0 < s_1 < \infty$ and $0 < \theta < 1$, and put $s_\theta := (1-\theta)s_0 + \theta s_1$. Then

$$[H_p^{s_0}, H_p^{s_1}]_\theta \doteq H_p^{s_\theta} , \qquad 1 < p < \infty , \tag{5.19}$$

and

$$[B_{p,q}^{s_0}, B_{p,q}^{s_1}]_\theta \doteq B_{p,q}^{s_\theta} , \qquad 1 \le p, q < \infty . \tag{5.20}$$

As for the real interpolation functors,

$$(H_p^{s_0}, H_p^{s_1})_{\theta,q} \doteq B_{p,q}^{s_\theta} , \qquad 1 < p < \infty , \quad 1 \le q \le \infty , \tag{5.21}$$

and

$$(B_{p,q_0}^{s_0}, B_{p,q_1}^{s_1})_{\theta,q} \doteq B_{p,q}^{s_\theta} , \qquad 1 \le p, q, q_0, q_1 \le \infty . \tag{5.22}$$

Suppose now that $(\mathcal{A}, \mathcal{B}) \in \mathbb{E}^1(\Omega, \delta, N)$ and $1 < p < \infty$, and let A be the L_p-realization of $(\mathcal{A}, \mathcal{B})$. It is an easy consequence of Theorem 4.1 that

$$D(A) \doteq W_{p,\mathcal{B}}^2$$

if $(\mathcal{A}, \mathcal{B}) \in \mathcal{E}^1(\Omega)$. In the following, we characterize the interpolation spaces

$$\big(L_p, D(A)\big)_\theta \doteq (L_p, W_{p,\mathcal{B}}^2)_\theta$$

for most values of $\theta \in (0,1)$, where $(\cdot,\cdot)_\theta$ denotes either the complex interpolation functor or any one of the real ones. For this we introduce the Banach spaces $H_{p,\mathcal{B}}^s$ and $B_{p,q,\mathcal{B}}^s$ as follows:

Suppose that $(\mathcal{A},\mathcal{B})\in\mathbb{E}(\Omega)$ and $S_p^s\in\{H_p^s, B_{p,q}^s\ ;\ 1\le q\le\infty\}$. Then

$$S_{p,\mathcal{B}}^s:=\begin{cases}\{u\in S_p^s\ ;\ \mathcal{B}u=0\}\ , & 1+1/p<s\le 2\ ,\\ \{u\in S_p^s\ ;\ (1-\delta)\gamma_\partial u=0\}\ , & 1/p<s<1+1/p\ ,\\ S_p^s\ , & 0\le s<1/p\ .\end{cases}\tag{5.23}$$

If $\delta=0$ (Dirichlet boundary operator: $\mathcal{B}=\gamma_\partial$), we extend this definition to include the value $s=1+1/p$ by putting

$$S_{p,\mathcal{B}}^s:=\{u\in S_p^s\ ;\ \gamma_\partial u=0\}\ ,\qquad 1/p<s\le 2\ .\tag{5.24}$$

Similarly, if $\delta=1$ (Neumann type boundary operator), we include the value $s=1/p$ by putting

$$S_{p,\mathcal{B}}^s:=S_p^s\ ,\qquad 0\le s<1+1/p\ .\tag{5.25}$$

To have a concise notation we define the set Σ_p^+ of 'singular positive values' by

$$\Sigma_p^+:=\begin{cases}\{1/p\} & \text{if } \delta=0\ ,\\ \{1+1/p\} & \text{if } \delta=1\ ,\\ \{1/p,1+1/p\} & \text{otherwise.}\end{cases}\tag{5.26}$$

Using these notations we have the following fundamental

5.2 Theorem *Suppose that $(\mathcal{A},\mathcal{B})\in\mathcal{E}^1(\Omega,\delta,N)$ and $1<p<\infty$. Then*

$$[L_p, W_{p,\mathcal{B}}^2]_\theta\doteq H_{p,\mathcal{B}}^{2\theta}$$

and

$$(L_p, W_{p,\mathcal{B}}^2)_{\theta,q}\doteq B_{p,q,\mathcal{B}}^{2\theta}\ ,\qquad 1\le q\le\infty\ ,$$

for $2\theta\in(0,2)\backslash\Sigma_p^+$.

The proof of this theorem is obtained by obvious modifications of the techniques of Grisvard [Gri69] and Seeley [See72] (cf. [Ama94]). Also cf. [Gui91b] for related results.

5.3 Remarks

(a) The space $S_{p,\mathcal{B}}^s$ is a closed linear subspace of S_p^s for $s\in[0,2]\backslash\Sigma_p^+$. If $0\le s<1+1/p$, it is independent of $(\mathcal{A},\mathcal{B})$. It depends only on δ if $1/p<s<1+1/p$.

(b) Given an interpolation couple (E_0,E_1), we define the continuous interpolation space $(E_0,E_1)_{\theta,\infty}^0$ to be the closure of $E_0\cap E_1$ in $(E_0,E_1)_{\theta,\infty}$ for $0<\theta<1$. By defining 'interpolated morphisms' in an obvious way by restriction, one obtains an exact interpolation functor of exponent θ, the 'continuous interpolation functor' $(\cdot,\cdot)_{\theta,\infty}^0$. It can be characterized by the 'continuous trace method' introduced by Da Prato and Grisvard [DaPG79] as has been shown in [DoF87]. This explains its name.

If $1 \leq p < \infty$ and $0 < s < \infty$ then $B^s_{p,\infty} \doteq N^s_p$, where $N^s_p := N^s_p(\Omega, \mathbb{K}^N)$ denote the **Nikolskii spaces** (e.g., [KuJF77]). For this reason, the spaces

$$n^s_p := n^s_p(\Omega, \mathbb{K}^N) := \text{closure of } C^\infty \text{ in } N^s_p \ , \qquad 1 \leq p < \infty \ , \quad s > 0 \ ,$$

are the **little Nikolskii spaces**. Similarly, the spaces

$$c^s := c^s(\overline{\Omega}, \mathbb{K}^N) := \text{closure of } C^\infty \text{ in } C^s$$

are the **little Hölder spaces**. It follows from (5.2), (5.5), (5.6), and (5.21) that

$$(L_p, W^k_p)^0_{\theta,\infty} \doteq n^{\theta k}_p \ , \qquad 1 < p < \infty \ , \quad k \in \dot{\mathbb{N}} \ , \quad 0 < \theta < 1 \ .$$

Consequently,

$$(L_p, W^2_{p,\mathcal{B}})^0_{\theta,\infty} \doteq n^{2\theta}_{p,\mathcal{B}} \ , \qquad 1 < p < \infty \ , \quad 2\theta \in (0,2)\backslash\Sigma^+_p \ .$$

(c) If one is interested in generalized Sobolev spaces of order s with $|s| \leq 2$ only, as is the case in connection with second order differential equations, everything said above remains valid if it is only assumed that Ω is of class C^2. More generally, if $\overline{\Omega}$ is a compact n-dimensional C^2–submanifold of $\mathbb{R}^n$, all the results (5.1)–(5.22) above remain valid if the orders of all spaces, that is, all the superscripts, are additionally restricted to belong to $[-2, 2]$. □

In the study of nonlinear problems it is often crucial to work in the category of real Banach spaces. This is, among other things, mainly due to the fact that — in the complex case — a map is (Fréchet) differentiable iff it is analytic. Thus differentiability is in the complex setting a condition which is by far too restrictive for most applications. This is the main reason for formulating all results in this paper for the general case $\mathbb{K} = \mathbb{R}$ or $\mathbb{C}$.

6 Abstract Interpolation-Extrapolation Scales

Let E_0 and E_1 be Banach spaces with $E_1 \overset{d}{\hookrightarrow} E_0$. Then (E_0, E_1) is said to be a **densely injected Banach couple**. By an **admissible interpolation functor** $(\cdot,\cdot)_\theta$, $0 < \theta < 1$, we mean an interpolation functor of exponent θ for the category of densely injected Banach couples (whose morphisms are the bounded linear operators, of course) such that

$$E_1 \overset{d}{\hookrightarrow} (E_0, E_1)_\theta$$

whenever (E_0, E_1) is such a couple. Note that each one of the following standard interpolation functors is admissible:

$$\begin{aligned} &\text{the complex interpolation functor } [\cdot,\cdot]_\theta \ , \\ &\text{the real interpolation functors } (\cdot,\cdot)_{\theta,q} \ , \qquad 1 \leq q < \infty \ , \\ &\text{the continuous interpolation functor } (\cdot,\cdot)^0_{\theta,\infty} \ . \end{aligned} \tag{6.1}$$

Let (E_0, E_1) be a densely injected Banach couple and denote by $\|\cdot\|_j$ the norm in E_j. Also assume that

$$A_0 \in \mathcal{H}(E_1, E_0) \ . \tag{6.2}$$

Then there exists $\omega \geq 0$ such that

$$\sigma(\omega + A_0) \subset [\mathrm{Re}\, z > 0] \ . \tag{6.3}$$

It is a consequence of (6.2) that $E_1 \doteq D(A_0)$. Thus it follows from (6.3) that

$$E_1 \doteq \big(\mathrm{dom}(A_0), \|(\omega + A_0)\cdot\|_0\big) \ .$$

Let A_1 be the E_1-realization of A_0 and let

$$E_2 := \big(\mathrm{dom}(A_1), \|(\omega + A_1)\cdot\|_1\big) \ .$$

Then (E_1, E_2) is a densely injected Banach couple and

$$A_1 \in \mathcal{H}(E_2, E_1)$$

with $\sigma(A_1) = \sigma(A_0)$.

We define now inductively a discrete scale of Banach spaces and operators $(E_k, A_k)_{k\in\mathbb{N}}$ by

$$E_{k+1} := (E_{k+1}, \|\cdot\|_{k+1}) := \big(\mathrm{dom}(A_k), \|(\omega + A_k)\cdot\|_k\big)$$

and

$$A_{k+1} := E_{k+1}\text{-realization of } A_k$$

for $k \in \dot{\mathbb{N}}$. Then

$$E_k \doteq D(A_0^k) \ , \qquad k \in \mathbb{N} \ ,$$

(E_k, E_{k+1}) is a densely injected Banach couple, and

$$A_k \in \mathcal{H}(E_{k+1}, E_k)$$

with $\sigma(A_k) = \sigma(A_0)$ for $k \in \mathbb{N}$.

On the other hand, we can recover A_0 and E_0 from A_1 and E_1 since $\big(E_1, \|(\omega + A_1)^{-1}\cdot\|_1\big)$ is a dense linear subspace of E_0 whose norm is equivalent to $\|\cdot\|_0$. Thus E_0 is a completion of the normed vector space $\big(E_1, \|(\omega + A_1)^{-1}\cdot\|_1\big)$ and $A_0 \in \mathcal{L}(E_1, E_0)$ is the continuous extension of $A_1 \in \mathcal{L}(E_2, E_1)$, or, considered as a linear operator in E_0, A_0 is the closure of A_1 in E_0. This observation suggests the following definition of a superspace E_{-1} of E_0 and an extension A_{-1} of A_0.

We put

$$E_{-1} := \big(E_0, \|(\omega + A_0)^{-1}\cdot\|\big)^{\sim} \ ,$$

where $^{\sim}$ denotes completion. Then

$$E_0 \overset{d}{\hookrightarrow} E_{-1} \ ,$$

that is, (E_{-1}, E_0) is a densely injected Banach couple. It follows that the closure A_{-1} of A_0 in E_{-1} is well defined and

$$A_{-1} \in \mathcal{H}(E_0, E_{-1})$$

with $\sigma(A_{-1}) = \sigma(A_0)$.

We fix now for each $\theta \in (0,1)$ an admissible interpolation functor $(\cdot,\cdot)_\theta$ arbitrarily and put

$$\mathfrak{F} := \big\{ (\cdot,\cdot)_\theta \ ; \ 0 < \theta < 1 \big\} \ .$$

Then we define

$$E_{k+\theta} := (E_{k+\theta}, \|\cdot\|_{k+\theta}) := (E_k, E_{k+1})_\theta$$

and

$$A_{k+\theta} := E_{k+\theta}\text{-realization of } A_k$$

for $k \in \mathbb{N} \cup \{-1\}$ and call

$$\{ (E_\alpha, A_\alpha) \ ; \ -1 \leq \alpha < \infty \} \tag{6.4}$$

the **interpolation-extrapolation scale (over $[-1, \infty)$) generated by A_0 and $\mathfrak{F}$.**

It follows that $(E_\alpha, E_{\alpha+1})$ is a densely injected Banach couple,

$$A_\alpha \in \mathcal{H}(E_{\alpha+1}, E_\alpha) \ , \quad \sigma(A_\alpha) = \sigma(A_0) \ , \qquad -1 \leq \alpha < \infty \ ,$$

and, given $-1 \leq \beta < \alpha < \infty$, the following diagram is commutative

$$\begin{array}{ccccc} E_{\alpha+1} & \xrightarrow{\ A_\alpha\ } & E_\alpha & \xrightarrow{\ e^{-tA_\alpha}\ } & E_\alpha \\ \big\downarrow d & & \big\downarrow d & & \big\downarrow d \\ E_{\beta+1} & \xrightarrow{\ A_\beta\ } & E_\beta & \xrightarrow{\ e^{-tA_\beta}\ } & E_\beta \end{array} \tag{6.5}$$

for $t \geq 0$. Observe that (6.5) shows that the operators A_α and A_β, as well as the corresponding semigroups $\{ e^{-tA_\alpha} \ ; \ t \geq 0 \}$ and $\{ e^{-tA_\beta} \ ; \ t \geq 0 \}$, respectively, are obtained from each other by natural restrictions and extensions, respectively.

Once the 'extrapolation space' E_{-1} and the extension A_{-1} of A_0 have been constructed, everything above follows easily from the basic facts about analytic semigroups and interpolation theory (cf. [Ama86d] and, in particular, [Ama94]). The above construction of E_{-1} and A_{-1} occurs perhaps for the first time in [Nag83] (also cf. [Wal86]). Without being aware of this it has been reintroduced in [Ama87] (also cf. [Ama86d]) and, again independently, in [Liu89]. In [DaPG84] there has been given a different, rather complicated construction of an extrapolation space $\widetilde{E}_{-1}$ and an extension $\widetilde{A}_{-1}$ of A_0. In [Ama88c, Remark 6.2] it has been shown that the Da Prato-Grisvard extrapolation pair $(\widetilde{E}_{-1}, \widetilde{A}_{-1})$ is equivalent to (E_{-1}, A_{-1}) (see also a corresponding statement in [DaP84]).

In [Nag83] and [Wal86] discrete scales $\{ (E_k, A_k) \ ; \ k \in \mathbb{Z}, \ k \geq -1 \}$ are considered only. In [Liu89] there is studied the 'scale of fractional power spaces' $E_\alpha := D(A_0^\alpha)$ for $\alpha \geq -1$, as is the case in [Ama87]. In addition, in [Liu89] the author also introduces at each point $\alpha \in [-1, \infty)$ inductive limit spaces $\bigcup_{\beta > \alpha} E_\beta$ and projective limit spaces $\bigcap_{\beta < \alpha} E_\beta$. (It should be mentioned that in all these papers the scales are not restricted below; see, however, Remark 6.1(a) below.) Lastly, in [DaPG84] the spaces E_α, $-1 < \alpha < 0$, have been defined by means of the continuous interpolation functors $(\cdot, \cdot)^0_{\theta,\infty}$, $0 < \theta < 1$.

Lastly, it should be noted that recently an extrapolation theory has been developed by B. Jawerth and M. Milman (cf. [JaM91]) which is of an entirely different nature than the one considered here.

6.1 Remarks

(a) Of course, we can replace in the above procedure A_0 by A_{-1} and construct the interpolation-extrapolation scale $\{(F_\alpha, B_\alpha)\ ;\ -1 \le \alpha < \infty\}$ over $[-1,\infty)$ generated by A_{-1} and $\mathfrak{F}$. Then $\{(F_{\alpha+1}, B_{\alpha+1})\ ;\ \alpha \ge -1\}$ is naturally isomorphic to $\{(E_\alpha, A_\alpha)\ ;\ \alpha \ge -1\}$. This means that we can extend the scale (6.4) to a scale over $[-2,\infty)$,

$$\{(E_\alpha, A_\alpha)\ ;\ -2 \le \alpha < \infty\} .$$

By repeating this argument, given any $m \in \mathbb{N}$, we can construct the interpolation–extrapolation scale over $[-m,\infty)$ generated by A_0 and $\mathfrak{F}$,

$$\{(E_\alpha, A_\alpha)\ ;\ -m \le \alpha < \infty\} .$$

Observe, however, that, thanks to the fact that we have to take a completion at each 'extrapolation step', we have to replace that part of the scale which has already been constructed in the previous step by an isomorphic image. Thus, in general, $E_{-\infty} := \bigcup_{k\ge 0} E_{-k}$ is not defined. (I am grateful to P. Koch Medina for pointing out this fact. It has been overlooked in my earlier papers — as well as in [Liu89], [Nag83], and [Wal86].) On the other hand, a 'universal superspace' $E_{-\infty}$ can be constructed by taking the inductive limit of the spaces $\{E_{-k}\ ;\ k \in \mathbb{N}\}$ with respect to the natural injections $j_k : E_{-k} \to E_{-k-1}$ induced by completions. In all of our applications of the interpolation-extrapolation scales only parameter values in $[-1,1]$ are relevant so that a universal superspace is not needed. For this reason we restrict ourselves to scales over $[-1,\infty)$.

(b) It is obvious that for the construction of an interpolation-extrapolation scale it is by no means necessary that A_0 be the negative infinitesimal generator of a semigroup. For more general situations we refer to [Ama94]. Moreover, the different choices of ω lead to equivalent norms on E_α. Thus we do not indicate the ω-dependence.

(c) Suppose that $-1 \le \beta < \alpha < \infty$. It follows from (6.5) that (E_β, E_α) is a densely injected Banach couple. Thus, given any $\theta \in (0,1)$ and any other admissible interpolation functor $((\cdot,\cdot))_\theta$, the interpolation space $((E_\beta, E_\alpha))_\theta$ is well defined. In general, $((E_\beta, E_\alpha))_\theta$ will not coincide with any one of the spaces E_γ, $\beta < \gamma < \alpha$. The following 'sandwich property' shows, however, that $((E_\beta, E_\alpha))_\theta$ can be arbitrarily closely approximated by spaces of the original interpolation-extrapolation scale. Namely,

$$E_\xi \hookrightarrow ((E_\beta, E_\alpha))_\theta \hookrightarrow E_\eta , \qquad \beta < \eta < (1-\theta)\beta + \theta\alpha < \xi < \alpha .$$

This is a consequence of an extension of the 'almost reiteration theorem' [Ama86c, Theorem 8.3] (cf. [Ama94]).

(d) There is an important case in which the scale $\{(E_\alpha, A_\alpha)\ ;\ -1 \le \alpha < \infty\}$ is stable under reiteration. Namely, suppose, in addition to (6.2) and (6.3), that $\omega + A_0$ has bounded imaginary powers and $(\cdot,\cdot)_\theta = [\cdot,\cdot]_\theta$ for $0 < \theta < 1$. Then

$$[E_\beta, E_\alpha]_\theta \doteq E_{(1-\theta)\beta+\theta\alpha} , \qquad -1 \le \alpha < \beta < \infty .$$

Moreover, in this case $E_\alpha \doteq D(A_0^\alpha)$ for $\alpha \in [-1,\infty)$, where $D(A_0^\alpha)$ denotes the completion of E_0 in the norm $\|(\omega + A_0)^\alpha \cdot\|_0$ if $\alpha < 0$.

This has been proven in [Ama87, Theorem 3.3].

(e) The following are equivalent:

(i) A_0 has a compact resolvent;

(ii) $E_\alpha \overset{cc}{\hookrightarrow} E_\beta$ for $-1 \le \beta < \alpha < \infty$;

(iii) There exist $\alpha, \beta \in [-1, \infty)$ with $\beta < \alpha$ such that $E_\alpha \overset{cc}{\hookrightarrow} E_\beta$.

This is an extension of [Ama86d, Theorem 8] for which we refer to [Ama94]. □

Suppose now that $(\mathcal{A}, \mathcal{B}) \in \mathcal{E}^1(\Omega)$, let A_0 be the L_p-realization of $(\mathcal{A}, \mathcal{B})$, where p is fixed in $(1, \infty)$, and put $E_0 := L_p$ and $E_1 := W^2_{p,\mathcal{B}}$. Then $A_0 \in \mathcal{H}(E_1, E_0)$ by Theorem 4.1. Hence we can construct the interpolation-extrapolation scale generated by A_0 and $\mathfrak{F}$. Thus if we restrict $(\cdot,\cdot)_\theta$ to any one of the choices (6.1), it follows from Theorem 5.2 that E_α is — except for equivalent norms — explicitly known, provided $2\alpha \in [0,2] \backslash \Sigma_p^+$. For this one has to observe that Theorem 5.2 implies

$$(L_p, W^2_{p,\mathcal{B}})^0_{\theta,\infty} \doteq n^{2\theta}_{p,\mathcal{B}} , \qquad 2\theta \in (0,2) \backslash \Sigma_p^+ . \tag{6.6}$$

On the other hand, the 'negative' spaces E_α, $-1 \le \alpha < 0$, are defined in a rather abstract way which does not seem to admit an easy concrete description. For applications of the above theory of interpolation-extrapolation spaces it is, of course, most important to possess a 'computable' description of the negative spaces E_α and the 'negative realizations' A_α of A_0, where $-1 \le \alpha < 0$. In the following, we shall show that this can be achieved by duality, provided we restrict the generality slightly.

We suppose now that

$$E_0 \text{ is a reflexive Banach space} \tag{6.7}$$

and

$$(\cdot,\cdot)_\theta \in \{ [\cdot,\cdot]_\theta, (\cdot,\cdot)^0_{\theta,\infty}, (\cdot,\cdot)_{\theta,q} \; ; \; 1 < q < \infty \} , \qquad 0 < \theta < 1 . \tag{6.8}$$

We put

$$(\cdot,\cdot)^\#_\theta := \begin{cases} [\cdot,\cdot]_\theta & \text{if } (\cdot,\cdot)_\theta = [\cdot,\cdot]_\theta , \\ (\cdot,\cdot)_{\theta,q'} & \text{if } (\cdot,\cdot)_\theta = (\cdot,\cdot)_{\theta,q} , \\ (\cdot,\cdot)_{\theta,1} & \text{if } (\cdot,\cdot)_\theta = (\cdot,\cdot)^0_{\theta,\infty} \end{cases} \tag{6.9}$$

for $0 < \theta < 1$, so that $(\cdot,\cdot)^\#_\theta$ is an admissible interpolation functor. We denote by $E^\#_0$ the dual space $(E_0)'$ of E_0 and by $A^\#_0$ the dual $(A_0)'$ of A_0. Moreover,

$$E^\#_1 := D(A^\#_0) .$$

Then it is known that

$$A^\#_0 \in \mathcal{H}(E^\#_1, E^\#_0) \quad \text{and} \quad e^{-tA^\#_0} = (e^{-tA_0})' , \qquad t \ge 0 .$$

Thus the **dual interpolation-extrapolation scale**

$$\{ (E^\#_\alpha, A^\#_\alpha) \; ; \; -1 \le \alpha < \infty \}$$

over $[-1, \infty)$ generated by $A^\#_0$ and $\mathfrak{F}^\# := \{ (\cdot,\cdot)^\#_\theta \; ; \; 0 < \theta < 1 \}$ is well defined.

6.2 Theorem *Let conditions* (6.7)–(6.9) *be satisfied. Then*

$$(E_\alpha)' \doteq E^\#_{-\alpha} \quad \textit{and} \quad (A_\alpha)' = A^\#_{-\alpha} , \qquad -1 \le \alpha \le 1 ,$$

with respect to the duality pairing naturally induced by $\langle \cdot,\cdot \rangle_E$. *If* $(\cdot,\cdot)_\theta \ne (\cdot,\cdot)^0_{\theta,\infty}$ *for* $0 < \theta < 1$ *then* E_α *is reflexive for* $-1 \le \alpha < \infty$.

This theorem has been proven in [Ama86d, Theorem 11], for the case that $(\cdot,\cdot)_\theta \neq (\cdot,\cdot)^0_{\theta,\infty}$ for $0<\theta<1$. The same proof extends to the case where the continuous interpolation functor is admitted too (cf. [Sit92, Theorem 2.3]). An analoguous result was proven earlier in [Ama87, Theorem 1.3] for the scale of fractional power spaces. That result was rediscovered in [Liu89, Theorem I.2.8].

7 Sobolev Interpolation-Extrapolation Scales

We extend now the definition of the Sobolev-Slobodeckii spaces to cover spaces of negative order by putting

$$W_p^{-s} := W_p^{-s}(\Omega,\mathbb{K}^N) := \begin{cases} H_p^{-s}\,, & s\in\mathbb{N}\,,\\ B_{p,p}^{-s}\,, & s\in\mathbb{R}^+\backslash\mathbb{N}\,,\end{cases}$$

for $1<p<\infty$. Henceforth we shall restrict our considerations to the **Sobolev-Slobodeckii and the Bessel potential scales**

$$W_p^s \quad\text{and}\quad H_p^s\,,\qquad s\in\mathbb{R}\,,$$

respectively, which are scales of Banach spaces of distributions.

We define the **formally dual boundary value problem** $(\mathcal{A}^\#,\mathcal{B}^\#)$ of $(\mathcal{A},\mathcal{B})\in\mathbb{E}(\Omega)$ by

$$\mathcal{A}^\# v := -\partial_j(a_{jk}^\#\partial_k v + a_j^\# v) + b_j^\#\partial_j v + a_0^\# v$$

and

$$\mathcal{B}^\# v := \delta\{\nu^j\gamma_\partial(a_{jk}^\#\partial_k v + a_j^\# v) + c^\#\gamma_\partial v\} + (1-\delta)\gamma_\partial v\,,$$

respectively, where

$$a_{jk}^\# := {}^t a_{kj}\,,\quad a_j^\# := {}^t b_j\,,\quad b_j^\# := {}^t a_j\,,\quad a_0^\# := {}^t a\,,\quad c^\# := {}^t c\,.$$

and ${}^t a\in X := \mathcal{L}(\mathbb{K}^N)$ is the transposed of $a\in X$.

Note that, given $(\mathcal{A},\mathcal{B})\in\mathbb{E}(\Omega)$ and $S\in\{H,W\}$, the set consisting of the two letters H and W,

$$\mathcal{B}\in\mathcal{L}(S_p^s, L_p(\partial\Omega,\mathbb{K}^N))\,,\qquad 1<p<\infty\,,\quad 1+1/p<s\leq 2\,,$$

and

$$\mathcal{B}^\#\in\mathcal{L}(S_{p'}^{2-t}, L_{p'}(\partial\Omega,\mathbb{K}^N))\,,\qquad 1<p<\infty\,,\quad 0\leq t<1/p\,.$$

If $\delta=0$ (Dirichlet boundary operator), we can admit $s\in(1/p,2]$ and $t\in[0,1+1/p)$. Hence the spaces

$$S_{p,\mathcal{B}}^s \quad\text{and}\quad S_{p',\mathcal{B}^\#}^{2-s}\,,\qquad p\in(1,\infty)\,,\quad s\in[0,2]\backslash\Sigma_p^+\,,$$

are well defined by (5.23) and closed linear subspaces of S_p^s and $S_{p'}^{2-s}$, respectively. Note that

$$S_{p',\mathcal{B}^\#}^s = S_{p',\mathcal{B}}^s\,,\qquad s\in[0,1+1/p')\backslash\Sigma_{p'}^+\,, \tag{7.1}$$

and that

$$S^s_{p',\mathcal{B}\#} \overset{d}{\hookrightarrow} L_{p'} \ , \qquad p \in (1,\infty) \ , \qquad s \in [0,2]\backslash\Sigma^+_{p'} \ .$$

Hence

$$L_p \overset{d}{\hookrightarrow} (S^s_{p',\mathcal{B}\#})' =: S^{-s}_{p,\mathcal{B}} \ , \qquad p \in (1,\infty) \ , \quad -s \in [-2,0]\backslash(-\Sigma^+_{p'}) \ .$$

with respect to the duality pairing naturally induced by $\langle\cdot,\cdot\rangle$ so that

$$\langle v,u\rangle_{S^{-s}_{p,\mathcal{B}}} = \langle v,u\rangle \ , \qquad v \in S^s_{p',\mathcal{B}\#} \ , \quad u \in L_p \ , \tag{7.2}$$

for $1 < p < \infty$, and $-s \in [-2,0]\backslash(-\Sigma^+_{p'})$. Thus, without fearing confusion, we simply write $\langle\cdot,\cdot\rangle$ for $\langle\cdot,\cdot\rangle_{S^{-s}_{p,\mathcal{B}}}$.

Observe that, thanks to (5.14) and (5.15),

$$S^s_{p,\mathcal{B}} = \begin{cases} \{\, u \in S^s_p \ ; \ \mathcal{B}u = 0 \,\} \ , & 1+1/p < s \le 2 \ , \\ \{\, u \in S^s_p \ ; \ (1-\delta)\gamma_\partial u = 0 \,\} \ , & 1/p < s < 1+1/p \ , \\ S^s_p \ , & -1+1/p < s < 1/p \ , \\ \{\, v \in S^{-s}_{p'} \ ; \ (1-\delta)\gamma_\partial v = 0 \,\}' \ , & -2+1/p < s < -1+1/p \ , \\ \{\, v \in S^{-s}_{p'} \ ; \ \mathcal{B}^\# v = 0 \,\}' \ , & -2 \le s < -2+1/p \ , \end{cases} \tag{7.3}$$

if $\delta \notin \{0,1\}$. If $\delta = 0$ then

$$S^s_{p,\mathcal{B}} = \begin{cases} \{\, u \in S^s_p \ ; \ \gamma_\partial u = 0 \,\} \ , & 1/p < s \le 2 \ , \\ S^s_p \ , & -2+1/p < s < 1/p \ , \quad s \ne -1+1/p \ , \\ \{\, v \in S^{-s}_{p'} \ ; \ \gamma_\partial v = 0 \,\}' \ , & -2 \le s \le -2+1/p \ , \end{cases} \tag{7.4}$$

whereas

$$S^s_{p,\mathcal{B}} = \begin{cases} \{\, u \in S^s_p \ ; \ \mathcal{B}u = 0 \,\} \ , & 1+1/p < s \le 2 \ , \\ S^s_p \ , & -1+1/p < s < 1+1/p \ , \\ (S^{-s}_{p'})' \ , & -2+1/p < s \le -1+1/p \ , \\ \{\, v \in S^{-s}_{p'} \ ; \ \mathcal{B}^\# v = 0 \,\}' \ , & -2 \le s < -2+1/p \ , \end{cases} \tag{7.5}$$

if $\delta = 1$. Moreover,

$$S^s_p = (S^{-s}_{p'})' \ , \qquad -1+1/p < s < 1/p \ . \tag{7.6}$$

It is convenient to put

$$\Sigma_p := \Sigma^+_p \cup (-\Sigma^+_{p'}) = \begin{cases} \{-1+1/p, 1/p\} & \text{if } \delta = 0 \ , \\ \{-2+1/p, 1+1/p\} & \text{if } \delta = 1 \ , \\ (\mathbb{Z}+1/p) \cap (-2,2) & \text{otherwise} \ . \end{cases} \tag{7.7}$$

Note that

$$S^s_{p,\mathcal{B}} \overset{d}{\hookrightarrow} S^t_{p,\mathcal{B}} \ , \qquad -2 \le t < s \le 2 \ , \quad s,t \notin \Sigma_p \ . \tag{7.8}$$

Also note that $S^s_{p,\mathcal{B}}$ is a space of distributions iff $s > -1 + 1/p$ and $\delta \neq 0$, or $s > -2 + 1/p$ and $\delta = 0$, thanks to the fact that the space of test functions is dense in $S^s_{p',\mathcal{B}\#}$ iff $s < 1/p'$ and $\delta \neq 0$, or $s < 1 + 1/p'$ and $\delta = 0$ (cf. [Tri78, Theorems 4.3.2/1 and 4.7.1]).

In the following,

$$H^s_{p,\mathcal{B}} \quad \text{and} \quad W^s_{p,\mathcal{B}} , \qquad p \in (1,\infty) , \quad s \in [-2,2] \backslash \Sigma_p ,$$

are said to be **Bessel potential** and **Sobolev-Slobodeckii scales induced by** $(\mathcal{A},\mathcal{B})$, respectively. Note that the 'center parts'

$$S^s_{p,\mathcal{B}} , \qquad p \in (1,\infty) , \quad s \in (-2 + 1/p, 1 + 1/p) \backslash \Sigma_p ,$$

are, despite of notations and names, independent of $(\mathcal{A},\mathcal{B})$. They are completely determined by the boundary characterization map δ. This fact will be of utmost importance for the study of quasilinear parabolic problems.

The following theorem shows that the spaces $S^s_{p,\mathcal{B}}$ characterize interpolation-extrapolation spaces generated by normally elliptic boundary value problems.

7.1 Theorem *Suppose that* $(\mathcal{A},\mathcal{B}) \in \mathcal{E}^1(\Omega,\delta,N)$ *and* $1 < p < \infty$. *Also suppose that*

$$(\cdot,\cdot)_\alpha \begin{cases} = [\cdot,\cdot]_\alpha & \text{if } \alpha = 1/2 , \\ \in \{ [\cdot,\cdot]_\alpha, (\cdot,\cdot)_{\alpha,p} \} & \text{if } \alpha \in (0,1) \backslash \{1/2\} , \end{cases} \tag{7.9}$$

and put $\mathfrak{F} := \{ (\cdot,\cdot)_\alpha \,;\, 0 < \alpha < 1 \}$. *Let* $\{ (E_\alpha, A_\alpha) \,;\, -1 \le \alpha < \infty \}$ *be the interpolation-extrapolation scale generated by the* L_p*-realization of* $(\mathcal{A},\mathcal{B})$ *and by* $\mathfrak{F}$. *Then*

$$E_\alpha \doteq \begin{cases} H^{2\alpha}_{p,\mathcal{B}} & \text{if } (\cdot,\cdot)_{\theta(\alpha)} = [\cdot,\cdot]_{\theta(\alpha)} \\ W^{2\alpha}_{p,\mathcal{B}} & \text{if } (\cdot,\cdot)_{\theta(\alpha)} = (\cdot,\cdot)_{\theta(\alpha),p} \end{cases}$$

for $2\alpha \in [-2,2] \backslash (\Sigma_p \cup \{0\})$, *where*

$$\theta(\alpha) := \begin{cases} \alpha & \text{if } \ 0 < \alpha < 1 , \\ \alpha + 1 & \text{if } -1 < \alpha < 0 . \end{cases}$$

This follows from Theorems 5.2 and 6.2 and the duality theory of normally elliptic boundary value problems (cf. [Ama90a] and [Ama94]).

Besides of the fact that the 'center parts' of the Bessel potential and Sobolev-Slobodeckii scales induced by $(\mathcal{A},\mathcal{B})$ have the property that they are independent of $(\mathcal{A},\mathcal{B})$, they enjoy the following important interpolation properties.

7.2 Theorem *Suppose that* $1 < p < \infty$ *and* $-2 + 1/p < s_o < s_1 < 1 + 1/p$. *Also suppose that* $0 < \theta < 1$ *and* $s_0, s_1, s_\theta := (1-\theta)s_0 + \theta s_1 \notin \Sigma_p$. *Then*

$$[H^{s_0}_{p,\mathcal{B}}, H^{s_1}_{p,\mathcal{B}}]_\theta \doteq H^{s_\theta}_{p,\mathcal{B}} , \tag{7.10}$$

and

$$(H^{s_0}_{p,\mathcal{B}}, H^{s_1}_{p,\mathcal{B}})_{\theta,p} \doteq W^{s_\theta}_{p,\mathcal{B}} , \qquad s_\theta \notin \mathbb{N} .$$

Moreover,

$$[W^{s_0}_{p,\mathcal{B}}, W^{s_1}_{p,\mathcal{B}}]_\theta \doteq W^{s_\theta}_{p,\mathcal{B}} , \qquad s_\theta \notin \mathbb{N} ,$$

and

$$(W^{s_0}_{p,\mathcal{B}}, W^{s_1}_{p,\mathcal{B}})_{\theta,p} \doteq W^{s_\theta}_{p,\mathcal{B}} , \qquad s_\theta \notin \mathbb{N} .$$

The proof of [Ama90a, Proposition 5.5] gives (7.10). The remaining assertions follow by reiteration, duality, and commutation arguments for the real and complex interpolation functors (cf. [Ama91b, Lemma 6.3] and [Ama94]).

7.3 Remark Let $\mathcal{A}_0 := (1-\Delta)1_{\mathbb{K}^N}$ and $\mathcal{B}_0 := \delta\partial_\nu + (1-\delta)\gamma_\partial$. Then the proof of Theorem 7.2 in the case that $s_0 < 0 < s_1$ relies on the fact that A_0, the L_p-realization of $(\mathcal{A}_0, \mathcal{B}_0)$, has bounded imaginary powers, and on Remark 6.1(d). The boundedness of A_0^{it} follows from a theorem of Seeley [See71] which requires a C^∞–domain. This is the only place in the whole paper where $\Omega \in C^\infty$ is essential. For all the other results, which do not employ Theorem 7.2 with $s_0 < 0 < s_1$, it suffices to assume that $\Omega \in C^2$. So far the boundedness of the imaginary powers of A_0, in the case where $\Omega \in C^2$, is only known if $\delta = 0$ and $N = 1$, thanks to [PrS93]. The proof of the latter paper does, however, not extend to Neumann boundary conditions. □

8 Extrapolated Elliptic Boundary Value Problems

Having obtained in Theorem 7.1 concrete descriptions of the interpolation-extrapolation spaces E_α for most values of $\alpha \in [-1, 1]$, we turn now to the problem of characterizing the operators $A_{\alpha-1}$ for $0 \le \alpha \le 1$ if A_0 is the L_p-realization of a normally elliptic boundary value problem.

We define $\delta_0 \in \mathbb{N}$ by

$$\delta_0 := \begin{cases} 0 & \text{if } \delta = 0 , \\ 1 & \text{otherwise} . \end{cases}$$

Suppose that $S \in \{H, W\}$. For dealing with quasilinear parabolic problems it will be of importance to know rather weak regularity conditions for the coefficients of $(\mathcal{A}, \mathcal{B})$ guaranteeing that, given $\delta_0 + 1/p < s \le 2$,

$$(\mathcal{A}, \mathcal{B}) \in \mathcal{L}(S^s_p, S^{s-2}_p \times L_p(\partial\Omega, \mathbb{K}^N))$$

and, given $0 \le t < 1 - \delta_0 + 1/p$,

$$(\mathcal{A}^\#, \mathcal{B}^\#) \in \mathcal{L}(S^{2-t}_{p'}, S^{-t}_{p'} \times L_{p'}(\partial\Omega, \mathbb{K}^N)) .$$

For this observe that, given $1 < p < \infty$ and $s \in \mathbb{R}$,

$$S^s_p \xrightarrow{\ \partial_k\ } S^{s-1}_p \xrightarrow{\ w \mapsto a_{jk} w\ } S^{s-1}_p \xrightarrow{\ \partial_j\ } S^{s-2}_p \tag{8.1}$$

implies, thanks to (5.10) and (5.11), that

$$[u \mapsto \partial_j(a_{jk}\partial_k u)] \in \mathcal{L}(S_p^s, S_p^{s-2}) ,$$

provided the middle arrow in (8.1) represents a continuous linear map, that is, if a_{jk} is a pointwise multiplier for S_p^{s-1}.

8.1 Remarks

(a) Let E be a Banach space of $\mathbb{K}^N$-valued distributions on Ω, that is, E is a Banach space such that $E \hookrightarrow \mathcal{D}'$. Then $a \in L_\infty(\Omega, X)$ is a (pointwise) **multiplier** for E if

$$[u \mapsto au] \in \mathcal{L}(E) . \tag{8.2}$$

We denote by $M(E)$ the set of all multipliers for E. It is a Banach space if the norm of $a \in M(E)$ is defined to be the norm of the linear map (8.2) (cf. [Tri83] or [Boud87] for more details).

(b) It is well known (e.g., [Proposition III.6][Boud87]) and easily seen that

$$M(L_p) = L_\infty(\Omega, X) , \qquad 1 \le p \le \infty ,$$

Multipliers on Besov and Triebel-Lizorkin spaces have been studied in detail in [Tri83], [Fra86], [Sic86], and [Sic93] (also cf. [MaS85] and [Ama91d]). In general, there is no easy description of $M(S_p^s)$, except for the case where $|s| > n/p$. Indeed,

$$M(S_p^s) = \begin{cases} H_p^{|s|}(\Omega, X) & \text{if } S = H , \\ W_p^{|s|}(\Omega, X) & \text{if } S = W , \end{cases}$$

for $1 < p < \infty$ and $|s| > n/p$. However,

$$C^\rho(\overline{\Omega}, X) \hookrightarrow M(S_p^s) , \qquad 1 < p < \infty , \quad s \in \mathbb{R} , \tag{8.3}$$

provided

$$\rho \begin{cases} > |s| & \text{if } s \notin \mathbb{Z} , \\ = |s| & \text{if } s \in \mathbb{Z} . \end{cases} \tag{8.4}$$

Thus Hölder spaces are kind of 'universal' multiplier spaces for S_p^s, independently of p. □

We fix now an increasing function $\sigma : [0,1] \to [0,1]$ satisfying

$$\sigma(0) = 0 , \qquad \sigma(t) > t , \quad 0 < t < 1 , \tag{8.5}$$

and put

$$\rho(\alpha) := \sigma(|2\alpha - 1|) , \qquad 0 \le \alpha \le 1 . \tag{8.6}$$

Moreover, we define $\mathbb{E}^\alpha(\Omega) := \mathbb{E}^\alpha(\Omega, N)$ by

$$\mathbb{E}^\alpha(\Omega) := \left[C^{\rho(\alpha)}(\overline{\Omega}, X)\right]^{n^2+n+n} \times L_\infty(\Omega, X) \times C^{\rho(\alpha)}(\partial\Omega, X) , \qquad 0 \le \alpha \le 1 .$$

Note that $\mathbb{E}^1(\Omega)$ is the space introduced in Section 4, that

$$\mathbb{E}^1(\Omega) \hookrightarrow \mathbb{E}^\alpha(\Omega) \hookrightarrow \mathbb{E}^\beta(\Omega) \hookrightarrow \mathbb{E}^{1/2}(\Omega) = \mathbb{E}(\Omega) , \qquad 1/2 < \beta < \alpha < 1 , \tag{8.7}$$

and that

$$\mathbb{E}^\alpha(\Omega) = \mathbb{E}^{1-\alpha}(\Omega) , \qquad 0 \le \alpha \le 1 . \tag{8.8}$$

Of course, $\mathbb{E}^\alpha(\Omega)$ depends on σ too for $\alpha \in (0,1)\backslash\{1/2\}$.

We choose a boundary characterization map δ and define **scales of boundary spaces**

$$\partial W_p^s , \qquad 1 < p < \infty , \quad 0 \le s \le 2 ,$$

by

$$\partial W_p^s := \partial W_p^s(\Omega, \delta, N) := \bigoplus_{\Gamma \in \boldsymbol{\Gamma}} \bigoplus_{r=1}^{N} W_p^{s-\delta_r(\Gamma)-1/p}(\Gamma, \mathbb{K}) ,$$

where $\boldsymbol{\Gamma}$ is the set of components of $\partial\Omega$ and $\delta_r(\Gamma) \in \{0,1\} \subset \mathbb{N}$ denotes the value of δ_r on $\Gamma \in \boldsymbol{\Gamma}$. Note that $\boldsymbol{\Gamma}$ is finite since $\boldsymbol{\Gamma}$ is an open covering of $\partial\Omega$ and $\partial\Omega$ is compact. We also put

$$\partial_1 W_p^{s-1} := \left\{ u \in W_p^{s-1-1/p}(\partial\Omega, \mathbb{K}^N) \ ; \ (1-\delta)u = 0 \right\}$$

and

$$\partial_0 W_p^s := \left\{ u \in W_p^{s-1/p}(\partial\Omega, \mathbb{K}^N) \ ; \ \delta u = 0 \right\} .$$

Then

$$\partial W_p^s = \partial_1 W_p^{s-1} \oplus \partial_0 W_p^s , \qquad 1 < p < \infty , \quad 0 \le s \le 2 . \tag{8.9}$$

It is an easy consequence of (8.3), (8.4), arguments of the type of (8.1), and the trace theorem that

$$(\mathcal{A}, \mathcal{B}) | S_p^{2\alpha} \in \mathcal{L}(S_p^{2\alpha}, S_p^{2\alpha-2} \times \partial W_p^{2\alpha}) , \qquad 1 < p < \infty , \quad \delta_0 + 1/p < 2\alpha \le 2 ,$$

whenever $(\mathcal{A}, \mathcal{B}) \in \mathbb{E}^\alpha(\Omega)$. In fact, the map

$$\mathbb{E}^\alpha(\Omega) \to \mathcal{L}(S_p^{2\alpha}, S_p^{2\alpha-2} \times \partial W_p^{2\alpha}) , \quad (\mathcal{A}, \mathcal{B}) \mapsto (\mathcal{A}, \mathcal{B}) | S_p^{2\alpha} \tag{8.10}$$

is linear and continuous for $1 < p < \infty$ and $\delta_0 + 1/p < 2\alpha \le 2$. Similarly, the map

$$\mathbb{E}^\beta(\Omega) \to \mathcal{L}(S_{p'}^{2-2\beta}, S_{p'}^{-2\beta} \times \partial W_{p'}^{2-2\beta}) , \quad (\mathcal{A}, \mathcal{B}) \mapsto (\mathcal{A}^\#, \mathcal{B}^\#) | S_{p'}^{2-2\beta} \tag{8.11}$$

is linear and continuous for $1 < p < \infty$ and $0 \le 2\beta < 1 - \delta_0 + 1/p$.

8.2 Remark The 'multiplier spaces' $\mathbb{E}^\alpha(\Omega)$ have been introduced to have relatively simple conditions guaranteeing (8.10) and (8.11). Of course, the corresponding regularity assumptions for the coefficients of $(\mathcal{A}, \mathcal{B})$ are by no means optimal. More general conditions can be obtained by using 'better' multiplier spaces, yet at the expense of more cumbersome formulations. □

With $(\mathcal{A},\mathcal{B}) \in \mathbb{E}(\Omega)$ we associate the **Dirichlet form** $\boldsymbol{a}$, defined by

$$\boldsymbol{a}(v,u) := \langle \partial_j v, a_{jk}\partial_k u + a_j u\rangle + \langle v, b_j \partial_j u + a_0 u\rangle + \langle \gamma_\partial v, c\gamma_\partial u\rangle_\partial \ .$$

It follows from Remark 8.1, the duality theorems (5.14) and (5.15), the imbeddings (5.5) and (5.6), and by arguments of the type of (8.1) that

$$\boldsymbol{a} \in \mathcal{L}(S_{p'}^{2\beta}, S_p^{2\alpha}; \mathbb{K}) \ , \tag{8.12}$$

in fact, that the map

$$\mathbb{E}^\alpha(\Omega) \to \mathcal{L}(S_{p'}^{2\beta}, S_p^{2\alpha}; \mathbb{K}) \ , \quad (\mathcal{A},\mathcal{B}) \mapsto \boldsymbol{a} \tag{8.13}$$

is linear and continuous for $1 < p < \infty$ and $1/p < 2\alpha \le 2$, $1/p' < 2\beta \le 2$ satisfying $\alpha + \beta \ge 1$.

With $(\mathcal{A},\mathcal{B}) \in \mathbb{E}(\Omega)$ we also associate two additional boundary operators defined by

$$\mathcal{C}u := (\delta - 1)\{\nu^j \gamma_\partial(a_{jk}\partial_k u + a_j u) + c\gamma_\partial u\} + \delta\gamma_\partial u$$

and

$$\mathcal{C}^\# v := (\delta - 1)\{\nu^j \gamma_\partial(a_{jk}^\# \partial_k v + a_j^\# v) + c^\# \gamma_\partial v\} + \delta\gamma_\partial v \ ,$$

respectively.

Suppose that $1 < p < \infty$. If $1 + 1/p < 2\alpha \le 2$ and $1/p' < 2\beta \le 2$ satisfy $\alpha + \beta \ge 1$, and if $(\mathcal{A},\mathcal{B}) \in \mathbb{E}^\alpha(\Omega)$, it is not difficult to verify that **Green's formula**

$$\langle v, \mathcal{A}u\rangle + \langle \delta\gamma_\partial v, \mathcal{B}u\rangle_\partial + \langle (\delta-1)\gamma_\partial v, \mathcal{C}u\rangle_\partial = \boldsymbol{a}(v,u) \ , \qquad (v,u) \in S_{p'}^{2\beta} \times S_p^{2\alpha} \ , \tag{8.14}$$

is valid. Similarly, if $(\mathcal{A},\mathcal{B}) \in \mathbb{E}^\beta(\Omega)$ and if $1/p < 2\alpha \le 2$ and $1 + 1/p' < 2\beta \le 2$ satisfy $\alpha + \beta \ge 1$, the **Green formula**

$$\langle \mathcal{A}^\# v, u\rangle + \langle \mathcal{B}^\# v, \delta\gamma_\partial u\rangle_\partial + \langle \mathcal{C}^\# v, (\delta-1)\gamma_\partial u\rangle_\partial = \boldsymbol{a}(v,u) \ , \qquad (v,u) \in S_{p'}^{2\beta} \times S_p^{2\alpha} \ , \tag{8.15}$$

is true. Lastly, if $(\mathcal{A},\mathcal{B}) \in \mathbb{E}^\alpha(\Omega)$ and if $1 + 1/p < 2\alpha \le 2$ and $1 + 1/p' < 2\beta \le 2$ satisfy $\alpha + \beta \ge 1$, the **second Green formula**

$$\langle v, \mathcal{A}u\rangle + \langle \mathcal{C}^\# v, \mathcal{B}u\rangle_\partial = \langle \mathcal{A}^\# v, u\rangle + \langle \mathcal{B}^\# v, \mathcal{C}u\rangle_\partial \ , \qquad (v,u) \in S_{p'}^{2\beta} \times S_p^{2\alpha} \ , \tag{8.16}$$

is valid.

Suppose that $1 < p < \infty$ and $(\mathcal{A},\mathcal{B}) \in \mathbb{E}^\alpha(\Omega)$. We define a unique

$$A_{\alpha-1} \in \mathcal{L}(S_{p,\mathcal{B}}^{2\alpha}, S_{p,\mathcal{B}}^{2\alpha-2}) \ ,$$

the **$S_{p,\mathcal{B}}^{2\alpha-2}$-realization of** $(\mathcal{A},\mathcal{B})$, as follows:

- If $\delta_0 + 1/p < 2\alpha \le 2$,

$$A_{\alpha-1} := \mathcal{A} | S_{p,\mathcal{B}}^{2\alpha} \ . \tag{8.17}$$

- If $1/p < 2\alpha < 1 + 1/p$ then $A_{\alpha-1}$ is the unique linear operator induced by the Dirichlet form $\boldsymbol{a}$ associated with $(\mathcal{A}, \mathcal{B})$, that is,
$$\langle v, A_{\alpha-1}u\rangle = \boldsymbol{a}(v,u) , \qquad (v,u) \in S^{2-2\alpha}_{p',\mathcal{B}\#} \times S^{2\alpha}_{p,\mathcal{B}} . \tag{8.18}$$
Observe that, thanks to (8.12) and (7.3)–(7.6), the operator $A_{\alpha-1}$ is well defined. Moreover, it follows from (8.13) that
$$\big[(\mathcal{A}, \mathcal{B}) \mapsto A_{\alpha-1}\big] \in \mathcal{L}\big(\mathbb{E}^{\alpha}(\Omega), \mathcal{L}(S^{2\alpha}_{p,\mathcal{B}}, S^{2\alpha-2}_{p,\mathcal{B}})\big) . \tag{8.19}$$
(Thanks to Green's formula (8.14), the two definitions (8.17) and (8.18) for $A_{\alpha-1}$ agree in the case that $\delta = 0$ and $1/p < 2\alpha < 1 + 1/p$.)
- Lastly, if $0 \leq 2\alpha < 1 - \delta_0 + 1/p$,
$$\langle v, A_{\alpha-1}u\rangle = \langle \mathcal{A}^{\#} v, u\rangle , \qquad (v,u) \in S^{2-2\alpha}_{p',\mathcal{B}\#} \times S^{2\alpha}_{p} , \tag{8.20}$$
that is,
$$A_{\alpha-1} := (\mathcal{A}^{\#} | S^{2-2\alpha}_{p',\mathcal{B}\#})' .$$
Note that $A_{\alpha-1}$ is well defined in this case too, thanks to (8.11) and (7.3) (and to the Green formula (8.15) if $\delta = 0$ and $1/p < 2\alpha < 1 + 1/p$).

Of course, the $S^{2\alpha-2}_{p,\mathcal{B}}$-realization $A_{\alpha-1}$ of $(\mathcal{A}, \mathcal{B})$ depends on $S \in \{H, W\}$ too. But we do not indicate this in order not to overburden the notation.

The following theorem justifies the notation $A_{\alpha-1}$ for the $S^{2\alpha-2}_{p,\mathcal{B}}$-realization of $(\mathcal{A}, \mathcal{B})$.

8.3 Theorem *Suppose that $(\mathcal{A}, \mathcal{B}) \in \mathcal{E}^1(\Omega)$, that $1 < p < \infty$, and that $\mathfrak{F}$ is defined by (7.9). Let $\big\{(E_\alpha, A_\alpha) \ ; \ -1 \leq \alpha < \infty\big\}$ be the interpolation-extrapolation scale generated by the L_p-realization of $(\mathcal{A}, \mathcal{B})$ and by $\mathfrak{F}$. Then, given $2\alpha \in [0,2]\backslash\Sigma_p$,*
$$A_{\alpha-1} = \begin{cases} H^{2\alpha-2}_{p,\mathcal{B}}\text{-realization of } (\mathcal{A}, \mathcal{B}) & \text{if } (\cdot,\cdot)_\alpha = [\cdot,\cdot]_\alpha , \\ W^{2\alpha-2}_{p,\mathcal{B}}\text{-realization of } (\mathcal{A}, \mathcal{B}) & \text{otherwise} . \end{cases}$$

For the proof of this theorem it suffices to observe that, thanks to (8.7), the first Green formula (8.14) implies
$$\langle v, A_0 u\rangle = \boldsymbol{a}(v,u) , \qquad (v,u) \in S^{2-2\alpha}_{p',\mathcal{B}\#} \times W^2_{p,\mathcal{B}} , \quad 1/p < 2\alpha < 1 + 1/p ,$$
and the second Green formula (8.16) gives
$$\langle v, A_0 u\rangle = \langle \mathcal{A}^{\#} v, u\rangle , \qquad (v,u) \in S^{2-2\alpha}_{p',\mathcal{B}\#} \times W^2_{p,\mathcal{B}} , \qquad 0 \leq 2\alpha < 1/p .$$
Now the assertion follows from Theorem 7.1, formulas (7.2) and (6.5), and an obvious density argument (cf. [Ama94] for more details).

8.4 Remarks

(a) Suppose that $(\mathcal{A}, \mathcal{B}) \in \mathcal{E}^1(\Omega)$. Since $W^2_p \overset{d}{\hookrightarrow} L_p$, it follows from Remark 6.1(e) and Theorem 8.3 that the $S^{2\alpha-2}_{p,\mathcal{B}}$-realization of $(\mathcal{A}, \mathcal{B})$ has a compact resolvent for $2\alpha \in [0,2]\backslash\Sigma_p$.

(b) Suppose that $(\mathcal{A},\mathcal{B})\in\mathcal{E}^1(\Omega)$. It is a consequence of Theorem 8.3 that

$$A_{\alpha-1}\in\mathcal{H}(S^{2\alpha}_{p,\mathcal{B}},S^{2\alpha-2}_{p,\mathcal{B}}) \tag{8.21}$$

and that

$$\sigma(A_{\alpha-1})=\sigma(A_0)$$

for $2\alpha\in[0,2]\backslash\Sigma_p$.

(c) Using (a) and the fact that the 'center parts' of the Bessel potential and Sobolev-Slobodeckii scales induced by $(\mathcal{A},\mathcal{B})$ are independent of $(\mathcal{A},\mathcal{B})$, we deduce from Remark 6.1(e) that

$$S^{2\alpha}_{p,\mathcal{B}}\overset{cc}{\hookrightarrow}S^{2\beta}_{p,\mathcal{B}}\ ,\qquad 1<p<\infty\ ,\quad 2\alpha,2\beta\in(-2+1/p,1+1/p)\backslash\Sigma_p\ ,$$

provided $2\alpha>2\beta$. □

Thanks to Remark 8.4(b) Theorem 8.3 implies generation theorems for analytic semigroups on the spaces $S^{2\alpha-2}_{p,\mathcal{B}}$ for the corresponding realizations of $(\mathcal{A},\mathcal{B})\in\mathcal{E}^1(\Omega)$. As will be shown in detail in Sections 9 and 11 below, these realizations are closely related to weak formulations of elliptic and parabolic boundary value problems of various 'degrees of weakness'. In this connection it is of importance to know that (8.21) is true if $(\mathcal{A},\mathcal{B})$ belongs to

$$\mathcal{E}^\alpha(\Omega):=\mathcal{E}^\alpha(\Omega,\delta,N):=\mathcal{E}(\Omega)\cap\mathbb{E}^\alpha(\Omega)\ ,\qquad 0\le\alpha\le1\ ,$$

only. Note that

$$\mathcal{E}^\alpha(\Omega)\text{ is open in }\mathbb{E}^\alpha(\Omega)\ ,\qquad 0\le\alpha\le1\ , \tag{8.22}$$

thanks to (4.8) and (8.7).

The following theorem contains the desired 'weak generation' result.

8.5 Theorem *Suppose that $1<p<\infty$ and $1\le2\alpha\le2$. If $(\mathcal{A},\mathcal{B})\in\mathcal{E}^\alpha(\Omega)$ then*

$$A_{\beta-1}\in\mathcal{H}(S^{2\beta}_{p,\mathcal{B}},S^{2\beta-2}_{p,\mathcal{B}})\ ,\qquad 2\beta\in[2-2\alpha,2\alpha]\backslash\Sigma_p\ , \tag{8.23}$$

and $A_{\beta-1}$ has a compact resolvent. Moreover, the map

$$\mathcal{E}^\alpha(\Omega)\to\mathcal{H}(S^{2\beta}_{p,\mathcal{B}},S^{2\beta-2}_{p,\mathcal{B}})\ ,\quad(\mathcal{A},\mathcal{B})\mapsto A_{\beta-1} \tag{8.24}$$

is analytic, provided either $1/p<2\beta<1+1/p$, or $\delta=0$ and $2\beta\in[0,2]\backslash\{1/p\}$.

The generation theorem (8.23) in the case $1/p<2\alpha<1+1/p$ has been proven in [Ama89a] under an additional hypothesis (cf. [Ama89a, Erratum]). The proof is based on Theorem 8.3. By a somewhat more sophisticated argument it can be shown that the additional hypothesis is not needed. Similar arguments can be used to prove (8.23) for $1+1/p<2\alpha<2$. The case $0\le2\alpha<1/p$ follows then by duality, of course. The compactness of the resolvent is an easy consequence of $S^{2\alpha}_{p,\mathcal{B}}\overset{cc}{\hookrightarrow}L_p\hookrightarrow S^{2\alpha-2}_{p,\mathcal{B}}$ which, in turn, follows from $W^s_p\overset{cc}{\hookrightarrow}L_p$ (also see Remark 8.4(c)). Lastly, (8.13) and (8.22) imply (8.24). (For details we refer to [Ama94].)

8.6 Remarks

(a) The regularity assumptions for the 'lower order coefficients' a_j, b_j, a_0, and c in

Theorem 8.5 can be considerably weakened (cf. [Ama89a] and [Ama94]). A particularly simple situation occurs if $p > n$ and $2\beta = 1$. Indeed, let

$$\mathbb{E}_p(\Omega) := C(\overline{\Omega}, X)^{n^2} \times L_p(\Omega, X)^n \times L_{p' \vee n}(\overline{\Omega}, X)^n \times L_q(\Omega, X) \times L_r(\partial\Omega, X) ,$$

where $1 \wedge pn/(n+p) \leq q < \infty$ and $1 \wedge p(n-1)/(p+n-1) \leq r < \infty$. Then

$$\mathcal{E}_p(\Omega) := \mathbb{E}_p(\Omega) \cap \mathcal{E}(\Omega)$$

is open in $\mathbb{E}(\Omega)$, and it is not difficult to verify that the map

$$\mathcal{E}_p(\Omega) \to \mathcal{H}(W^1_{p,\mathcal{B}}, W^{-1}_{p,\mathcal{B}}) , \quad (\mathcal{A}, \mathcal{B}) \mapsto A_{-1/2}$$

is well defined and analytic.

(b) Analoguous 'weak' generation theorems are valid for normally elliptic systems of higher order (cf. [Ama94]).

(c) It is a consequence of (3.1) and (8.23) that there exist constants $\kappa \geq 1$ and $\omega \geq 0$ such that

$$|\lambda| \, \|u\|_{S^{2\beta-2}_{p,\mathcal{B}}} + \|u\|_{S^{2\beta}_{p,\mathcal{B}}} \leq \kappa \, \|(\lambda + A_{\beta-1})u\|_{S^{2\beta-2}_{p,\mathcal{B}}} , \qquad u \in S^{2\beta}_{p,\mathcal{B}} ,$$

for all $\lambda \in [\mathrm{Re}\, z \geq \omega]$ and $2\beta \in [2-2\alpha, 2\alpha] \backslash \Sigma_p$.

(d) If we restrict our consideration to values of β with $1 + 1/p < 2\beta \leq 2$, Theorem 8.5 remains true for second order normally elliptic boundary value problems of the general form

$$\mathcal{A}u := -a_{jk}\partial_j\partial_k u + a_j\partial_j u + a_0 u$$

and

$$\mathcal{B}u := \delta\{b_j\gamma_\partial\partial_j u + c\gamma_\partial u\} + (1-\delta)du ,$$

provided $a_{jk} \in C^\rho(\overline{\Omega}, X)$ for some $\rho > 2 - 2\beta$, and $a_j, a_0 \in L_\infty(\Omega, X)$, and where $b_j, c \in C^1(\partial\Omega, X)$ and $d \in C^2(\partial\Omega, X)$ (cf. Remark 4.2(b) and observe that every regular boundary operator of order at most one can be written in the above form). For a proof we refer to [Ama94]. □

In the very special situation that $\alpha = 1/2$, it follows from Theorem 8.5 and Remark 8.6(a), in particular, that the Dirichlet form

$$(v, u) \mapsto \langle \partial_j v, a_{jk}\partial_k u + a_j u \rangle + \langle v, b_j\partial_j u + a_0 u \rangle + \langle \gamma_\partial v, c\gamma_\partial u \rangle_\partial$$

with continuous top-order coefficients a_{jk} and lower-order coefficients belonging to L_∞ (for example) defines a negative generator $A_{-1/2}$ of an analytic semigroup on $W^{-1}_{p,\mathcal{B}}$ with domain $W^1_{p,\mathcal{B}}$, provided $(\mathcal{A}, \mathcal{B}) \in \mathcal{E}(\Omega)$. If A is uniformly strongly elliptic and $\mathcal{B} = \gamma_\partial$ is the Dirichlet boundary operator, this has also been proven (for continuous coefficients and systems of order $2m$) in [Ves89] by a different technique. If $N = 1$ and $\mathcal{A}$ is uniformly strongly elliptic with continuous coefficients, the fact that $A_{-1/2} \in \mathcal{H}(W^1_{p,\mathcal{B}}, W^{-1}_{p,\mathcal{B}})$ has again been shown in [LuV91] (also cf. [Ves91]). In the latter paper, spaces of continuous and Hölder continuous functions are considered too (also cf. [Ves88]). The methods of proofs in these papers do not extend to general normally elliptic boundary value problems of the class $\mathcal{E}(\Omega)$ (or to $\alpha \neq 1/2$) since they rely (implicitly or explicitly) on the coercivity of the quadratic form $u \mapsto \boldsymbol{a}(\overline{u}, u)$ on $H^1_{\mathcal{B}} := H^1_{2,\mathcal{B}}$, which is not available in the general case.

Note that the Dirichlet form $\boldsymbol{a}$ is a continuous bilinear form on $W^1_{p',\mathcal{B}\#} \times W^1_{p,\mathcal{B}}$ if it is only assumed that it has L_∞-coefficients. If $p = 2$ and the associated quadratic form $u \mapsto \boldsymbol{a}(\overline{u}, u)$ is coercive, it is well known that $A_{-1/2} \in \mathcal{H}(H^1_{\mathcal{B}}, H^{-1}_{\mathcal{B}})$, since $A_{-1/2}$ is 'regularly accretive' (e.g., [Tan79, Theorem 3.6.1]). In [GrR89] (also cf. [Grö89], [Grö92]) this result is extended — under weak conditions on $\partial\Omega$ and without requiring that Dirichlet and Neumann boundary parts do not touch — to values of p different from 2 but close to 2.

Given $(\mathcal{A}, \mathcal{B}) \in \mathbb{E}^\alpha(\Omega) = \mathbb{E}^{1-\alpha}(\Omega)$, we define

$$A^{\#}_{-\alpha} \in \mathcal{L}(S^{2-2\alpha}_{p',\mathcal{B}\#}, S^{-2\alpha}_{p',\mathcal{B}\#}) ,$$

the $S^{-2\alpha}_{p',\mathcal{B}\#}$**-realization** of $(\mathcal{A}^{\#}, \mathcal{B}^{\#})$, as follows:

- If $0 \le 2\alpha < 1 - \delta_0 + 1/p$,

$$A^{\#}_{-\alpha} := \mathcal{A}^{\#} | S^{2-2\alpha}_{p',\mathcal{B}\#} .$$

- If $1/p < 2\alpha < 1 + 1/p$,

$$\langle A^{\#}_{-\alpha} v, u \rangle := \boldsymbol{a}(v, u) , \qquad (v, u) \in S^{2-2\alpha}_{p',\mathcal{B}\#} \times S^{2\alpha}_{p,\mathcal{B}} .$$

- If $\delta_0 + 1/p < 2\alpha \le 2$,

$$\langle A^{\#}_{-\alpha} v, u \rangle := \langle v, \mathcal{A} u \rangle , \qquad (v, u) \in S^{2-2\alpha}_{p',\mathcal{B}\#} \times S^{2\alpha}_{p,\mathcal{B}} .$$

Similarly as above, one verifies the unambiguity of this definition.

Let $(\mathcal{A}, \mathcal{B}) \in \mathcal{E}^\alpha(\Omega)$. Then it follows from the fact that $A_{-\alpha}$ has a nonempty resolvent set, from Theorem 8.5, and by Green's formulas that

$$(A_{\alpha-1})' = A^{\#}_{1-\alpha} := S^{2-2\alpha}_{p',\mathcal{B}\#}\text{-realization of } A^{\#}_{-\alpha} \tag{8.25}$$

for $2\alpha \in [0, 2] \backslash (\mathbb{N} + 1/p)$.

For simplicity of presentation, we have restricted ourselves to the 'classical scales' H^s_p and W^s_p. It should be clear to the reader that all the results presented so far extend to the scales of Besov spaces $B^s_{p,q}$ with $1 < q < \infty$. For this it suffices to admit the interpolation functors $(\cdot,\cdot)_{\theta,q}$ with $1 < q < \infty$. If one considers the continuous interpolation functors $(\cdot,\cdot)^0_{\theta,\infty}$ too, it is possible to extend (suitable modifications of) the above results to include the scales of little Nikolskii spaces, $n^s_{p,\mathcal{B}}$, induced by $(\mathcal{A}, \mathcal{B})$ (cf. [Sit92], [Sit93a]; also see [Gui91b] for a study of elliptic and parabolic boundary value problems in little Nikolskii spaces of positive orders).

Let now $\mathbb{K} = \mathbb{R}$. We always equip $\mathbb{R}^N$ with its natural, that is, componentwise, order, so that $\mathbb{R}^N$ is a Banach lattice with positive cone $(\mathbb{R}^N)^+ := (\mathbb{R}^+)^N$. The natural order of L_p, $1 \le p \le \infty$, is induced by the positive cone

$$L_p^+ := \{ u \in L_p \ ; \ u \ge 0 \text{ a.e.} \} .$$

Then L_p is a Banach lattice. Note that $L_{p'}^+ = (L_p^+)'$ for $1 < p < \infty$.

Suppose that $S \in \{H, W\}$, that $1 < p < \infty$, and that $s \in \dot{\mathbb{R}}^+$. Then $S_p^s \hookrightarrow L_p$ so that S_p^s is an ordered Banach space with respect to the natural order canonically induced by L_p. By mollifying, for example, it is easily seen that $(S_p^s)^+ = S_p^s \cap L_p^+$, the positive cone of S_p^s, is total.

If $1 < p < \infty$ and $s \in \mathbb{R}^+$, the natural order of $W_p^1(\partial\Omega, \mathbb{R}^N)$ is defined similarly. It follows that the positive cone $W_p^s(\partial\Omega, \mathbb{R}^N)^+$ of $W_p^s(\Omega, \mathbb{R}^N)$ is total. Thus $W_p^{-s}(\partial\Omega, \mathbb{R}^N)$ is an ordered Banach space with respect to the canonical dual order and, given $u \in W_p^{-s}(\partial\Omega, \mathbb{R}^N)$, it follows that

$$u \ge 0 \quad \text{iff} \quad \langle v, u \rangle_\partial \ge 0 \ , \quad v \in W_{p'}^s(\partial\Omega, \mathbb{R}^N)^+ \ .$$

Consequently, the spaces

$$\partial W_p^s \ , \qquad 1 < p < \infty \ , \quad 0 \le s \le 2 \ ,$$

are ordered Banach spaces with respect to the obvious natural order.

Suppose now that $N = 1$. If $1 < p < \infty$ and $0 < s < 1 + 1/p$, it follows from [BouM91] and [BouK92] that S_p^s is a sublattice of L_p and that there exists a constant c such that

$$\| \, |u| \, \|_{S_p^s} \le c \|u\|_{S_p^s} \ , \qquad u \in S_p^s \ . \tag{8.26}$$

Let $1/p < s < 1 + 1/p$. By means of local coordinates and the fact that the translation group is strongly continuous on $S_p^s(\mathbb{R}^n, \mathbb{R}^N)$, it is not difficult to see that

$$C^2_{(1-\delta)\gamma_\partial} := \{ u \in C^2 \ ; \ (1-\delta)\gamma_\partial u = 0 \}$$

is dense in $S_{p,\mathcal{B}}^s$ (cf. the proof of [Ada75, Theorem 3.18]). Thus, given $u \in S_{p,\mathcal{B}}^s$, there exists a sequence (u_j) in $C^2_{(1-\delta)\gamma_\partial}$ such that $u_j \to u$ in S_p^s. From (8.26) and the reflexivity of S_p^s we deduce the existence of $v \in S_p^s$ and a subsequence, again denoted by (u_j), such that $|u_j| \rightharpoonup v$ in S_p^s, where $\rightharpoonup$ means 'weak convergence'. Since $|u_j| \to |u|$ in L_p, we see that $v = |u|$, that is, $|u_j| \rightharpoonup |u|$. Note that $(1-\delta)\gamma_\partial$ is continuous, whence weakly continuous, from S_p^s into ∂W_p^s, and that $(1-\delta)\gamma_\partial \, |u_j| = |(1-\delta)\gamma_\partial u_j| = 0$. Consequently $(1-\delta)\gamma_\partial \, |u| = 0$, that is, $|u| \in S_{p,\mathcal{B}}^s$. This shows that

$$S_{p,\mathcal{B}}^s \text{ is a sublattice of } L_p \quad \text{if} \quad 0 < s < 1 + 1/p \text{ and } s \notin \Sigma_p \ .$$

Note, however, that $S_{p,\mathcal{B}}^s$ is not a Banach lattice since the positive cone is not normal. But, of course, the positive cone $(S_{p,\mathcal{B}}^s)^+ = (S_{p,\mathcal{B}}^s) \cap L_p^+$ is total in $S_{p,\mathcal{B}}^s$. Hence the dual positive cone

$$(S_{p,\mathcal{B}}^{-s})^+ := \big((S_{p',\mathcal{B}\#}^s)^+ \big)' \ , \qquad -2 + 1/p < -s < 0 \ , \quad s \notin \Sigma_p \ ,$$

is a proper cone in $S_{p,\mathcal{B}}^{-s}$ inducing the natural order. Thus, given $u \in S_{p,\mathcal{B}}^{-s}$, it follows that

$$u \ge 0 \quad \text{iff} \quad \langle v, u \rangle \ge 0 \ , \quad v \in (S_{p',\mathcal{B}\#}^s)^+ \ . \tag{8.27}$$

Suppose now that $2\alpha \in [1,2]\backslash\Sigma_p$ and that $(\mathcal{A},\mathcal{B}) \in \mathcal{E}^\alpha(\Omega)$. Given $\beta \in [2-2\alpha, 2\alpha]\backslash\Sigma_p$, let $A_{\beta-1}$ be the $S^{2\beta-2}_{p,\mathcal{B}}$-realization of $(\mathcal{A},\mathcal{B})$. If $(\mathcal{A},\mathcal{B}) \in \mathbb{E}^1(\Omega)$, it follows from the maximum principle for uniformly elliptic equations that A_0 is resolvent positive (cf. [Ama83]). Hence $A_0^\# = (A_0)'$ is resolvent positive too, thanks to $L^+_{p'} = (L^+_p)'$. Thus, given $2\beta \in (1/p, 2]\backslash\Sigma_p$ and $(u,v) \in (S^{2\beta-2}_{p,\mathcal{B}})^+ \times (S^{2-2\beta}_{p',\mathcal{B}\#})^+$, it follows from Theorem 6.2 and (8.27) that

$$\langle v, (\lambda + A_{\beta-1})^{-1}u\rangle = \langle (\lambda + A^\#_{1-\beta})^{-1}v, u\rangle = \langle (\lambda + A^\#_0)^{-1}v, u\rangle \geq 0 \tag{8.28}$$

for all sufficiently large $\lambda > 0$. Thus $A_{\beta-1}$ is resolvent positive for $2\beta \in (1/p, 2]\backslash\Sigma_p$, provided $(\mathcal{A},\mathcal{B}) \in \mathcal{E}^1(\Omega)$.

Suppose now that $1 \leq 2\alpha < 2$. Since $\rho(\alpha) > 2\alpha - 1$ for $2\alpha > 1$ and since $C^s \hookrightarrow c^t$ and, similarly, $C^s(\partial\Omega,\mathbb{R}) \hookrightarrow c^t(\partial\Omega,\mathbb{R})$ for $0 < t < s < 1$ and for $s = t = 0$, we can assume, by replacing $\rho(\alpha)$ by a slightly smaller number if $2\alpha > 1$, that

$$(\mathcal{A},\mathcal{B}) \in \left[c^{\rho(\alpha)}(\overline{\Omega},\mathbb{R})\right]^{n^2+n+n} \times L_\infty(\Omega,\mathbb{R}) \times c^{\rho(\alpha)}(\partial\Omega,\mathbb{R}) .$$

Note that $\mathbb{E}^1(\Omega)$ is dense in the latter space. Hence, thanks to (8.22), there exists a sequence $((\mathcal{A}_j,\mathcal{B}_j)) \in \mathcal{E}^1(\Omega)$ such that $(\mathcal{A}_j,\mathcal{B}_j) \to (\mathcal{A},\mathcal{B})$ in $\mathcal{E}^\alpha(\Omega)$. Now it follows from (8.24) and the continuity of the inversion map that, denoting by $(A_j)_{\beta-1}$ the $S^{2\beta-2}_{p,\mathcal{B}}$-realization of $(\mathcal{A}_j,\mathcal{B}_j)$,

$$\left(\lambda + (A_j)_{\beta-1}\right)^{-1} \to (\lambda + A_{\beta-1})^{-1} \qquad \text{in } \mathcal{L}(S^{2\beta-2}_{p,\mathcal{B}})$$

for each sufficiently large $\lambda > 0$, provided $2\beta \in [2-2\alpha, 2\alpha] \cap (1/p, 1+1/p)$. Since

$$(\lambda + A_{\gamma-1})^{-1} \subset (\lambda + A_{\beta-1})^{-1} \qquad \text{for} \quad \beta < \gamma \leq 1 ,$$

it is clear that $A_{\beta-1}$ is resolvent positive for each $2\beta \in (1/p, 2\alpha] \cap [2-2\alpha, 2\alpha]\backslash\Sigma_p$.

Since $\mathbb{E}^\alpha(\Omega) = \mathbb{E}^{1-\alpha}(\Omega)$, it follows that

$$A^\#_{-\beta} \text{ is resolvent positive for } \quad 2\beta \in [2-2\alpha, 2\alpha] \cap [2-2\alpha, 1+1/p)\backslash\Sigma_p .$$

Note that the positive cone $(S^{-2\beta}_{p',\mathcal{B}\#})^+$ is total, thanks to the fact that $L_{p'}$ is dense in $S^{-2\beta}_{p',\mathcal{B}\#}$ and $L_{p'}$ is a lattice so that $L_{p'} = L^+_{p'} - L^+_{p'}$. Choose $\lambda > 0$ such that $(\lambda + A^\#_{-\beta})^{-1}$ is a positive continuous isomorphism from $S^{-2\beta}_{p',\mathcal{B}\#}$ onto $S^{2-2\beta}_{p',\mathcal{B}\#}$. Then it follows that $(S^{2-2\beta}_{p',\mathcal{B}\#})^+ = S^{2-2\beta}_{p',\mathcal{B}\#} \cap L^+_{p'}$ is total too. Hence the dual positive cone

$$(S^{2\beta-2}_{p,\mathcal{B}})^+ := ((S^{2-2\beta}_{p',\mathcal{B}\#})^+)'$$

is a proper cone defining the natural order of $S^{2\beta-2}_{p,\mathcal{B}}$ for $2\beta \in [2-2\alpha, 2\alpha] \cap [2-2\alpha, 1/p)$. Now the argument (8.28) applies again for these values of 2β to show that $A_{\beta-1}$ is resolvent positive.

Lastly, if $2\beta \in [2-2\alpha, 2\alpha]\backslash\Sigma_p$, it follows that there exists $\varepsilon_0 > 0$ such that

$$(1 + \varepsilon A_{\beta-1})^{-1}(S^{2\beta-2}_{p,\mathcal{B}})^+ \subset (S^{2\gamma}_{p,\mathcal{B}})^+ , \qquad 2\gamma \in [2\beta-2, 2\beta \wedge 2\alpha]\backslash\Sigma_p , \quad 0 < \varepsilon < \varepsilon_0 .$$

Since $(1 + \varepsilon A_{\beta-1})^{-1}$ converges strongly to the identity in $S^{2\beta-2}_{p,\mathcal{B}}$ as $\varepsilon \to 0$, we see that $(S^{2\gamma}_{p,\mathcal{B}})^+$ is dense in $(S^{2\beta}_{p,\mathcal{B}})^+$ if $2\alpha - 2 \leq 2\beta \leq 2\gamma \leq 2\alpha$ and $2\gamma, 2\beta \notin \Sigma_p$. This proves

8.7 Theorem *Suppose that $\mathbb{K} = \mathbb{R}$ and $N = 1$. Also suppose that $1 < p < \infty$, that $1 \leq 2\alpha \leq 2$, and that $(\mathcal{A}, \mathcal{B}) \in \mathcal{E}^\alpha(\Omega)$. Then each one of the Banach spaces*

$$S^{2\beta}_{p,\mathcal{B}} , \qquad 2\beta \in [2-2\alpha, 2\alpha]\backslash\Sigma_p ,$$

is an ordered Banach space with the natural order, which is canonically induced by L_p^+ if $2\beta \geq 0$ and which is the canonical dual order otherwise. If $2\alpha - 2 \leq 2\beta \leq 2\gamma \leq 2\alpha$ with $2\beta, 2\gamma \notin \Sigma_p$, the injection $S^{2\gamma}_{p,\mathcal{B}} \hookrightarrow S^{2\beta}_{p,\mathcal{B}}$ is positive and

$$(S^{2\gamma}_{p,\mathcal{B}})^+ \text{ is dense in } (S^{2\beta}_{p,\mathcal{B}})^+ .$$

Lastly, each one of the operators

$$A_{\beta-1} \in \mathcal{H}(S^{2\beta}_{p,\mathcal{B}}, S^{2\beta-2}_{p,\mathcal{B}}) , \qquad 2\beta \in [2-2\alpha, 2\alpha]\backslash\Sigma_p , \tag{8.29}$$

is resolvent positive.

Observe that the above theorem is a generalization of the classical naximum principle for strongly uniformly elliptic equations. It appears here for the first time in print.

8.8 Remark It should be noted that for the validity of the resolvent positivity of (8.29) the regularity assumptions for the 'lower order coefficients' of $A_{\beta-1}$ can be considerably weakened. In particular, if $p > n$, the operators

$$A_{-1/2} \in \mathcal{H}(W^1_{p,\mathcal{B}}, W^{-1}_{p,\mathcal{B}}) , \quad (\mathcal{A}, \mathcal{B}) \in \mathcal{E}_p(\Omega)$$

are resolvent positive. This follows from the fact that $\{ (\mathcal{A}, \mathcal{B}) \in \mathcal{E}^1(\Omega) \ ; \ a_0 \in C(\overline{\Omega}, X) \}$ is dense in $\mathcal{E}_p(\Omega)$, from Remark 8.6(a), and from the proof of Theorem 8.7. □

9 Nonhomogeneous Elliptic Boundary Value Problems

Suppose that $1 < p < \infty$ and $2\alpha \in [1,2]\backslash\Sigma_p$. Given

$$(\mathcal{A}, \mathcal{B}) \in \mathbb{E}^\alpha(\Omega) := \mathbb{E}^\alpha(\Omega, \delta, N) \tag{9.1}$$

and

$$(\lambda, f, g) \in \mathbb{K} \times S^{2\alpha-2}_{p,\mathcal{B}} \times \partial W^{2\alpha}_p , \tag{9.2}$$

we consider the linear boundary value problem

$$(\lambda + \mathcal{A})u = f \quad \text{in } \Omega , \qquad \mathcal{B}u = g \quad \text{on } \partial\Omega . \tag{9.3}$$

If $\delta_0 + 1/p < 2\alpha \leq 2$ then u is said to be a **strong $S^{2\alpha}_p$-solution** of (9.3) if $u \in S^{2\alpha}_{p,\mathcal{B}}$ and

$(\lambda + \mathcal{A}, \mathcal{B})u = (f, g)$ (in the sense of distributions).

If $1/p < 2\alpha < 1 + 1/p$ then u is a **weak $S_p^{2\alpha}$-solution** if $u \in S_p^{2\alpha}$ and satisfies

$$\begin{aligned} &(1-\delta)\gamma_\partial u = (1-\delta)g \\ &\text{and} \\ &\boldsymbol{a}_\lambda(v, u) = \langle v, f\rangle + \langle \gamma_\partial v, \delta g\rangle_\partial , \qquad v \in S_{p'\mathcal{B}}^{2-2\alpha} , \end{aligned} \tag{9.4}$$

where

$$\boldsymbol{a}_\lambda := \boldsymbol{a} + \lambda\langle\cdot,\cdot\rangle$$

is the Dirichlet form associated with $(\lambda + \mathcal{A}, \mathcal{B})$.

Lastly, if $0 \le 2\alpha < 1 - \delta_0 + 1/p$ then u is a **very weak $S_p^{2\alpha}$-solution** of (9.3) if $u \in S_p^{2\alpha}$ and

$$\langle(\lambda + \mathcal{A}^\#)v, u\rangle = \langle v, f\rangle + \langle \mathcal{C}^\# v, g\rangle_\partial , \qquad v \in S_{p',\mathcal{B}^\#}^{2-2\alpha} . \tag{9.5}$$

In general, $u \in S_p^{2\alpha}$ is said to be an **$S_p^{2\alpha}$-solution** of (9.3) provided it is

- a strong $S_p^{2\alpha}$-solution if $\delta_0 + 1/p < 2\alpha \le 2$,
- a weak $S_p^{2\alpha}$-solution if $1/p < 2\alpha < 1 + 1/p$,
- a very weak $S_p^{2\alpha}$-solution if $0 \le 2\alpha < 1 - \delta_0 + 1/p$.

The following remarks show that the above definitions are natural and consistent.

9.1 Remarks

(a) Let (9.1) and (9.2) be satisfied. If u is an $S_p^{2\alpha}$-solution of (9.3), it is an $S_p^{2\beta}$-solution for each $2\beta \in [2\alpha - 2, 2\alpha]\backslash(\mathbb{N} + 1/p)$.

This is an easy consequence of (5.5), (7.4), (8.7), (8.8), and the Green formulas (8.14)–(8.16).

(b) Suppose that $\delta = 0$ and $1/p < 2\alpha < 1 + 1/p$. In this case it is an easy consequence of Green's formulas that the three concepts: strong, weak, and very weak $S_p^{2\alpha}$-solution all coincide.

(c) The concept of a strong $S_p^{2\alpha}$-solution is, of course, meaningful for general second order boundary value problems which are not necessarily in divergence form. □

Suppose now that

$$(\mathcal{A},\mathcal{B}) \in \mathcal{E}^\alpha(\Omega)\ . \tag{9.6}$$

If $1+1/p<2\alpha\le 2$, it is a consequence of the normal ellipticity and the trace theorem that

$$\mathcal{B}|S_p^{2\alpha} \in \mathcal{L}(S_p^{2\alpha},\partial W_p^{2\alpha}) \tag{9.7}$$

is a retraction. Hence we can choose a coretraction

$$\mathcal{T}_\alpha := (\mathcal{B}|S_p^{2\alpha})^c \in \mathcal{L}(\partial W_p^{2\alpha},S_p^{2\alpha}) \tag{9.8}$$

(cf. [Ama90a, Appendix B], for example). Then

$$\mathcal{S}_{\alpha-1,\lambda} := -(\lambda+\mathcal{A})\mathcal{T}_\alpha \in \mathcal{L}(\partial W_p^{2\alpha},S_p^{2\alpha-2})\ ,\qquad \lambda\in\mathbb{K}\ . \tag{9.9}$$

Moreover, given (9.2), u is an $S_p^{2\alpha}$-solution of (9.3) iff $u=v+\mathcal{T}_\alpha g$ and v is an $S_p^{2\alpha}$-solution of

$$(\lambda+\mathcal{A})v = f+\mathcal{S}_{\alpha-1,\lambda}g \quad \text{in } \Omega\ ,\qquad \mathcal{B}v=0 \quad \text{on } \partial\Omega\ .$$

Hence we see from (8.17) that (9.3) is equivalent to

$$\begin{aligned}&\text{find } v\in S_{p,\mathcal{B}}^{2\alpha} \text{ satisfying}\\ &(\lambda+A_{\alpha-1})v = f+\mathcal{S}_{\alpha-1,\lambda}g\\ &\text{and put } u:=v+\mathcal{T}_\alpha g\ ,\end{aligned} \tag{9.10}$$

provided $1+1/p<2\alpha\le 2$.

Suppose that $1/p<2\alpha<1+1/p$. Then $(1-\delta)\gamma_\partial \in \mathcal{L}(S_p^{2\alpha},\partial_0 W_p^{2\alpha})$ is a retraction by the trace theorem. Choose a coretraction

$$\mathcal{T}_\alpha := \big[(1-\delta)\gamma_\partial|S_p^{2\alpha}\big]^c \in \mathcal{L}(\partial_0 W_p^{2\alpha},S_p^{2\alpha})\ . \tag{9.11}$$

It follows from (8.12) that there is a unique

$$\mathcal{S}_{\alpha-1,\lambda} \in \mathcal{L}(\partial_0 W_p^{2\alpha},S_{p,\mathcal{B}}^{2\alpha-2})\ ,\qquad \lambda\in\mathbb{K}\ , \tag{9.12}$$

satisfying

$$\langle v,\mathcal{S}_{\alpha-1,\lambda}w\rangle = -\boldsymbol{a}_\lambda(v,\mathcal{T}_\alpha w)\ ,\qquad (v,w)\in S_{p',\mathcal{B}\#}^{2-2\alpha}\times\partial_0 W_p^{2\alpha}\ . \tag{9.13}$$

Moreover, $\delta\gamma_\partial \in \mathcal{L}(S_{p',\mathcal{B}\#}^{2-2\alpha},\partial_1 W_{p'}^{2-2\alpha})$ and, consequently,

$$\mathcal{R}_{\alpha-1} := (\delta\gamma_\partial|S_{p',\mathcal{B}\#}^{2-2\alpha})' \in \mathcal{L}(\partial_1 W_p^{2\alpha-1},S_{p,\mathcal{B}}^{2\alpha-2}) \tag{9.14}$$

is well defined, thanks to (7.1) and the fact that $(\partial_1 W_{p'}^{2-2\alpha})' \doteq \partial_1 W_p^{2\alpha-1}$, as follows from (5.15). Now, given (9.2), we deduce that u is an $S_p^{2\alpha}$-solution of (9.3) iff $u=v+\mathcal{T}_\alpha(1-\delta)g$ and v is an $S_p^{2\alpha}$-solution of

$$(\lambda+\mathcal{A})v = f+\mathcal{S}_{\alpha-1,\lambda}(1-\delta)g \quad \text{in } \Omega\ ,\qquad \mathcal{B}v=\delta g \quad \text{on } \partial\Omega\ .$$

We see from (9.4) that this is the case iff $v\in S_{p,\mathcal{B}}^{2\alpha}$ and

$$\boldsymbol{a}_\lambda(w,v) = \big\langle w,f+\mathcal{S}_{\alpha-1,\lambda}(1-\delta)g\big\rangle + \langle\gamma_\partial w,\delta g\rangle_\partial = \big\langle w,f+\mathcal{S}_{\alpha-1,\lambda}(1-\delta)g+\mathcal{R}_{\alpha-1}\delta g\big\rangle\ .$$

for $w \in S^{2-2\alpha}_{p',\mathcal{B}^\#}$. Thus, thanks to (8.18), it follows that (9.4) is equivalent to

$$\begin{aligned}&\text{find } v \in S^{2\alpha}_{p,\mathcal{B}} \text{ satisfying}\\ &(\lambda + A_{\alpha-1})v = f + \mathcal{S}_{\alpha-1,\lambda}(1-\delta)g + \mathcal{R}_{\alpha-1}\delta g\\ &\text{and put } u := v + \mathcal{T}_\alpha(1-\delta)g\end{aligned} \tag{9.15}$$

provided $1/p < 2\alpha < 1 + 1/p$.

Lastly, suppose that $0 \le 2\alpha < 1/p$. Then

$$\mathcal{S}_{\alpha-1} := (\mathcal{C}^\# | S^{2-2\alpha}_{p',\mathcal{B}^\#})' \in \mathcal{L}(\partial W^{2\alpha}_p, S^{2\alpha-2}_{p,\mathcal{B}}) \tag{9.16}$$

and it is clear that (9.5) is equivalent to

$$\begin{aligned}&\text{find } u \in S^{2\alpha}_{p,\mathcal{B}} \text{ satisfying}\\ &(\lambda + A_{\alpha-1})u = f + \mathcal{S}_{\alpha-1}g\end{aligned} \tag{9.17}$$

provided $0 \le 2\alpha < 1/p$.

Thanks to (9.10), (9.15), and (9.17) the problem of the $S^{2\alpha}_p$-solvability of the boundary value problem (9.3) has been reduced to the study of a linear operator equation of the form

$$(\lambda + A_{\alpha-1})u = F\ , \tag{9.18}$$

where $F \in S^{2\alpha-2}_{p,\mathcal{B}}$ depends linearly and continuously upon (f,g). Since $A_{\alpha-1}$ has a compact resolvent by Theorem 8.5, provided $(\mathcal{A},\mathcal{B}) \in \mathcal{E}^\alpha(\Omega)$, the Riesz-Schauder theory, (8.25), and Theorem 8.5 imply the following 'Fredholm alternative theorem' (for details see [Ama94]).

9.2 Theorem *Suppose that $1 < p < \infty$ and $2\alpha \in [0,2]\backslash(\mathbb{N} + 1/p)$. Also suppose that $(\mathcal{A},\mathcal{B}) \in \mathcal{E}^\alpha(\Omega)$. Then the normally elliptic boundary value problem*

$$(\lambda + \mathcal{A})u = f \quad \text{in } \Omega\ , \qquad \mathcal{B}u = g \quad \text{on } \partial\Omega \tag{9.19}$$

has for each

$$(f,g) \in W^{2\alpha-2}_{p,\mathcal{B}} \times \partial W^{2\alpha}_p \tag{9.20}$$

a unique $W^{2\alpha}_p$-solution u, provided $\lambda \notin \sigma(A_{\alpha-1})$. The set $\sigma(A_{\alpha-1})$ is a nonempty countable subset of a sector of the form $\omega + \big([\,|\arg z| \le \varphi] \cup \{0\}\big)$ for some $\omega \ge 0$ and $\varphi := \varphi(\mathcal{A},\mathcal{B}) \in [0,\pi/2)$, without finite limit point. If $(\mathcal{A},\mathcal{B})$ is formally self-dual, that is, if $(\mathcal{A},\mathcal{B}) = (\mathcal{A}^\#,\mathcal{B}^\#)$, then $\varphi = 0$.

For $\lambda \in \sigma(A_{\alpha-1})$ the homogeneous problems

$$(\lambda + \mathcal{A})u = 0 \quad \text{in } \Omega\ , \qquad \mathcal{B}u = 0 \quad \text{on } \partial\Omega \tag{9.21}$$

and

$$(\lambda + \mathcal{A}^\#)v = 0 \quad \text{in } \Omega\ , \qquad \mathcal{B}^\# v = 0 \quad \text{on } \partial\Omega \tag{9.22}$$

possess the same finite positive number of linearly independent $W^{2\alpha}_p$-solutions and $W^{2-2\alpha}_{p'}$-solutions, respectively. Moreover, the inhomogeneous problem (9.19) is solvable iff the following 'orthogonality conditions' are satisfied:

- *if* $1+1/p<2\alpha\le 2$,
$$\langle w,f\rangle-\langle w,(\lambda+\mathcal{A})\mathcal{T}_\alpha g\rangle=0 \tag{9.23}$$
- *if* $1/p<2\alpha<1+1/p$,
$$\boldsymbol{a}_\lambda\big(w,\mathcal{T}_\alpha(1-\delta)g\big)=\langle w,f\rangle+\langle\gamma w,\delta g\rangle_\partial \tag{9.24}$$
- *if* $0\le 2\alpha<1/p$,
$$\langle w,f\rangle+\langle\mathcal{C}^\# w,g\rangle_\partial=0 \tag{9.25}$$

for each $W_{p'}^{2-2\alpha}$*-solution* w *of the homogeneous boundary value problem* (9.22).

Lastly, suppose that $2\beta\in[2-2\alpha,2\alpha]\backslash(\mathbb{N}+1/p)$. *Then*

$$\sigma(A_{\beta-1})=\sigma(A_{\alpha-1})$$

and, given (9.20), u *is a* $W_p^{2\beta}$*-solution of* (9.19) *iff it is a* $W_p^{2\alpha}$*-solution.*

9.3 Remarks

(a) Suppose that $(\mathcal{A},\mathcal{B})\in\mathcal{E}^1(\Omega)$. Then $\sigma(A_{\alpha-1})=\sigma(A_0)$ for $2\alpha\in[0,2]\backslash(\mathbb{N}+1/p)$. Moreover, thanks to the regularity assertions contained in the last part of Theorem 9.2, it follows that we can restrict w in (9.23)–(9.25) to belong to $\ker(\lambda+A_0^\#)$. Then each one of the orthogonality conditions (9.23)–(9.25) takes the form

$$\langle w,f\rangle+\langle\mathcal{C}^\# w,g\rangle_\partial=0$$

for $w\in W_{p'}^2$ satisfying $(\lambda+\mathcal{A}^\#)w=0$ in Ω and $\mathcal{B}^\# w=0$ on $\partial\Omega$, as follows from (9.8), (9.11), Green's formulas and a density argument.

(b) If $\delta=0$ [resp. $\delta=1$], the exceptional value $2\alpha=1+1/p$ [resp. $2\alpha=1/p$] can be admitted too (cf. (5.24) [resp. (5.25)]).

(c) Using the fact that

$$W_{p,\mathcal{B}}^{2\alpha}\hookrightarrow W_{q,\mathcal{B}}^{2\alpha}\quad\text{and}\quad\partial W_p^{2\alpha}\hookrightarrow\partial W_q^{2\alpha}\ ,\qquad 1<q<p<\infty\ ,$$

it is obvious that, given (9.20), each $W_q^{2\alpha}$-solution, where $1<q<p<\infty$, is a $W_p^{2\alpha}$-solution.

(d) It is an easy consequence of Remark 8.6(c) and (9.9)–(9.17) that there exists a constant c such that the 'a priori estimate'

$$\|u\|_{W_p^{2\alpha}}\le c\big(\|f\|_{W_{p,\mathcal{B}}^{2\alpha-2}}+\|g\|_{\partial W_p^{2\alpha}}+\|u\|_{W_{p,\mathcal{B}}^{2\alpha-2}}\big)$$

is true.

(e) Theorem (9.2) and (a) remain true if the Sobolev-Slobodeckii scale W_p^s is replaced by the Bessel potential scale H_p^s.

(f) If $1+1/p<2\alpha\le 2$, the first part of Theorem 9.2 is valid for general second order normally elliptic boundary value problems (cf. Remark 8.6(d)). □

It may be worthwhile to elucidate the general results by means of the following simple

9.4 Example Suppose that $a_{jk} = a_{kj} \in C^1(\overline{\Omega}, \mathbb{R})$ and that $a := [a_{jk}] \in \mathbb{R}^{n\times n}$ is uniformly positive definite. Also suppose that $a_0 \in L_\infty(\Omega, \mathbb{R})$ and put

$$\mathcal{A}u = -\partial_j(a_{jk}\partial_k u) + a_0 u$$

so that $\mathcal{A}$ is a uniformly strongly elliptic differential operator. Let ∂_{ν_a} be the conormal derivative with respect to a and suppose that $\delta \in \{0,1\}$. Put

$$\mathcal{B}u := \delta\partial_{\nu_a}u + (1-\delta)\gamma_\partial u \ .$$

Then $(\mathcal{A}, \mathcal{B}) \in \mathcal{E}^1(\Omega)$ with $N = 1$ (scalar case), and $(\mathcal{A}, \mathcal{B})$ is formally self-dual.

Suppose that $1 < p < \infty$ and $s \in [0,2]\backslash\{\delta + 1/p\}$. Also suppose that

$$(f,g) \in W^{s-2}_{p,\mathcal{B}} \times W^{s-\delta-1/p}_p(\partial\Omega) \ .$$

Then the scalar linear elliptic boundary value problem

$$(\lambda + \mathcal{A})u = f \quad \text{in } \Omega\ , \qquad \mathcal{B}u = g \quad \text{on } \partial\Omega \tag{9.26}$$

has for each $\lambda \notin \sigma(A_0)$ a unique W^s_p-solution. The spectrum $\sigma(A_0)$ of the $L_p(\Omega)$-realization of $(\mathcal{A}, \mathcal{B})$ consists of a countably infinite sequence of real eigenvalues of finite multiplicity possessing ∞ as its only limit point. If $\lambda \in \sigma(A_0)$ then (9.26) is solvable iff

$$\langle v, f\rangle + \langle\gamma_\partial v, \delta g\rangle_\partial - \big\langle\partial_{\nu_a}v, (1-\delta)g\big\rangle_\partial = 0$$

for each $v \in W^2_{p'}$ satisfying

$$(\lambda + \mathcal{A})v = 0 \quad \text{in } \Omega\ , \qquad \mathcal{B}v = 0 \quad \text{on } \partial\Omega\ .$$

Suppose that u is a W^s_p-solution of (9.26). Then

- $u \in W^s_p$ and $(\lambda + \mathcal{A})u = f$ in Ω in the sense of distributions and $\mathcal{B}u = g$ on $\partial\Omega$ in the sense of traces, provided $\delta + 1/p < s \leq 2$.

- $u \in W^s_p$ and
$$(1-\delta)\gamma_\partial u = (1-\delta)g$$
and
$$\langle\partial_j v, a_{jk}\partial_k u\rangle + \big\langle v, (\lambda + a_0)u\big\rangle = \langle v, f\rangle + \langle\gamma_\partial v, \delta g\rangle_\partial \tag{9.27}$$
for $v \in W^{2-s}_{p',\mathcal{B}\#}$, provided $1/p < s < 1 + 1/p$.

- $u \in W^s_p$ and
$$\big\langle -\partial_j(a_{jk}\partial_k v) + (\lambda + a_0)v, u\big\rangle = \langle v, f\rangle + \langle\gamma_\partial v, \delta g\rangle_\partial - \big\langle\partial_{\nu_a}v, (1-\delta)g\big\rangle_\partial$$
for $v \in W^{2-s}_{p',\mathcal{B}\#}$, provided $0 \leq s < \delta + 1/p$.

Observe that

$$\langle v, f\rangle + \langle\gamma_\partial v, \delta g\rangle_\partial - \langle\partial_{\nu_a} v, (1-\delta)g\rangle_\partial = \int_\Omega vf\,dx + \delta\int_{\partial\Omega} vg\,d\sigma - (1-\delta)\int_{\partial\Omega}\partial_{\nu_a} vg\,d\sigma$$

for $v \in W^2_{p'}$, if $(f,g) \in L_p \times L_p(\partial\Omega)$. On the other hand, if $2-s > n/p'$, it follows from $W^{2-s}_{p'} \hookrightarrow C$ that $\langle v, f\rangle$ is well defined for $v \in W^{2-s}_{p'}$ if f is a Radon measure on $\overline{\Omega}$. Similarly, $\langle\gamma_\partial v, g\rangle_\partial$ and $\langle\partial_{\nu_a} v, g\rangle_\partial$ are well defined for $v \in W^{2-s}_{p'}$ if g is a Radon measure on $\partial\Omega$ provided $2-s > n/p'$ and $1-s > n/p'$, respectively. Thus (9.26) can be solved for rather singular data, provided we choose p sufficiently close to 1. □

9.5 Remarks

(a) Presuppose the hypotheses of Example 9.4 and consider the Neumann problem

$$-\partial_j(a_{jk}\partial_k u) = f \quad \text{in } \Omega\ , \qquad \partial_{\nu_a} u = g \quad \text{on } \partial\Omega\ . \tag{9.28}$$

Suppose that $1 < p < \infty$ and $s \in [0,2]\backslash\{1+1/p\}$. Given

$$(f,g) \in W^{s-2}_{p,\mathcal{B}} \times W^{s-1-1/p}_p(\partial\Omega)\ ,$$

it follows from Example 9.4 that (9.28) has a W^s_p-solution iff

$$\langle \mathbf{1}, f\rangle + \langle \mathbf{1}, g\rangle_\partial = 0\ , \tag{9.29}$$

where $\mathbf{1}(x) := 1$ for $x \in \overline{\Omega}$, thanks to the fact that $\ker(A_{\alpha-1}) = \ker(A_0) = \mathbb{K}\mathbf{1}$, where $\alpha := s/2$. Since 0 is a simple eigenvalue of A_0, whence of $A_{\beta-1}$, as follows from the maximum principle, for example, the spectral theory for operators with compact resolvents implies the existence of a subspace F_β of $W^{2\beta}_{p,\mathcal{B}}$ such that

$$W^{2\beta}_{p,\mathcal{B}} = \mathbb{K}\mathbf{1} \oplus F_\beta\ , \qquad 2\beta \in [-2,2]\backslash\{-2+1/p, 1+1/p\}\ ,$$

and that this decomposition reduces the linear operator $A_{\beta-1}$. Hence we deduce from Theorem 8.3 that

$$B_{\alpha-1} := A_{\alpha-1}|F_{\alpha-1} \in \mathcal{L}\mathrm{is}(F_\alpha, F_{\alpha-1})\ . \tag{9.30}$$

Let

$$\mathcal{R}_{\beta-1} := (\gamma_\partial|W^{2-2\beta}_{p',\mathcal{B}\#})' \in \mathcal{L}\big(W^{2\beta-1-1/p}_p(\partial\Omega), W^{2\beta-2}_{p,\mathcal{B}}\big) \tag{9.31}$$

for $0 \le 2\beta < 1+1/p$ and note that $\mathcal{S}_{\alpha-1}$ as defined in (9.16) equals $\mathcal{R}_{\alpha-1}$ if $0 \le 2\alpha < 1/p$ and $\delta = 1$. Thus, if u is a $W^{2\alpha}_p$-solution of (9.28), we deduce from (9.10)–(9.17) that

$$u = \xi\mathbf{1} + \begin{cases} B^{-1}_{\alpha-1}(f + \mathcal{S}_{\alpha-1,0}g) + \mathcal{T}_\alpha g & \text{if } 1+1/p < 2\alpha \le 2\ ,\\ B^{-1}_{\alpha-1}(f + \mathcal{R}_{\alpha-1}g) & \text{if } \qquad 0 \le 2\alpha < 1+1/p\ ,\end{cases}$$

where $\xi \in \mathbb{K}$, provided (9.29) is satisfied. Consequently, it follows from (9.8), (9.30), and (9.31) that there exists a constant c such that

$$\|u - \langle\mathbf{1}, u\rangle\mathbf{1}\|_{W^s_p} \le c\big(\|f\|_{W^{s-2}_{p,\mathcal{B}}} + \|g\|_{W^{s-1-1/p}_p(\partial\Omega)}\big)\ , \tag{9.32}$$

whenever (9.29) is satisfied and u is a solution of (9.28).

In [vW90] von Wahl proved (among other things) by estimating directly integral operators on $\partial\Omega$ that every solution of the three-dimensional Neumann problem (i.e., $n = 3$)

$$-\Delta u = 0 \quad \text{in } \Omega\ , \qquad \partial_\nu u = g \quad \text{on } \partial\Omega$$

satisfies the a priori estimate

$$\|\nabla u\|_{L_p(\Omega)} \le c\,\|g\|_{W_p^{-1/p}(\partial\Omega)}\ ,$$

which is a very special case of (9.32).

(b) Let the assumptions of Example 9.4 be satisfied and suppose, for example, that

$$f \in L_{p,\mathrm{loc}}\ .$$

Then, given any W_p^s-solution u of (9.26),

$$u \in W_{p,\mathrm{loc}}^2\ .$$

Indeed, given $\varphi \in \mathcal{D}$, it follows that φu is a W_p^s-solution of

$$(\lambda + \mathcal{A})v = f_\varphi \quad \text{in } \Omega\ , \qquad \mathcal{B}v = 0 \quad \text{on } \partial\Omega\ , \tag{9.33}$$

where (using formal notation)

$$f_\varphi := \varphi f - \partial_j(a_{jk}u\partial_k\varphi) - \partial_j\varphi a_{jk}\partial_k u \in W_{p,\mathcal{B}}^{(s-1)\wedge 0}\ .$$

From this we deduce that φu is a W_p^t-solution of (9.33), where $t := 2 \wedge (s+1)$. By repeating this 'boot strapping argument', if necessary, the assertion follows.

(c) Suppose that $s = 1$, $p = 2$, and $\delta = 0$ in (9.27). Then (9.27) is the usual weak formulation of the Dirichlet problem

$$(\lambda + \mathcal{A})u = f \quad \text{in } \Omega\ , \qquad u = 0 \quad \text{on } \partial\Omega$$

in $\mathring{H}^1 := W_{2,\mathcal{B}}^1$. Since $\mathcal{A}$ is a scalar uniformly strongly elliptic operator, Gårding's inequality implies the coerciveness of the quadratic form $u \mapsto a(\overline{u}, u)$ on $\mathring{H}^1$. Hence the assertions are in this particular case well known consequences of the Lax-Milgram lemma and the compact imbedding $\mathring{H}^1 \overset{cc}{\hookrightarrow} L_2$ (e.g., [DauL90, Chapter VII]). More generally, the case $p = 2$ and $2\alpha = 1$ of Theorem 9.2 follows from the Lax-Milgram lemma and from $W_{2,\mathcal{B}}^1 \overset{cc}{\hookrightarrow} L_2$ whenever the quadratic form $u \mapsto a_\lambda(\overline{u}, u)$ is coercive on $W_{2,\mathcal{B}}^1$ for some $\lambda \ge 0$. This is, however, in general not the case for $(\mathcal{A}, \mathcal{B}) \in \mathcal{E}(\Omega)$ if $N > 1$. (We refer to [deF63], [LiM72], and [Neč67], for example, for discussions of the coercivity of Dirichlet forms.) Thus, even in the 'classical' case $p = 2$ and $2\alpha = 1$, the weak solvability of (9.19) is not accessible by 'variational techniques' if $N > 1$. □

In the literature there is only one result concerning the weak $S_p^{2\alpha}$-solvability of problem (9.13), namely [Sim72]. Simader studies the weak W_p^1-solvability of the Dirichlet problem (9.14), (9.15) for Dirichlet forms with continuous coefficients and arbitrary $p \in (1, \infty)$, but in the case of a single equation ($N = 1$), by a completely different technique. (As mentioned earlier, we restrict ourselves here to the case of second order problems. However, our techniques extend straightforwardly to normally elliptic boundary value problems of order $2m$ (cf. [Ama94]). Hence our results also cover the higher order case treated by Simader.)

Nonhomogeneous elliptic boundary value problems of arbitrary order, but for smooth coefficients and in the case $N = 1$, have been studied in the L_p-setting in detail by Lions and Magenes in [LiM60]. An extension of the Lions-Magenes theory to systems is given in [Gey65]. The simpler L_2-case is treated in [LiM72]. There is, however, an important difference between the Lions-Magenes theory and the theory exposed in this section. Namely, in the Lions-Magenes theory the inhomogeneity f in (9.19) is restricted to belong to a space of distributions (which cannot be 'optimally' described), whereas we admit inhomogeneities f from the space $W_{p,\mathcal{B}}^{2\alpha-2}$ which is, in general, not a space of distributions. However, $W_{p,\mathcal{B}}^{2\alpha-2}$ has a natural description as the dual of $W_{p',\mathcal{B}^\#}^{2-2\alpha}$ which makes it a very reasonable and optimal choice. Thus the above theory can be considered to be an extension and simplification of the Lions-Magenes theory, even in the case of smooth coefficients, for the class of systems considered here.

The results of this section occur here for the first time in print. However, special cases have already been presented by the author in lecture series at TH Delft in the fall of 1989 and at Tokyo University in the spring of 1990.

10 Abstract Linear Parabolic Equations

Let E be a Banach space and let J be a perfect subinterval of $\mathbb{R}^+$ containing 0. For each $t \in J$ let $A(t)$ be a linear operator in E, and let $f : J \to E$. By a **solution** of the **linear evolution equation**

$$\dot{u} + A(t)u = f(t) \ , \qquad t \in \dot{J} := J \setminus \{0\} \ , \tag{10.1}$$

where $\dot{u} := \partial_t u$, we mean a function $u \in C(J, E) \cap C^1(\dot{J}, E)$ such that $u(t) \in \operatorname{dom}(A(t))$ for $t \in \dot{J}$ and u satisfies (10.1) pointwise. If $u(0) = x$ then u is a solution of the **linear Cauchy problem** (in E)

$$\dot{u} + A(t)u = f(t) \ , \quad t \in \dot{J} \ , \qquad u(0) = x \ . \tag{10.2}$$

Lastly, u is a **strict solution** of (10.2) if $u \in C^1(J, E)$ and $u(0) \in \operatorname{dom}(A(0))$.

The linear evolution equation (10.1) and the linear Cauchy problem (10.2) are said to be **parabolic** if $-A(t)$ is the infinitesimal generator of a strongly continuous analytic semigroup on E for each $t \in J$.

In the following, $(\cdot, \cdot)_\theta$ is for each $\theta \in (0, 1)$ an admissible interpolation functor and $E_\theta := (E_0, E_1)_\theta$, if (E_0, E_1) is a densely injected Banach couple.

As for the solvability of parabolic linear Cauchy problems we have the following fundamental

10.1 Theorem *Suppose that (E_0, E_1) is a densely injected Banach couple and*

$$\big(x, (A, f)\big) \in E_0 \times C^{\rho}\big(J, \mathcal{H}(E_1, E_0) \times E_0\big)$$

for some $\rho \in (0,1)$. Then the linear parabolic Cauchy problem (10.2) *possesses a unique solution $u := u(\cdot, x, A, f)$, and*

$$u \in C^{\rho}(\dot{J}, E_1) \cap C^{1+\rho}(\dot{J}, E_0) \ . \tag{10.3}$$

If $0 \leq \beta \leq \alpha \leq 1$ and $x \in E_\alpha$,

$$u \in C^{\alpha-\beta}(J, E_\beta) \ .$$

If $x \in E_1$ then u is a strict solution.

The existence of a unique solution, which is strict if $x \in E_1$, has been proven by Sobolevskii [Sob66] and, independently, by Tanabe [Tan60]. A different proof has more recently been given by Acquistapace and Terreni [AcT85]. The regularity result (10.3) is contained in [Sob66] and [AcT85]. The continuity assertion involving interpolation spaces is due to the author (cf. [Ama86b] and [Ama94], also cf. [DaK92, Section 5]).

It is an important property of parabolic evolution equations that they 'regularize'. In fact, (10.3) shows that the solution is more regular than its initial value, in general. The following 'automatic regularity theorem' implies that the solution of (10.2) has even better regularity properties if f is more regular.

10.2 Theorem *Let (E_0, E_1) be a densely injected Banach couple. Suppose that $0 < \gamma < \rho < 1$ and $0 < \varepsilon \leq 1 - \gamma$, and that*

$$(x, A) \in E_0 \times C^{\rho}\big(J, \mathcal{H}(E_1, E_0)\big)$$

and

$$f \in C^{\varepsilon}(J, E_\gamma) + C(J, E_{\gamma+\varepsilon}) \ .$$

Then the Cauchy problem (10.2) *has a unique solution u, and*

$$u \in C(J, E_0) \cap C(\dot{J}, E_1) \cap C^1(\dot{J}, E_\gamma) \ .$$

It is also a solution of the parabolic equation

$$\dot{v} + A_\gamma(t)v = f(t) \ , \qquad t \in \dot{J} \ , \tag{10.4}$$

in E_γ, where A_γ is the E_γ-realization of A. If $x \in \operatorname{dom}\big(A_\gamma(0)\big)$ then u is a strict solution of (10.4).

This theorem has been proven in [Ama88c, Theorem 8.2].

It should be noted that the domains, $\operatorname{dom}(A_\gamma(t))$, of the linear operators $A_\gamma(t)$ depend on $t \in J$, in general, whereas the operators $A(t),\ t \in J$, have 'constant domains'. Also observe that the assertion that u is a solution of (10.4) implies $u(t) \in \operatorname{dom}(A_\gamma(t))$ for $t \in \dot{J}$. This means that $u(t)$ has, in general, better 'spacial' regularity than follows from $u(t) \in E_1$.

Theorem 10.2 guarantees the unique solvability of the abstract Cauchy problem for operators 'with variable domains but a constant interpolation space'. Related recent results can be found in [AcT86] and [Yag88], [Yag90].

Suppose that $\mu \in \mathbb{R}^+$ and $\rho \in (0,1)$. Put

$$\|u\|_{C^\rho_{\rho,\mu}} := \sup_{t \in \dot{J}} t^\mu \|u(t)\| + \sup_{2\varepsilon \in \dot{J}} \varepsilon^{\rho+\mu} [u]_{\rho,[\varepsilon,2\varepsilon]}$$

and denote by

$$BUC^\rho_{\rho,\mu}(J,E)$$

the subspace of $C^\rho(\dot{J},E)$ consisting of all u satisfying $\|u\|_{C^\rho_{\rho,\mu}} < \infty$ and

$$\lim_{\varepsilon \to 0} \varepsilon^{\rho+\mu} [u]_{\rho,[\varepsilon,2\varepsilon]} = 0 \ , \tag{10.5}$$

equipped with the norm $\|\cdot\|_{C^\rho_{\rho,\mu}}$. It is not difficult to see that $BUC^\rho_{\rho,\mu}(J,E)$ is a Banach space, a 'singular Hölder space'. Moreover,

$$BUC^\rho(J,E) \hookrightarrow BUC^\rho_{\rho,\mu}(J,E) \hookrightarrow BUC^\sigma_{\sigma,\nu}(J,E) \hookrightarrow C^\sigma(\dot{J},E)$$

for $0 < \sigma \le \rho < 1$ and $0 \le \mu \le \nu < \infty$. Also put

$$BUC^{1+\rho}_{\rho,\mu}(J,E) := \left(\{\, u \in BUC^\rho_{\rho,\mu}(J,E) \ ; \ \partial u \in BUC^\rho_{\rho,\mu}(J,E) \,\},\ \|\cdot\|_{C^{1+\rho}_{\rho,\mu}} \right) ,$$

where

$$\|u\|_{C^{1+\rho}_{\rho,\mu}} := \|u\|_{C^\rho_{\rho,\mu}} + \|\partial u\|_{C^\rho_{\rho,\mu}} \ .$$

Lastly, $BUC^\rho_\rho := BUC^\rho_{\rho,0}$. Note that the elements of $BUC^\rho_{\rho,\mu}(J,E)$ are allowed to 'become singular' as $t \to 0$ at a controlled rate.

10.3 Theorem *Let (E_0,E_1) be a densely injected Banach couple, let J be bounded, and let $\rho \in (0,1)$ and $\mu \in [0,1)$. Put*

$$(\mathbb{E}_0,\mathbb{E}_1) := \left(BUC^\rho_{\rho,\mu}(J,E_0),\ BUC^\rho_{\rho,\mu}(J,E_1) \cap BUC^{1+\rho}_{\rho,\mu}(J,E_0) \right)$$

and

$$E_{1-\mu} := (E_0,E_1)^0_{1-\mu,\infty} \ .$$

Denote by γ_0 the 'trace operator at $t=0$', that is, $\gamma_0 u = u(0)$. Then

$$(\partial + A, \gamma_0) \in \mathcal{L}\mathrm{is}(\mathbb{E}_1, \mathbb{E}_0 \times E_{1-\mu}) \ ,$$

provided $A \in C^\rho_\rho(J, \mathcal{H}(E_1,E_0))$.

Theorem 10.3 is a slight extension of some results contained in [AcT87] and [Lun87]. The spaces $BUC^{\rho}_{\rho,\mu}$ — but without the requirement (10.5) — are the spaces $Z_{\mu,\rho}$ in [AcT87]. Related spaces also occur in [Sob64]. On the other hand, the spaces BUC^{ρ}_{ρ} correspond to the spaces z^{ρ} in [Lun87].

Theorem 10.3 is a 'maximal regularity' result. There are other maximal regularity theorems: in spaces of continuous functions, due to Da Prato and Grisvard [DaPG79], and in Sobolev spaces, due to Dore and Venni [DoV87]. All these maximal regularity results require additional restrictions either for the class of admissible initial values (as in Theorem 10.3), or for the class of admissible Banach spaces E_0, E_1, and the space of initial values (as in the Da Prato-Grisvard case), or — in addition — for the class of admissible operators (as in the case of the Dore-Venni result). In contrast, there are no restrictions of this type in Theorems 10.1 and 10.2. But, of course, the above theorems do not give maximal regularity. Since maximal regularity plays a minor rôle in this paper, we refer the interested reader to the literature (e.g., [DaPG79], [CH*87], [Lun87], [DoV87], or [Ama94]).

11 Nonhomogeneous Linear Parabolic Boundary Value Problems

Let J be a perfect subinterval of $\mathbb{R}^+$ containing 0, and let $1 < p < \infty$ and $1 \leq 2\alpha \leq 2$. Also suppose that $\rho \in (0,1)$ and

$$(\mathcal{A}, \mathcal{B}) \in C^{\rho}\big(J, \mathcal{E}^{\alpha}(\Omega)\big) . \tag{11.1}$$

Given $S \in \{H, W\}$, we write

$$S^s_{p,\mathcal{B}} = S^s_{p,\mathcal{B}(t)}$$

to mean that the spaces $S^s_{p,\mathcal{B}(t)}$ are independent of $t \in J$. Note that this is the case, for example, if one of the following conditions is satisfied:

- $\mathcal{B}(t)$ is independent of $t \in J$ and $s \in [-2,2]\backslash\Sigma_p$;
- $s \in (-2+1/p, 1+1/p)\backslash\Sigma_p$.

Recall that

$$S^s_{p,\mathcal{B}} = S^s_p , \qquad -1+1/p < s < 1/p .$$

Also assume that 2β satisfies

$$2\alpha - 2 \leq 2\beta \leq 2\alpha \tag{11.2}$$

and

$$2\beta \in \begin{cases} [0,2]\backslash\Sigma_p & \text{if } S^{2\beta}_{p,\mathcal{B}} = S^{2\beta}_{p,\mathcal{B}(t)} , \\ (1/p,2]\backslash\Sigma_p & \text{otherwise} . \end{cases} \tag{11.3}$$

Lastly, assume that $\sigma \in [0,1)$ and

$$\big(u^0, (f,g)\big) \in S^{2\beta-2}_{p,\mathcal{B}} \times C^{\sigma}(J, S^{2\beta-2}_{p,\mathcal{B}} \times \partial W^{2\beta}_p) . \tag{11.4}$$

Then we consider the nonhomogeneous linear **parabolic boundary value problem**

$$\begin{aligned} \partial_t u + \mathcal{A}(t)u &= f(t) && \text{in } \Omega \times \dot{J} , \\ \mathcal{B}(t)u &= g(t) && \text{on } \partial\Omega \times \dot{J} , \\ u(\cdot,0) &= u^0 && \text{on } \Omega . \end{aligned} \tag{11.5}$$

Similarly as in the case of elliptic boundary value problems, we introduce several types of generalized solutions of (11.5). Namely, u is said to be

- a **strong $S_p^{2\beta}$-solution** if $1+1/p<2\beta\leq 2$ and
$$u\in C(J,S_p^{2\beta-2})\cap C(\dot{J},S_p^{2\beta})\cap C^1(\dot{J},S_p^{2\beta-2})\ ,$$
and if it satisfies (11.5) pointwise (in the obvious sense);
- a **weak $S_p^{2\beta}$-solution** if $1/p<2\beta<1+1/p$ and
$$u\in C(J,S_{p,\mathcal{B}}^{2\beta-2})\cap C(\dot{J},S_p^{2\beta})\cap C^1(\dot{J},S_{p,\mathcal{B}}^{2\beta-2})\ ,$$
and if it satisfies $u(0)=u^0$,
$$(1-\delta)\gamma_\partial u(t)=(1-\delta)g(t)\ ,\qquad t\in\dot{J}\ ,$$
and
$$\langle v,\dot{u}(t)\rangle+\mathfrak{a}(t)\bigl(v,u(t)\bigr)=\langle v,f(t)\rangle+\langle\gamma_\partial v,\delta g(t)\rangle_\partial$$
for $t\in\dot{J}$ and $v\in S_{p',\mathcal{B}\#}^{2-2\beta}$;
- a **very weak $S_p^{2\beta}$-solution** if $0\leq 2\beta<1/p$ and $S_{p,\mathcal{B}}^{2\beta}$ and $S_{p',\mathcal{B}\#}^{2-2\beta}$ are independent of t, and if
$$u\in C(J,S_{p,\mathcal{B}}^{2\beta})$$
and u satisfies
$$\int_J\langle-\dot{\varphi}+\mathcal{A}^\#\varphi,u\rangle\,dt=\int_J\bigl\{\langle\varphi,f\rangle+\langle\mathcal{C}^\#\varphi,g\rangle_\partial\bigr\}\,dt+\langle\varphi(0),u^0\rangle$$
for all
$$\varphi\in C(J,S_{p',\mathcal{B}\#}^{2-2\beta})\cap C^1(J,S_{p'}^{-2\beta})$$
having compact support in $J\backslash\{\sup J\}$.

In each one of the above cases u is simply said to be an **$S_p^{2\beta}$-solution** of (11.5).

11.1 Remarks

(a) Every strong $S_p^{2\beta}$-solution is a weak $S_p^{2\gamma}$-solution if $1/p<2\gamma<1+1/p$. Similarly, if u is a weak $S_p^{2\gamma}$-solution for some $2\gamma\in(1/p,1+1/p)$ and if $0\leq 2\beta<1/p$ and $S_{p,\mathcal{B}}^{2\beta}$ and $S_{p',\mathcal{B}\#}^{2-2\beta}$ are independent of t, it is a very weak $S_p^{2\beta}$-solution.

This is an easy consequence of Green's formulas and obvious integrations.

(b) Of course, it suffices to require
$$(\mathcal{A},\mathcal{B})\in C\bigl(J,\mathcal{E}(\Omega)\bigr)$$
for the definition of $S_p^{2\beta}$-solutions to be meaningful.

(c) In the definition of a very weak solution we require u to be continuous on J into $S^{2\beta}_{p,\mathcal{B}}$ and not on $\dot{J}$ as may be expected. Thus is done to guarantee the existence of the occurring integrals. □

Consider now first the case $1/p < 2\beta < 1 + 1/p$. Suppose that

$$\begin{aligned}&\text{there exists}\\ &u_g \in C^{\sigma}(J, S^{2\beta}_{p}) \cap C^{1+\sigma}(J, S^{2\beta-2}_{p,\mathcal{B}})\\ &\text{such that } (1-\delta)\gamma_{\partial} u_g = (1-\delta) g \ .\end{aligned} \tag{11.6}$$

Define $v_g(t) \in S^{2\beta-2}_{p,\mathcal{B}}$ by

$$\langle v, v_g(t)\rangle := \mathfrak{a}(t)(v, u_g(t)) \ , \qquad t \in J \ , \quad v \in S^{2-2\beta}_{p',\mathcal{B}\#} \ ,$$

and note that (11.6), (8.7), and (8.13) imply

$$v_g \in C^{\rho\wedge\sigma}(J, S^{2\beta-2}_{p,\mathcal{B}}) \ .$$

Consequently,

$$F_{\beta-1} := f + \mathcal{R}_{\beta-1}\delta g - v_g - \dot{u}_g \in C^{\rho\wedge\sigma}(J, S^{2\beta-2}_{p,\mathcal{B}}) \ . \tag{11.7}$$

Also note that, thanks to (11.1) and (8.24),

$$A_{\beta-1} \in C^{\rho}\big(J, \mathcal{H}(S^{2\beta}_{p,\mathcal{B}}, S^{2\beta-2}_{p,\mathcal{B}})\big) \ . \tag{11.8$_\beta$}$$

It is now obvious that

$$\begin{aligned}&u \text{ is an } S^{2\beta}_{p,\mathcal{B}}\text{-solution of (11.5) iff}\\ &u = v + u_g \text{ and } v \text{ is a solution of the Cauchy problem}\\ &\dot{v} + A_{\beta-1}(t)v = F_{\beta-1}(t) \ , \quad t \in \dot{J} \ , \qquad v(0) = u^0 - u_g(0)\\ &\text{in } S^{2\beta-2}_{p,\mathcal{B}} \ ,\end{aligned} \tag{11.9}$$

provided $1/p < 2\beta < 1 + 1/p$ and (11.6) is satisfied.

Clearly, Theorem 10.1 guarantees the unique solvability of the Cauchy problem in (11.9) provided $\sigma > 0$. On the other hand, suppose that

$$\sigma = 0 \quad \text{and} \quad 2 - 2\alpha < 2\beta < 2\alpha \ . \tag{11.10}$$

Fix $\gamma \in (1/p, \beta)$ and put $E_0 := S^{2\gamma-2}_{p,\mathcal{B}}$ and $E_1 := S^{2\gamma}_{p,\mathcal{B}}$. Note that

$$A_{\gamma-1} \in C^{\rho}\big(J, \mathcal{H}(E_1, E_0)\big) \ .$$

Choose $\theta \in (0, \beta - \gamma)$, select an admissible interpolation functor $(\cdot,\cdot)_\theta$, put $E_\theta := (E_1, E_0)_\theta$, and observe that

$$S^{2\gamma-2}_{p,\mathcal{B}} \hookrightarrow E_\theta \hookrightarrow E_0 \ .$$

Hence it follows from (11.7) that

$$F_{\beta-1} \in C(J, E_\theta) \ .$$

Consequently, we can apply Theorem 10.2 to deduce that the Cauchy problem in (11.9) has a unique solution, considered as a Cauchy problem in E_0. These arguments prove the following

11.2 Theorem *Let* (11.1), (11.2), (11.4), *and* (11.6) *be satisfied with* $S := W$, *and suppose that* $1/p < 2\beta < 1 + 1/p$. *Then the parabolic boundary value problem* (11.5) *has a unique weak* $W_p^{2\beta}$*-solution, provided* $\sigma > 0$. *If* $\sigma = 0$ *and* $2 - 2\alpha < 2\beta < 2\alpha$, *it has u unique weak* $S_p^{2\gamma}$*-solution for any* γ *satisfying* $1/p < 2\gamma < 2\beta$.

Next we consider the case

$$1 + 1/p < 2\beta \leq 2\alpha \leq 2 , \tag{11.11}$$

and assume that $\sigma \in [0, 1)$ and

$$\begin{aligned}&\text{there exists}\\ &u_g \in C^\sigma(J, S_p^{2\beta}) \cap C^{1+\sigma}(J, S_p^{2\beta-2})\\ &\text{such that } \mathcal{B}u_g = g .\end{aligned} \tag{11.12}$$

Then

$$F_{\beta-1} := f - \mathcal{A}u_g - \dot{u}_g \in C^{\rho\wedge\sigma}(J, S_p^{2\beta-2}) ,$$

thanks to (11.1) and (8.10). Note that Theorem 8.5 implies

$$A_{\beta-1}(t) \in \mathcal{H}(S_{p,\mathcal{B}(t)}^{2\beta}, S_p^{2\beta-2}) , \qquad t \in J .$$

It is obvious that (11.9) is true in this case too.

Suppose that $2\rho > 2\beta - 1 - 1/p$ and fix γ such that $1/p < 2\gamma < 1 + 1/p$ and $\beta - \gamma < \rho$. Let $E_0 := H_{p,\mathcal{B}}^{2\gamma-2}$ and $E_1 := H_{p,\mathcal{B}}^{2\gamma}$ and note that Theorem 7.2 implies $S_p^{2\beta-2} \doteq (E_0, E_1)_{\beta-\gamma}$, provided $(\cdot,\cdot)_{\beta-\gamma} := [\cdot,\cdot]_{\beta-\gamma}$ if $S = H$, and $(\cdot,\cdot)_{\beta-\gamma} := (\cdot,\cdot)_{\beta-\gamma,p}$ if $S = W$ and $2\beta \neq 1$. Also note that $(11.8)_\gamma$ is true. Now, as an easy consequence of Theorem 10.2, we obtain the following result.

11.3 Theorem *Let assumptions* (11.1), (11.4), (11.11), *and* (11.12) *be satisfied with* $S := W$, *and suppose that* $2\rho > 2\beta - 1 - 1/p$. *If* $\sigma > 0$, *the parabolic boundary value problem* (11.5) *has a unique strong* $W_p^{2\beta}$*-solution. If* $\sigma = 0$ *and* $\beta < \alpha$, *it has a unique strong* $W_p^{2\gamma}$*-solution for each* γ *satisfying* $1 + 1/p < 2\gamma < 2\beta$.

Lastly, we consider the case of very weak solutions. Thus we assume that $0 \leq 2\beta < 1/p$ and $S_{p,\mathcal{B}}^{2\beta}$ and $S_{p',\mathcal{B}^\#}^{2-2\beta}$ are independent of t. We also presuppose (11.2).

Note that

$$F_{\beta-1} := f + \mathcal{S}_{\beta-1}g \in C^{\rho\wedge\sigma}(J, S_{p,\mathcal{B}}^{2\beta-2}) ,$$

thanks to (8.7), (8.8), (8.10), (8.11), (11.1), and the definition of $\mathcal{S}_{\beta-1}$ in (9.16). Also note that $(11.8)_\beta$ is satisfied, as follows from Theorem 8.5, the definition of $A_{\beta-1}(t)$ in (8.20), and (8.11) and (11.1).

We associate with (11.5) the Cauchy problem

$$\dot{u} + A_{\beta-1}(t)u = F_{\beta-1}(t) , \quad t \in \dot{J} , \qquad u(0) = u^0 \tag{11.13}$$

in $S^{2\beta-2}_{p,\mathcal{B}}$. Suppose that u is a strict $S^{2\beta}_p$-solution of (11.13). Then an obvious integration argument implies that u is a very weak $S^{2\beta}_p$-solution of (11.5). Conversely, let u be a very weak $S^{2\beta}_p$-solution of (11.5) satisfying

$$u \in C(J, S^{2\beta}_p) \cap C^1(J, S^{2\beta-2}_{p,\mathcal{B}}) .$$

Then it follows from (8.25) and the arguments of the proof of [Ama88b, Theorem 10.2] that u is a strict solution of (11.13).

Using these facts and similar arguments as above one obtains the following

11.4 Theorem *Let* (11.1) *and* (11.2) *be satisfied, where* $0 \leq 2\beta < 1/p$, *and let* $S^{2\beta}_{p,\mathcal{B}}$ *and* $S^{2-2\beta}_{p',\mathcal{B}^\#}$ *be independent of* t. *Also suppose that* $u^0 \in W^{2\beta}_p$ *and*

$$(f, g) \in C^\sigma(J, W^{2\beta-2}_{p,\mathcal{B}} \times \partial W^{2\beta}_p)$$

for some $\sigma \in [0, 1)$. *If* $\sigma > 0$, *the parabolic boundary value problem* (11.5) *has a unique very weak* $S^{2\beta}_p$*-solution* u *satisfying*

$$u \in C(J, W^{2\beta}_p) \cap C^1(J, W^{2\beta-2}_{p,\mathcal{B}}) . \tag{11.14$_\beta$}$$

If $\sigma = 0$ *and* $2 - 2\alpha < 2\beta < 2\alpha$ *then, given any* $\gamma \in [0, \beta)$, *problem* (11.5) *has a unique very weak* $S^{2\gamma}_p$*-solution satisfying* (11.14)$_\gamma$.

11.5 Remarks

(a) The hypotheses (11.6) and (11.12) imply additional restrictions for g, in general. They are trivially satisfied if (11.4) is true and g is independent of t. Note that (11.6) and (11.12) are necessary conditions, provided $\sigma = 0$ and we are interested in strict $S^{2\beta}_p$-solutions of (11.5), that is, in $S^{2\beta}_p$-solutions satisfying

$$u \in C(J, S^{2\beta}_p) \cap C^1(J, S^{2\beta-2}_{p,\mathcal{B}}) .$$

In this case u^0 has to belong to $S^{2\beta}_p$ too, of course.

We do not discuss conditions (11.6) and (11.12) here in more detail, but refer to [Ama94]. Of course, it is easy to give (nonoptimal) sufficient conditions for their validity (along the lines of [Ama90a, Appendix B], for example).

(b) It is obvious that Theorems 11.2–11.4 remain valid if W is replaced by H.

(c) The arguments preceding Theorem 11.3 show that the spaces $H^s_{p,\mathcal{B}}$ have to be considered too, even if one is interested in the Sobolev-Slobodeckii spaces $W^s_{p,\mathcal{B}}$ only. This stems from the fact that L_p and $W^1_{p,\mathcal{B}}$ cannot be obtained from Besov spaces by real or complex interpolation if $p \neq 2$.

(d) Suppose that $1/p < 2\beta < 1 + 1/p$. Then u is a strict $S_p^{2\beta}$-solution of (11.5) in the sense of (a) iff

$$u \in C(J, S_p^{2\beta})$$

and u satisfies

$$(1-\delta)\gamma_\partial u = (1-\delta)g$$

and

$$\int_J \{-\langle\dot\varphi, u\rangle + \boldsymbol{a}(\varphi, u)\}\, dt = \int_J \{\langle\varphi, f\rangle + \langle\gamma_\partial\varphi, \delta g\rangle_\partial\}\, dt + \langle\varphi(0), u^0\rangle$$

for all $\varphi \in C(J, S_{p',\mathcal{B}\#}^{2-2\beta}) \cap C^1(J, S_{p',\mathcal{B}\#}^{-2\beta})$ having compact support in $J\backslash\{\sup J\}$.

This follows by the arguments of the proof of [Ama88b, Theorem 10.3].

(e) The above considerations show that the parabolic boundary value problem (11.5) is equivalent to a Cauchy problem of the form

$$\dot v + A_{\beta-1}(t)v = F_{\beta-1}(t)\ , \quad t \in \dot J\ , \qquad v(0) = v^0 \tag{11.15}$$

in $S_{p,\mathcal{B}}^{2\beta-2}$, provided assumption (11.6) or (11.12) is satisfied if 2β belongs to the corresponding interval. Hence it is easy to deduce from Theorem 10.1 additional smoothness properties of the solutions provided u^0 is sufficiently regular. Moreover, by applying abstract maximal regularity theorems to (11.15) (e.g., Theorem 10.3), it is possible to obtain maximal regularity results for (11.5) (cf. [Ama94]).

(f) Of course, Theorems 11.2–11.4 allow bootstrapping arguments (in the obvious way) to improve the (local) regularity of the solutions if the data enjoy better (local) regularity properties (cf. Remark 9.5(b) above). Moreover, if the data are smooth, it can be shown (by employing classical results about linear parabolic boundary value problems in Hölder spaces; see [LadSU68], for example) that the solutions constructed above are smooth classical solutions of (11.5). For details we refer to [Ama94].

(g) It is obvious that Theorem 11.3 remains valid for general normally elliptic boundary value problems $(\mathcal{A}, \mathcal{B})$ which are not necessarily in divergence form (cf. Remark 8.6(d)).

(h) Nonhomogeneous (nonlocal) boundary data possessing little regularity occur naturally in control problems for parabolic equations (cf. [Ama89b] and [Koc92], for example, as well as the references given there). In this connection it is of particular interest to admit Radon measures as boundary data. Observe that this is possible in the weak settings studied in Theorems 11.2 and 11.4, provided p is sufficiently close to 1 (cf. Remark 9.5(b)).

(i) Suppose, for simplicity, that $(\mathcal{A}, \mathcal{B}) \in \mathcal{E}^1(\Omega)$, that $\sigma(A_0) \subset [\operatorname{Re} z > 0]$, and that $\delta = 1$. Consider the 'semi-autonomous' Neumann type parabolic boundary value problem

$$\begin{aligned} \partial_t u + \mathcal{A}u &= f(t) && \text{in } \ \Omega \times \dot{J} \ , \\ \mathcal{B}u &= g(t) && \text{on } \partial\Omega \times \dot{J} \ , \\ u(\cdot,0) &= u^0 && \text{on } \ \Omega \ . \end{aligned} \tag{11.16}$$

Let $W^2_{p,\mathcal{A}} := \ker(\mathcal{A})$. It is not difficult to see that

$$W^2_p = W^2_{p,\mathcal{B}} \oplus W^2_{p,\mathcal{A}} \tag{11.17}$$

and that

$$\mathcal{B}^c := (\mathcal{B}|W^2_{p,\mathcal{A}})^{-1} \in \mathcal{L}\mathrm{is}\big(W^{1-1/p}_p(\partial\Omega,\mathbb{R}^N), W^2_{p,\mathcal{A}}\big) \ .$$

By decomposing a solution u of (11.16) according to (11.17) in the form $u = v + w$, it is easily verified that $w = \mathcal{B}^c g$ and v satisfies

$$\dot{v} + A_0 v = f(t) - (\mathcal{B}^c g)^{\cdot} \ , \quad t \in \dot{J} \ , \qquad v(0) = u_0 - w(0) \ . \tag{11.18}$$

Since $A_0 \in \mathcal{H}(W^2_{p,\mathcal{B}}, L_p)$, the solution of (11.18) can be represented by the variation-of-constants formula. Using the fact that $(e^{-tA_0})^{\cdot} = -A_0 e^{-tA_0}$ for $t > 0$, a formal partial integration then leads to the representation

$$u(t) = u^0 + \int_0^t e^{-(t-\tau)A_0}\big\{f(\tau) + A_0\mathcal{B}^c g(\tau)\big\}\,d\tau \ , \qquad t \in J \ , \tag{11.19}$$

where we also used the fact that a semigroup commutes with its generator. Unfortunately, formula (11.19) is not meaningful since, in general, $\mathcal{B}^c g(\tau)$ will not be contained in $\mathrm{dom}(A_0)$. However, if we replace A_0 in (11.19) by $A_{\alpha-1}$ for $2\alpha \in (1/p, 1+1/p)$, we obtain the **generalized varation-of-constants formula**

$$u(t) = u^0 + \int_0^t e^{-(t-\tau)A_{\alpha-1}}\big\{f(\tau) + A_{\alpha-1}\mathcal{B}^c g(\tau)\big\}\,d\tau \ , \qquad \tau \in J \ , \tag{11.20}$$

for solutions of (11.16), which is meaningful, thanks to

$$\mathrm{im}(\mathcal{B}^c) = W^2_{p,\mathcal{A}} \hookrightarrow W^2_p \hookrightarrow W^{2\alpha}_p = W^{2\alpha}_{p,\mathcal{B}} \ .$$

Moreover, it follows from Theorem 9.2 that $\mathcal{B}^c$ has a unique continuous extension to a map

$$\mathcal{B}^c_\alpha \in \mathcal{L}(\partial W^{2\alpha}_p, W^{2\alpha}_p) \ .$$

Hence (11.20) is well defined for $(f,g) \in C(J, W^{2\alpha-2}_{p,\mathcal{B}} \times \partial W^{2\alpha}_p)$.

Note that u of (11.20) is a 'mild' solution of the Cauchy problem

$$\dot{u} + A_{\alpha-1}u = f + A_{\alpha-1}\mathcal{B}^c_\alpha g \quad \text{in } \dot{J} \ , \qquad u(0) = u^0$$

in $W^{2\alpha-2}_{p,\mathcal{B}}$. Using Green's formula (8.14) and a density argument, it is not difficult to see that

$$A_{\alpha-1}\mathcal{B}^c_\alpha = \mathcal{R}_{\alpha-1}$$

where $\mathcal{R}_{\alpha-1}$ is defined in (9.14).

The above approach to the nonhomogeneous linear parabolic boundary value problem (11.6) has been proposed by the author in [Ama86a] and extended to nonautonomous (linear and semilinear) problems in [Ama88b] (also cf. [Ama86d] for a heuristic explanation of this method). Later we discovered that the representation formula (11.19) had already been introduced by Balakrishnan [Bal76] and exploited by Lasiecka and others (e.g., [Las80], [Trig80]) for problems in control theory. These authors work in a Hilbert space setting which is crucial for their approach.

It should be mentioned that our investigations around the generalized variation-of-constants formula (11.20) eventually led us to introduce the interpolation-extrapolation scales described in Section 6 above and to our approach to quasilinear parabolic systems to be described in the following sections. □

Suppose now that $\mathbb{K}=\mathbb{R}$ and $N=1$, and endow all spaces with their respective natural orders as in Theorem 8.7. Also suppose that

$$g=0 \qquad \text{if} \quad 1+1/p<2\beta\le 2\alpha\le 2 \tag{11.21}$$

and

$$(1-\delta)g=0 \qquad \text{if} \quad 2-2\alpha\le 2\beta<1+1/p\ . \tag{11.22}$$

Then it follows from (11.9), (11.12), and (11.13), respectively, that the unique $S_p^{2\beta}$-solution u of (11.5) is the solution of the Cauchy problem (11.13), where $A_{\beta-1}$ satisfies $(11.8)_\beta$ and $F_{\beta-1}=f+\mathcal{R}_{\beta-1}\delta g$. Thus u is given by

$$u(t)=U_{\beta-1}(t,0)u^0+\int_0^t U_{\beta-1}(t,\tau)F_{\beta-1}(\tau)\,d\tau\ , \qquad t\in J\ , \tag{11.23}$$

where $U_{\beta-1}$ is the parabolic evolution operator for $A_{\beta-1}$ (cf. [DaK92] or [Ama94], for example). Note that $\mathcal{R}_{\beta-1}\ge 0$, thanks to (9.14) and the positivity of the trace operator. Hence $F_{\beta-1}(t)\ge 0$ if $\big(f(t),g(t)\big)\ge 0$. By Theorem 8.7 we know that $A_{\beta-1}(t)$ is resolvent positive for each $t\in J$. Hence [DaK92, Corollary 10.10] implies that $U_{\beta-1}(t,s)\ge 0$ for $s,t\in J$ with $s\le t$. Now the following theorem is an obvious consequence of (11.23).

11.6 Theorem *Suppose that $\mathbb{K}=\mathbb{R}$ and $N=1$, and let conditions* (11.21) *and* (11.22), *respectively, be satisfied. If $u^0\ge 0$ and $\big(f(t),g(t)\big)\ge 0$ for $t\in J$ then $u(t)\ge 0$ for $t\in J$, where u is the unique $S_p^{2\beta}$-solution of Theorems* 11.2–11.4, *respectively.*

As always, the regularity assumptions for the 'lower order coefficients' of $(\mathcal{A},\mathcal{B})$ can be weakened. For later purposes we exhibit the following simple special case.

11.7 Remark Suppose that $p>n$ and $\rho\in(0,1)$. Also suppose that

$$(\mathcal{A},\mathcal{B})\in C^\rho\big(J,\mathcal{E}_p(\Omega)\big)$$

and

$$\big(u^0,(f,g)\big)\in W_{p,\mathcal{B}}^{-1}\times C^\rho(J,W_{p,\mathcal{B}}^{-1}\times\partial W_p^1)$$

with $(1-\delta)g = 0$. Then problem (11.5) has a unique weak W_p^1-solution u. If $\mathbb{K} = \mathbb{R}$ and $N = 1$ then $u^0 \geq 0$ and $\big(f(t), g(t)\big) \geq 0$ for $t \in J$ imply $u(t) \geq 0$ for $t \in J$.

This follows from Remarks 8.6 and 8.8 and the above considerations. □

The results of the present section occur in the given generality in this paper for the first time in print. The basic observation of Remark 11.5(e) is due to the author. It has been used — in the case $1/p < 2\beta < 1 + 1/p$ and under the assumption that $(1-\delta)g = 0$ — implicitly or explicitly in many of the papers of the author cited in the references, but almost all the times under more restrictive assumptions (see Remark 11.5(h)).

Nonhomogeneous linear parabolic boundary value problems in W_p^2 (in fact, in W_p^{2m} for systems of order $2m$, which we do not consider here) have been studied by Solonnikov in great generality (cf. [Sol65] and [LadSU68]), where optimal regularity results are given. The case $p = 2$ is also considered in [AgV64]. A different approach to W_p^2-solutions has been proposed by Terreni [Ter89], who obtained optimal regularity results in spaces of functions which are Hölder continuous in t with values in W_p^2. Terreni's technique is inspired by the 'operator theoretic' approach of Grisvard [Gri69], based on Dunford integral representations. For further results in this direction using spaces of continuous and Hölder continuous functions and Nikolskii spaces we refer to [LuSW92], [Lun89], and [Gui90], respectively, as well as to the references in those papers (also see [Gui91a]).

Whereas the papers cited above all deal with strong W_p^2-solutions (or even more regular ones), there are very few results for nonhomogeneous parabolic problems with nonregular data. Lions and Magenes have extended their 'elliptic' theory, referred to in Section 9, to nonhomogeneous linear parabolic problems in [LiM72, Chapter 4]. However, due to the considerable 'technical' difficulties they encounter, an L_2-theory has been developed only (see the relevant remarks in [LiM72, vol. II pp. 85–88]). In addition, the Lions-Magenes theory requires rather smooth coefficients and differs from our theory (in the case $p = 2$, of course) in that we do not necessarily require that f be a distribution, whereas f has to belong to a suitable (nonoptimal) space of distributions on Ω in their theory.

As already mentioned in Remark 11.5(h), Lasiecka [Las80] and Triggiani [Trig80] used a semigroup approach to autonomous problems with irregular data, based on Balakrishnan's representation formula (11.19). However, their theory is a Hilbert space theory which uses fractional powers in an essential way. For these reasons it is neither applicable to problems with nonautonomous principal parts nor to W_p^s-spaces with $p \neq 2$.

Lastly, in [LuV91] the authors prove Hölder regularity results for weak W_p^1-solutions in the case of a single equation ($N = 1$).

12 Abstract Quasilinear Parabolic Problems

Let J be a perfect subinterval of $\mathbb{R}^+$ containing 0, let E be a Banach space, and let V be a nonempty subset of E. Suppose that $A(t,x)$ is for each $(t,x) \in J \times V$ a linear operator in E and that $f : J \times V \to E$.

By a **solution** u on $\dot{J}_u$ of the **quasilinear evolution equation**

$$\dot{u} + A(t,u)u = f(t,u) , \qquad t \in J , \tag{12.1}$$

we mean a function

$$u \in C(J_u, V) \cap C^1(\dot{J}_u, E) ,$$

where J_u is a perfect subinterval of J containing 0, such that $u(t) \in \operatorname{dom} A\big(t, u(t)\big)$ and

$$\dot{u}(t) + A\big(t, u(t)\big)u(t) = f\big(t, u(t)\big) , \qquad t \in \dot{J}_u .$$

If, in addition, $u(0) = x$ then u is a solution on J_u of the **quasilinear Cauchy problem**

$$\dot{u} + A(t, u)u = f(t, u) \quad \text{in } \dot{J}_u , \qquad u(0) = x . \tag{12.2}$$

The evolution equation (12.1) and the Cauchy problem (12.2) are said to be **parabolic** if $A(t, x)$ is for each $(t, x) \in J \times V$ a negative infinitesimal generator of an analytic semigroup on E.

Given $u : J_u \to V$, put

$$(A_u, f_u)(t) := \big(A(t, u(t)), f(t, u(t))\big) , \qquad t \in J .$$

Then u is a solution of (12.2) iff u is a solution of the linear Cauchy problem

$$\dot{v} + A_u(t)v = f_u(t) , \quad t \in \dot{J}_u , \qquad v(0) = x . \tag{12.3}$$

It is a **strict** solution of the Cauchy problem (12.2) on J_u if u is a strict solution of (12.3). Lastly, a solution u of (12.1) is **maximal** if there does not exist a solution of (12.1) which is a proper extension of u.

Let X be a metric space and let $t^+ : X \to (0, \infty]$. Put

$$\mathfrak{X} := \bigcup_{x \in X} [0, t^+(x)) \times \{x\}$$

and, given $\varphi : \mathfrak{X} \to X$, let $\varphi^t(x) := \varphi(t, x)$. Then φ is a (local) **semiflow** on X if

(i) $\mathfrak{X}$ is open in $\mathbb{R}^+ \times X$;

(ii) $\varphi \in C(\mathfrak{X}, X)$;

(iii) $\varphi^0 = \mathrm{id}_X$;

(iv) if $x \in X$ and $0 \leq s < t^+(x)$, and if $0 \leq t < t^+\big(\varphi^s(x)\big)$ then $s + t < t^+(x)$ and $\varphi^{s+t}(x) = \varphi^t \circ \varphi^s(x)$.

If $t^+(x) = \infty$ for each $x \in X$ then φ is a **global** semiflow on X. The set

$$\gamma^+(x) := \{\, \varphi^t(x) \; ; \; 0 \leq t < t^+(x) \,\}$$

is the (positive semi-) **orbit** through x, and $t^+(x)$ is the (positive) **escape time** of x.

Throughout the remainder of this section we assume that

$$(E_0, E_1) \text{ is a densely injected Banach couple} ,$$

and $(\cdot,\cdot)_\theta$ is for each $\theta \in (0,1)$ an admissible interpolation functor. Moreover,

$$E_\theta := (E_0, E_1)_\theta \ , \qquad 0 < \theta < 1 \ ,$$

and, given $X \subset E_\theta$ for some $\theta \in (0,1)$, we put $X_\eta := X \cap E_\eta$ for $\eta \in [\theta, 1]$, equipped with the topology induced by E_η, as explained in Section 3.

After these preparations we can formulate the basic existence, uniqueness, and continuity theorem for quasilinear parabolic Cauchy problems.

12.1 Theorem *Suppose that $0 < \gamma \leq \beta < \alpha < 1$, that X_β is open in E_β, and that*

$$(A, f) \in C^{1-}\big(X_\beta, \mathcal{H}(E_1, E_0) \times E_\gamma\big) \ .$$

Then the autonomous quasilinear parabolic Cauchy problem

$$\dot{u} + A(u)u = f(u) \ , \quad 0 < t < \infty \ , \qquad u(0) = x \tag{12.4}$$

possesses for each $x \in X_\alpha$ a unique maximal solution

$$u(\cdot, x) \in C\big([0, t^+(x)), X_\alpha\big) \cap C^\alpha\big([0, t^+(x)), E_0\big) \ .$$

The map $(t,x) \mapsto u(t,x)$ is a semiflow on X_α.

This theorem is due to the author. A proof is contained implicitly in [Ama86c]. A detailed derivation has also been given in the lecture series at TH Delft in the fall of 1989, referred to already in Section 9 (cf. [Li92, Chapter IV, Appendix A], where this proof has been written up for the particular case $f = 0$). A complete proof will be given in [Ama94]. It relies on a careful study of linear parabolic evolution equations (cf. [Ama86c], [Ama88a]) and on Banach's fixed point theorem. (In [Ama88a] a more complicated version (containing some unnecessary restrictions) has been proven which was needed to obtain the results of [Ama90a] since at that time the generation Theorem 8.5, which simplifies the study of quasilinear parabolic boundary value problems considerably, was not yet available.)

The first existence theorem for the quasilinear Cauchy problem (12.4) is due to Sobolevskii [Sob66]. Roughly speaking, Sobolevskii uses fractional power spaces $D\big(A^\alpha(0)\big)$ and $D\big(A^\beta(0)\big)$ instead of the interpolation spaces E_α and E_β, respectively. However, fractional power spaces are not particularly well suited for quasilinear problems and their use causes considerable technical problems. In particular, Sobolevskii did not prove that

$$u(\cdot, x) \in C\big([0, t^+(x)), D(A^\alpha(0))\big) \ , \qquad x \in D\big(A^\alpha(0)\big) \ .$$

The continuity of the solution is only shown with respect to a topology which is weaker than the topology of the space of initial values. Hence the results of [Sob66] do not guarantee that the solutions generate a semiflow.

The use of interpolation spaces simplifies the theory considerably (cf. [DaK92] for an exposition of this fact in the semilinear case). Moreover, it has the advantage that in concrete applications the spaces E_θ often can be identified with well known function spaces as is shown in this article. In contrast to this, the spaces $D\big(A^\alpha(0)\big)$, $0 < \alpha < 1$, are not explicitly known, in general, or only if restrictive regularity assumptions are satisfied, which is of little value for quasilinear parabolic boundary value problems.

More recently, Angenent [Ang90] has studied autonomous quasilinear parabolic Cauchy problems in 'spaces of maximal regularity' in the sense of Da Prato and Grisvard [DaPG79] (also cf. [Sit92], [Sit93a], [Sit93b] for some extensions and complements). The maximal regularity setting requires, however, serious restrictions for the Banach couple (E_0, E_1). In particular, E_0 (whence E_1) cannot be reflexive if $E_0 \neq E_1$, as has been shown in [Bai80].

It should be noted that Theorem 12.1 has the advantage of being of great generality and flexibility. Namely, there are no restrictions for the Banach spaces, except for $E_1 \overset{d}{\hookrightarrow} E_0$, and there are practically no restrictions for the interpolation functors, except for the density requirements $E_1 \overset{d}{\hookrightarrow} (E_0, E_1)_\theta$.

12.2 Remarks

(a) It is easy to deduce from the 'autonomous' Theorem 12.1 an existence, uniqueness, and continuity theorem for parameter-dependent nonautonomous Cauchy problems. For this we suppose that Λ is an open subset of some Banach space F and put

$$\boldsymbol{E}_\theta := \mathbb{R} \times E_\theta \times F , \qquad 0 \leq \theta \leq 1 .$$

Note that $(\boldsymbol{E}_0, \boldsymbol{E}_1)$ is a densely injected Banach couple and $(\boldsymbol{E}_0, \boldsymbol{E}_1)_\theta \doteq \boldsymbol{E}_\theta$ for $0 < \theta < 1$. We also assume that $0 < \gamma \leq \beta < \alpha < 1$ and that

$$(A, f) \in C^{1-}(J \times X_\beta \times \Lambda, \mathcal{H}(E_1, E_0) \times E_\gamma) . \tag{12.5}$$

Then we consider the parameter-dependent nonautonomous quasilinear parabolic Cauchy problem

$$\dot{u} + A(t, u, \lambda)u = f(t, u, \lambda) , \quad t \in \dot{J} , \qquad u(0) = x^0 . \tag{12.6}$$

Put $\boldsymbol{X}_\beta := \mathbb{R} \times X_\beta \times \Lambda$, fix any $T \in \dot{J}$, and extend $(A, f)|([0, T] \times X_\beta \times \Lambda)$ over all of $\boldsymbol{X}_\beta$ in the class C^{1-}. Denoting this extension again by (A, f), define

$$(\boldsymbol{A}, \boldsymbol{f}) \in C^{1-}(\boldsymbol{X}_\beta, \mathcal{H}(\boldsymbol{E}_1, \boldsymbol{E}_0) \times \boldsymbol{E}_\gamma)$$

by

$$(\boldsymbol{A}, \boldsymbol{f}) := \left(\begin{bmatrix} 0 & 0 & 0 \\ 0 & A & 0 \\ 0 & 0 & 0 \end{bmatrix}, \begin{bmatrix} 1 \\ f \\ 0 \end{bmatrix} \right) .$$

Then (12.6) is equivalent to the autonomous parabolic Cauchy problem

$$\dot{\boldsymbol{u}} + \boldsymbol{A}(\boldsymbol{u})\boldsymbol{u} = \boldsymbol{f}(\boldsymbol{u}) , \quad 0 < t < \infty , \qquad \boldsymbol{u}(0) = \boldsymbol{x} , \tag{12.7}$$

where $\boldsymbol{u} := (t, u, \lambda)$ and $\boldsymbol{x} := (0, x, \lambda)$. Of course, if $\boldsymbol{u} := (t, u, \lambda)$ is a solution to (12.7) on $[0, t^+)$ then $u|[0, T] \cap [0, t^+)$ is a solution to (12.6) on this interval. Since $T \in \dot{J}$ was arbitray, we obtain a maximal solution to (12.5) by uniqueness.

However, by extending the proof of Theorem 12.1 to deal with (12.6) directly, it is possible to weaken the continuity requirements for A and f with respect to t and λ. This is important for some applications. For a proof we refer to [Ama94] (also cf. [Ama88a]).

(b) Of course, the above results also apply to the **semilinear Cauchy problem**

$$\dot{u} + A(t)u = f(t,u) \ , \quad t \in \dot{J} \ , \qquad u(0) = x \tag{12.8}$$

(and to the parameter-dependent analogue), where

$$A \in C^{\rho}\big(J, \mathcal{H}(E_1, E_0)\big) \tag{12.9}$$

for some $\rho \in (0,1)$. In this case standard arguments based upon the integral equation deduced from the variation-of-constants formula guarantee the solvability of (12.8) also in E_1 if f maps into E_γ for some $\gamma \in (0,1)$. More precisely, suppose that

$$X_\alpha \text{ is an open subset of } E_\alpha \text{ for some } \alpha \in [0,1]$$

and

$$f \in C^{1\text{-}}(J \times X_\alpha, E_\gamma) \qquad \text{for some } \gamma \in [0,\alpha] \quad \text{with } \alpha - \gamma < 1 \ .$$

Then, given $x \in X_\alpha$, there exists a unique maximal solution $u(\cdot,x)$ of (12.8) satisfying

$$u(\cdot,x) \in C\big([0,t^+(x)), X_\alpha\big) \cap C^\alpha\big([0,t^+(x)), X_0\big) \ .$$

Moreover, $u(\cdot,x)$ depends continuously upon $x \in X_\alpha$ in the topology of X_α. In particular, if (12.8) is autonomous, $(t,x) \mapsto u(t,x)$ is a semiflow on X_α. Lastly, $u(\cdot,x)$ is the unique solution of the Volterra integral equation

$$u(t) = U(t,0)x + \int_0^t U(t,\tau) f\big(\tau, u(\tau)\big)\, d\tau \ , \qquad t \in J \ ,$$

in E_α, where U is the unique parabolic fundamental solution for (12.9) (cf. [Ama84] and [Ama94]). Recall that $U(t,\tau) = e^{-(t-\tau)A}$ if $A(t) := A \in \mathcal{H}(E_1, E_0)$ for $t \in J$. □

For simplicity we restrict our considerations in the following to the autonomous quasilinear Cauchy problem (12.4) and we leave it to the reader to extend the results below to the nonautonomous case (along the lines of Remark 12.2(a), for example; cf. [Ama94]). We also presuppose the hypotheses of Theorem 12.1 for what follows.

The fact that $(t,x) \mapsto u(t,x)$ is a semiflow on X_α has important consequences for the long time behavior of solutions to (12.4). For example, that the ω-limit set $\omega(x)$ of $x \in X_\alpha$, defined by

$$\omega(x) := \{\, y \in X_\alpha \ ; \text{ there exists a sequence } t_j \to \infty \text{ such that } u(t_j, x) \to y \,\} \ ,$$

is a nonempty compact connected subset of X_α, provided the orbit $\gamma^+(x)$ is relatively compact. In particular, $t^+(x) = \infty$, that is, the solution to problem (12.4) exists globally, if $\gamma^+(x)$ is relatively compact. (We refer to [Ama90b] for the elementary theory of semiflows.)

It is an important consequence of Theorem 12.1 that the semiflow $(t,x) \mapsto u(t,x)$ is smoothing. In fact, $u(\cdot,x) \in C\big((0,t^+(x)),E_1\big)$ for $x \in X_\alpha$. Thus, if we are interested in the long time behavior of $u(\cdot,x)$, it is no restriction to assume that $x \in X_1$, which we will do in the following (since it will simplify the presentation somewhat).

Now suppose that $E_1 \overset{cc}{\hookrightarrow} E_0$ and that $\gamma^+(x)$ is bounded in E_δ for some $\delta \in (\alpha,1)$. Then it follows from $E_\delta \overset{cc}{\hookrightarrow} E_\alpha$ (cf. Remark 6.1(e)) that $\gamma^+(x)$ is relatively compact in X_α, provided it is bounded away from ∂X_α. Thus $t^+(x) = \infty$ in this case. In fact, one can do better if one is interested in global existence only. Namely, by analyzing the proof of Theorem 12.1 carefully and using the smoothing properties of the semiflow one proves the following

12.3 Theorem *Suppose that $E_1 \overset{cc}{\hookrightarrow} E_0$ and there exists $\delta \in (\beta,\alpha)$ such that*

$$u(\cdot,x)[0,T] := u(\cdot,x)\big|\big[0,t^+(x)\big) \cap [0,T]$$

is bounded in E_δ and bounded away from ∂X_β for each $T > 0$. Then $t^+(x) = \infty$.

12.4 Remark Theorem 12.3 guarantees the global existence of $u(\cdot,x)$ for a given x, provided it is possible to establish **a priori bounds** for $u(\cdot,x)$ in the norm of E_δ for some $\delta > \beta$. This shows that one should try to choose β as close to zero as possible since then one has to establish a priori bounds in relatively 'weak norms' only. □

Theorem 12.3 is closely related to the following 'blow up' result for solutions having a finite 'lifetime' only.

12.5 Theorem *Suppose that $E_1 \overset{cc}{\hookrightarrow} E_0$ and $t^+(x) < \infty$. Then either*

$$u(t,x) \to \partial X_\beta \quad \text{or} \quad \|u(t,x)\|_\delta \to \infty \qquad \text{as} \quad t \to t^+(x)$$

for each $\delta \in (\beta,1]$. If $u(\cdot,x)$ is bounded in E_δ for some $\delta \in (\beta,1]$ and uniformly continuous from $\big[0,t^+(x)\big)$ to E_β then $u(\cdot,x)$ converges towards a single point of ∂X_β.

Lastly, we give a sufficient condition for a bounded orbit to be relatively compact.

12.6 Theorem *Suppose that $E_1 \overset{cc}{\hookrightarrow} E_0$ and that $\gamma^+(x)$ is bounded away from ∂X_β for some $x \in X_\alpha$. Also suppose that*

$$u \in BUC^\rho\big(\big[0,t^+(x)\big),E_\delta\big)$$

for some $\delta \in (\beta,\alpha]$. Then $\gamma^+(x)$ is relatively compact in X_α, whence $t^+(x) = \infty$. Moreover, $\big\{\, u(t,x) \;;\; t \geq \tau \,\big\}$ is bounded in E_ε for $\varepsilon \in [0,1)$ and $\tau > 0$.

Proofs of the above results will be found in [Ama94], where the case that E_1 is not compactly injected in E_0 will be considered too.

Suppose now that
$$(A,f) \in C^k\big(X_\beta, \mathcal{H}(E_1,E_0) \times E_\gamma\big)$$
for some $k \in \dot{\mathbb{N}} \cup \{\infty,\omega\}$, where C^ω means 'analytic'. Then it can be shown that the semiflow $(t,x) \mapsto u(t,x)$ belongs to the class C^k too for $t > 0$. Moreover, considering $\partial_2 u(\cdot,x)$, for example, it follows that, given any $h \in E_\alpha$, the function $\partial_2 u(\cdot,x)h$ is the unique solution of the linearized Cauchy problem
$$\begin{aligned} \dot v + A\big(u(t,x)\big)v + \partial A\big(u(t,x)\big)\big[v,u(t,x)\big] &= \partial f\big(u(t,x)\big)v\ , \qquad 0 < t < t^+(x)\ ,\\ v(0) &= h\ . \end{aligned}$$

Thus, roughly speaking, the results known from the theory of ordinary differential equations hold in the case of abstract quasilinear parabolic equations too. In fact, this analogy extends further: the principle of linearized stability is true; given suitable hypotheses, there are stable, unstable, and center manifolds possessing properties similar to the ones for ordinary differential equations; Hopf bifurcation theorems are valid; etc.

A qualitative theory analoguous to the one for ordinary differential equations has first been established by D. Henry in [Hen81] for the case of semilinear parabolic evolution equations. In the last years such a theory has also been developed for 'fully nonlinear parabolic problems' by Da Prato and Lunardi (jointly, and separately by Lunardi; cf. [Lun91] for a survey). By using maximal regularity too, Angenent [Ang90] and Simonett [Sit92], [Sit93a], [Sit93b] developed aspects of a qualitative theory for quasilinear parabolic evolution equations. The use of maximal regularity results greatly simplifies this task. In fact, given maximal regularity, the proofs follow pretty closely the corresponding ones of the theory of ordinary differential equations (modulo 'technical' difficulties, of course). As explained already, maximal regularity requires restrictive assumptions which we try to avoid. Without maximal regularity proofs are somewhat harder and technically more involved since one has to handle a 'loss of regularity'. This can be seen by comparing, for example, the 'differentiability proof' in the presence of maximal regularity assumptions in [Ang90] or [Sit93b] with the corresponding proof without maximal regularity hypotheses in [Ama88a]. In fact, a center-manifold theory has not yet been established without using maximal regularity.

We shall not go into further detail here but refer the interested reader to [Ama94]. However, it should be mentioned that the applicability of the qualitative theory of quasilinear parabolic evolution equations to quasilinear parabolic systems, to reaction-diffusion systems, for example — though of great importance — is somewhat limited due to the great complexity of those problems. This should be clear to any one knowing a bit about the great complexity of a single semilinear parabolic equation (cf. [Ama91c] and [AmR93] for explicit formulations of simple looking open problems which have to be solved for obtaining deeper qualitative information).

'Fully nonlinear parabolic problems' are of the form
$$\dot u + \varphi(u) = 0\ , \quad t > 0\ , \qquad \varphi(0) = x\ , \tag{12.10}$$
where $\varphi \in C^2(X_1, E_0)$ is such that
$$\partial\varphi \in C^1\big(X_1, \mathcal{H}(E_1,E_0)\big)\ . \tag{12.11}$$

The investigations of this class of problems are all based on 'maximal regularity properties' either in the sense of [DaPG79] (namely in [DaPG79]) and [Ang90], for example) or in the sense of singular Hölder spaces (e.g., [Lun87]). In fact, the following theorem is a consequence of [Lun87, Theorem 2].

12.7 Theorem *Let condition* (12.11) *be satisfied. Then the Cauchy problem* (12.10) *possesses for each $x \in X_1$ a unique maximal solution*

$$u(\cdot, x) \in C\big([0, t^+(x)), X_1\big) \cap C^1\big([0, t^+(x)), E_0\big) \ .$$

The map $(t, x) \mapsto u(t, x)$ is a semiflow on X_1.

This theorem can, of course, also be applied to quasilinear Cauchy problems. In general, it is, however, not advisable to use this approach since it requires the initial values to belong to E_1 so that the solutions do not exhibit any smoothing effect. Of course, there are situations where no smoothing does occur and one has to have recourse to this approach (cf. Section 17 below).

13 General Quasilinear Parabolic Boundary Value Problems

We show now how the abstract results of Section 12 can be combined with the theory developed above to prove that quasilinear parabolic boundary value problems generate semiflows on suitable phase spaces. In this section we consider the general case where the nonlinearities are not supposed to be local. Thus we study problems of the form

$$\begin{aligned} \partial_t u + \mathcal{A}(u)u &= F(u) && \text{in} \quad \Omega \times (0,\infty) \ , \\ \mathcal{B}(u)u &= G(u) && \text{on} \ \partial\Omega \times (0,\infty) \ , \\ u(\cdot, 0) &= u^0 && \text{on} \quad \Omega \ . \end{aligned} \tag{13.1}$$

We start by considering weak solutions. For this we suppose that

$$1 \le \sigma < 1 + 1/p < 2 \quad \text{and} \quad \sigma - 2 < \tau \le r < s < \sigma \tag{13.2}$$

such that $\tau, r, s \notin \Sigma_p$. We also suppose that

$$\sigma \le 2\widehat{\alpha} \le 2 \ ,$$

that

$$V^r_{p,\mathcal{B}} \text{ is a nonempty open subset of } W^r_{p,\mathcal{B}}$$

and that

$$(\mathcal{A}, \mathcal{B}) \in C^{1\text{-}}(V^r_{p,\mathcal{B}}, \mathcal{E}^{\widehat{\alpha}}(\Omega)) \tag{13.3}$$

and

$$(F, G) \in C^{1\text{-}}(V^r_{p,\mathcal{B}}, W^\tau_{p,\mathcal{B}} \times \partial W^{\tau+2}_p) \ . \tag{13.4}$$

Lastly, we assume that

$$(1 - \delta)G = 0 \ . \tag{13.5}$$

Let $S \in \{H, W\}$ and suppose that $\rho \in (r, 1+1/p) \backslash \Sigma_p$ satisfies $2-2\widehat{\alpha} \leq \rho \leq 2\widehat{\alpha}$. Then u is said to be a **weak S_p^ρ-solution** on J of (13.1) if J is a perfect subinterval of $\mathbb{R}^+$ containing 0, if

$$u \in C(J, V_{p,\mathcal{B}}^r) , \tag{13.6}$$

and u is a weak S_p^ρ-solution of the linear parabolic boundary value problem

$$\begin{aligned} \partial_t v + \mathcal{A}(u(t))v &= F(u(t)) && \text{in } \Omega \times \dot{J} , \\ \mathcal{B}(u(t))v &= G(u(t)) && \text{on } \partial\Omega \times \dot{J} , \\ u(\cdot, 0) &= u^0 && \text{on } \Omega . \end{aligned} \tag{13.7}$$

Note that (13.3)–(13.6) imply that the concept of weak S_p^ρ-solutions of (13.7) is well defined.

Choose $\sigma_1 \in [1, 1+1/p)$ such that

$$\sigma_1 = 1 \quad \text{if } 2\widehat{\alpha} = 1 \qquad \text{and} \qquad \sigma - 1 < \sigma_1 - 1 < \rho(\widehat{\alpha}) \wedge (\tau + 1) \quad \text{otherwise} ,$$

where $\rho(\widehat{\alpha})$ is the Hölder exponent (8.6). Put $E_0 := H_{p,\mathcal{B}}^{\sigma_1 - 2}$ and $E_1 := H_{p,\mathcal{B}}^{\sigma_1}$ and define $\mathfrak{F} := \{ (\cdot,\cdot)_\theta \ ; \ 0 < \theta < 1 \}$ by

$$(\cdot,\cdot)_\theta := \begin{cases} [\cdot,\cdot]_\theta & \text{if } \theta \in \{(3-\sigma_1)/2, (2-\sigma_1)/2\} , \\ (\cdot,\cdot)_\theta & \text{otherwise} . \end{cases} \tag{13.8}$$

Then (E_0, E_1) is a densely injected Banach couple and Theorem 7.2 implies

$$E_\theta := (E_0, E_1)_\theta \doteq W_{p,\mathcal{B}}^{\sigma_1 - 2 + 2\theta} , \qquad 0 < \theta < 1 , \quad \sigma_1 - 2 + 2\theta \notin \Sigma_p . \tag{13.9}$$

Let

$$\alpha := 1 - (\sigma_1 - s)/2 , \quad \beta := 1 - (\sigma_1 - r)/2 , \quad \gamma := 1 - (\sigma_1 - \tau)/2 .$$

Then $0 < \gamma \leq \beta < \alpha < 1$ and

$$E_\alpha \doteq W_{p,\mathcal{B}}^s , \quad E_\beta \doteq W_{p,\mathcal{B}}^r , \quad E_\gamma \doteq W_{p,\mathcal{B}}^\tau .$$

Hence $X_\beta := V_{p,\mathcal{B}}^r$ is a nonempty open subset of E_β. Let $A := A_{(\sigma_1/2)-1}$ and note that (13.3) and Theorem 8.5 imply

$$A \in C^{1\text{-}}(X_\beta, \mathcal{H}(E_1, E_0)) .$$

Also note that

$$f := F + \mathcal{R}_{\tau/2} G \in C^{1\text{-}}(X_\beta, E_\gamma) ,$$

thanks to (13.4), where $\mathcal{R}_{\tau/2}$ has been defined in (9.14).

It is an obvious consequence of (11.9) that u is a weak $H_p^{\sigma_1}$-solution of (13.1) on J iff u is a solution of the quasilinear Cauchy problem

$$\dot{u} + A(u)u = f(u) , \quad 0 < t < \infty , \qquad u(0) = u^0 \tag{13.10}$$

in E_0. Thanks to Theorem 12.1 the Cauchy problem (13.10) generates a local semiflow on

$$X_\alpha := V_{p,\mathcal{B}}^s := V_{p,\mathcal{B}}^r \cap W_p^s$$

consisting of weak $H_p^{\sigma_1}$-solutions of (13.1). It is obvious that each weak $H_p^{\sigma_1}$-solution of (13.1) is a weak W_p^σ-solution too.

Suppose now that $2-2\widehat{\alpha} \le s \le 2\widehat{\alpha}$ and let $\gamma^+(u^0)$ be bounded in X_α and bounded away from ∂X_β for some $u^0 \in X_\alpha$. Since $E_\alpha \overset{cc}{\hookrightarrow} E_\beta$, it follows that $\gamma^+(u^0)$ is relatively compact in X_β. Hence $(A,f)\big(\gamma^+(u^0)\big)$ is relatively compact in $\mathcal{L}(W^s_{p,\mathcal{B}}, W^{s-2}_{p,\mathcal{B}}) \times W^{s-2}_{p,\mathcal{B}}$, whence bounded. Thus we deduce from (13.10) that

$$\|\dot{u}(t,u^0)\|_{W^{s-2}_{p,\mathcal{B}}} \le c , \qquad t \in \big[0, t^+(u^0)\big) .$$

Consequently,

$$u(\cdot,u^0) \in B\big([0,t^+(u^0)), W^s_{p,\mathcal{B}}\big) \cap BUC^{1\text{-}}\big([0,t^+(u^0)), W^{s-2}_{p,\mathcal{B}}\big) . \tag{13.11}$$

Fix $s_\delta \in (r,s)\backslash\mathbb{N}$ and note that Theorem 7.2 implies

$$E_\delta \doteq W^{s_\delta}_{p,\mathcal{B}} = (W^{s-2}_{p,\mathcal{B}}, W^s_{p,\mathcal{B}})_{\theta,p} , \qquad \theta := 1-(s-s_\delta)/2 .$$

From this and from (13.11) we easily deduce that

$$u(\cdot,u^0) \in BUC^{1-\theta}\big([0,t^+(u^0)), E_\delta\big) .$$

Since $\theta \in (0,1)$ and $\delta \in (\beta,\alpha)$, we can now invoke Theorem 12.6. This shows that the following fundamental existence, uniqueness, and continuity theorem is true, where the last part of the assertion follows from Theorem 12.3.

13.1 Theorem *The quasilinear parabolic boundary value problem (13.1) possesses for each $u^0 \in V^s_{p,\mathcal{B}}$ a unique maximal weak W_p^σ-solution $u(\cdot,u^0)$, and $(t,u^0) \mapsto u(t,u^0)$ is a semiflow on $V^s_{p,\mathcal{B}}$. If $2-2\widehat{\alpha} \le s \le 2\widehat{\alpha}$, bounded orbits which are bounded away from $\partial V^r_{p,\mathcal{B}}$ are relatively compact in $V^s_{p,\mathcal{B}}$. If there exists $\rho \in (r,s)$ such that $u(\cdot,u^0)|[0,T]$ is bounded in W^ρ_p and bounded away from $\partial V^r_{p,\mathcal{B}}$ for each $T>0$ then $t^+(u^0)=\infty$.*

13.2 Remarks

(a) The assumption (13.5) that the Dirichlet part of the boundary conditions has to be homogeneous can easily be substituted by the condition

$$(1-\delta)G = (1-\delta)g^0$$

for some fixed $g^0 \in \partial W_p^\sigma$. Then Theorem 13.1 remains valid provided $V^r_{p,\mathcal{B}}$ is replaced by a nonempty open subset of the closed affine subspace

$$\big\{\, u \in W^r_p \ ;\ (1-\delta)\gamma_\partial u = (1-\delta)g^0 \,\big\}$$

of W^r_p.

(b) From (13.9) it should be clear why we have to consider both scales, the Sobolev-Slobodeckii scale and the Bessel potential scale, even if we are interested in the Sobolev-Slobodeckii scale only. Namely, if we would restrict ourselves to the latter scale only, we would have to put $E_1 := W^{\sigma_1}_{p,\mathcal{B}}$ and $E_0 := W^{\sigma_1-2}_{p,\mathcal{B}}$. But then we would not obtain the — in many cases most natural — space $W^1_{p,\mathcal{B}}$ by interpolating between E_0 and E_1. □

By imposing additional restrictions it is easy to get better regularity properties for a given weak solution $u(\cdot,u^0)$ of (13.1). Indeed, the following theorem is an easy consequence of Theorem 11.3 and the fact that $u(\cdot,u^0)$ is a weak W_p^σ-solution of the linear problem (13.7).

13.3 Theorem *Let $u := u(\cdot,u^0)$ and $t^+ := t^+(u^0)$ for some $u^0 \in V_{p,\mathcal{B}}^s$. Suppose that $G = 0$ and $1+1/p < 2\alpha \le 2$ and that*

$$\big(\mathcal{A}(u(\cdot)),\mathcal{B}(u(\cdot))\big) \in C^\rho\big((0,t^+),\mathbb{E}^\alpha(\Omega)\big)$$

and

$$F\big(u(\cdot)\big) \in C^\mu\big((0,t^+),W_p^{2\alpha-2}\big)\ ,$$

where $2\rho > 2\alpha - 1 - 1/p$ and $\mu > 0$. Then

$$u \in C\big((0,t^+),W_p^{2\alpha}\big) \cap C^1\big((0,t^+),W_p^{2\alpha-2}\big)$$

and u satisfies

$$\begin{aligned} \partial_t u + \mathcal{A}(u)u &= F(u) \qquad &&\text{in } \ \Omega\times(0,t^+)\ ,\\ \mathcal{B}(u)u &= 0 &&\text{on } \partial\Omega\times(0,t^+) \end{aligned}$$

in the strong sense.

We turn now to the particular situation that

$$\begin{aligned} &\text{there exists } (\mathcal{A}_0,\mathcal{B}_0) \in \mathcal{E}^1(\Omega) \text{ such that}\\ &S_{p,\mathcal{B}_0}^\rho = S_{p,\mathcal{B}(v)}^\rho\ , \qquad 1+1/p<\rho\le 2\ , \quad v \in \operatorname{dom}\big(\mathcal{A}(\cdot),\mathcal{B}(\cdot)\big)\ , \quad S \in \{H,W\}\ , \end{aligned} \tag{13.12}$$

that is, to the case where the Bessel potential and the Sobolev-Slobodeckii scales induced by $\big(\mathcal{A}(v),\mathcal{B}(v)\big)$ are independent of v for $\rho \in [0,2]\backslash\Sigma_p$. (By $\operatorname{dom}\big(\mathcal{A}(\cdot),\mathcal{B}(\cdot)\big)$ we mean the domain of the map $v \mapsto \big(\mathcal{A}(v),\mathcal{B}(v)\big)$.)

13.4 Examples

(a) $\delta = 0$, that is, $\mathcal{B} = \gamma_\partial$, the Dirichlet boundary operator. Then we can choose for $(\mathcal{A}_0,\mathcal{B}_0)$ the (diagonal) normally elliptic boundary value problem $(-\Delta,\gamma_\partial)$.

(b) $\delta \in \{0,1\}$ and $(\mathcal{A},\mathcal{B})$ is of separated divergence form with $a_j = 0,\ j = 1,\dots,n$, and $c = 0$. In this case we can put $(\mathcal{A}_0,\mathcal{B}_0) := \big(-\Delta,\delta\partial_{\nu_\alpha} + (1-\delta)\gamma_\partial\big)$, where we use the notations of Example 4.3(e).

(c) $\big(\mathcal{A}(\cdot),\mathcal{B}(\cdot)\big) = (\mathcal{A}_0,\mathcal{B}_0)$ is independent of v, and $(\mathcal{A}_0,\mathcal{B}_0) \in \mathcal{E}^1(\Omega)$. □

Given condition (13.12), we replace assumption (13.2) by

$$1+1/p<\sigma\le 2 \quad \text{and} \quad \sigma-2<\tau\le r<s<\sigma \tag{13.13}$$

with $\tau,r,s \notin \Sigma_p$. In this case we fix $\sigma_1 \in (1+1/p,2]$ such that

$$\sigma_1 = 2 \ \text{ if } \ \sigma = 2 \qquad \text{and} \qquad \sigma-1<\sigma_1-1<\rho(\widehat{\alpha})\wedge(\tau+1) \ \text{ otherwise}\ ,$$

and we define $\mathfrak{F}$ again by (13.8).

Suppose that $S \in \{H, W\}$ and that $\rho \in (r, 2] \backslash \Sigma_p$ satisfies $2 - 2\widehat{\alpha} \le \rho \le 2\widehat{\alpha}$. Then a **strong S_p^ρ-solution** of the quasilinear parabolic boundary value problem

$$\begin{aligned} \partial_t u + \mathcal{A}(u)u &= F(u) && \text{in} \quad \Omega \times (0, \infty) , \\ \mathcal{B}(u)u &= 0 && \text{on} \quad \partial\Omega \times (0, \infty) , \\ u(\cdot, 0) &= u^0 && \text{on} \quad \Omega \end{aligned} \tag{13.14}$$

on J is a function u satisfying (13.6) such that u is a strong S_p^ρ-solution of the linear parabolic boundary value problem (13.7) (with $G = 0$, of course).

It follows that u is a strong $H_p^{\sigma_1}$-solution of (13.14) on J iff it is a solution on J of the quasilinear Cauchy problem

$$\dot{u} + A(u)u = F(u) , \quad 0 < t < \infty , \qquad u(0) = u^0$$

in E_0. An obvious modification of the arguments leading to Theorem 13.1, based on the 'almost reiteration theorem' of Remark 6.1(c), proves the following

13.5 Theorem *Let (13.3) and (13.4) be satisfied. Then the assertions of Theorem 13.1 remain valid for strong solutions, except for the following modification: the map $(t, u^0) \mapsto u(t, u^0)$ is a semiflow on*

$$V_\alpha := (H_p^{\sigma_1 - 2}, H_{p,\mathcal{B}}^{\sigma_1})_{\alpha,p} \cap V_{p,\mathcal{B}}^r .$$

13.6 Remarks

(a) It follows from Remark 6.1(c) and from Theorem 7.1 that

$$V_{p,\mathcal{B}}^{s_1} \hookrightarrow V_\alpha \hookrightarrow V_{p,\mathcal{B}}^{s_2} ,$$

provided $r \le s_2 < s < s_1 \le 2$ and $s_1, s_2 \notin \Sigma_p$.

(b) It is obvious that Theorem 13.5 remains valid if $(\mathcal{A}(v), \mathcal{B}(v))$ is for each $v \in V_{p,\mathcal{B}}^r$ a second order normally elliptic boundary value problem, which is not necessarily of divergence form, provided suitable regularity conditions are satisfied. This is a consequence of Remark 8.6(d). □

Lastly, we turn to the study of very weak solutions. For this we impose the assumption that

$$\begin{aligned} &\text{there exists } (\mathcal{A}_0, \mathcal{B}_0) \in \mathcal{E}^1(\Omega) \text{ such that} \\ &S_{p',\mathcal{B}_0^\#}^{2-\rho} = S_{p',\mathcal{B}^\#(v)}^{2-\rho} , \qquad 0 \le \rho < 1/p , \quad v \in \operatorname{dom}(\mathcal{A}(\cdot), \mathcal{B}(\cdot)) , \quad S \in \{H, W\} . \end{aligned} \tag{13.15}$$

Then we replace assumption (13.2) by

$$0 \le \sigma < 1/p \quad \text{and} \quad \sigma - 2 < \tau \le r < s < \sigma \tag{13.16}$$

such that $\tau, r, s \notin \Sigma_p$.

Suppose that $\rho \in (r, 1/p)$ satisfies $2 - 2\widehat{\alpha} \leq \rho \leq 2\widehat{\alpha}$. Then a **very weak S_p^ρ-solution** of the quasilinear parabolic boundary value problem (13.1) on J is a function u satisfying (13.6) such that u is a very weak S_p^ρ-solution of the linear problem (13.7).

Assume that $\sigma \geq 2 - 2\widehat{\alpha}$, fix $\sigma_1 \in \big(\sigma, (\tau + 2) \wedge 1/p\big)$, and define $\mathfrak{F}$ by

$$(\cdot,\cdot)_\theta := \begin{cases} [\cdot,\cdot]_\theta & \text{if } \theta \in \big\{(2-\sigma_1)/2, (1-\sigma_1)/2\big\} , \\ (\cdot,\cdot)_{\theta,p} & \text{otherwise} . \end{cases}$$

Note that (13.3) implies

$$S_\gamma := (\mathcal{C}^\# | W_{p',\mathcal{B}_0^\#}^{-\tau})' \in C^{1\text{-}}(V_{p,\mathcal{B}}^r, \mathcal{L}(\partial W_p^{\tau+2}, W_p^\tau))$$

and, consequently and thanks to (13.4),

$$f := F + S_\gamma G \in C^{1\text{-}}(V_{p,\mathcal{B}}^r, E_\gamma) . \tag{13.17}$$

Then u is a very weak $H_p^{\sigma_1}$-solution of (13.1) on J iff u is a solution of (13.10) on J (where now f is given by (13.17), of course). Hence, similarly as above, we obtain

13.7 Theorem *Let (13.15) and (13.16) be satisfied. Then the assertions of Theorem 13.1 are true for very weak solutions, except for the following modification: the map $(t, u^0) \mapsto u(t, u^0)$ is a semiflow on*

$$V_\alpha := (H_{p,\mathcal{B}}^{\sigma_1 - 2}, H_p^{\sigma_1})_{\alpha,p} \cap V_{p,\mathcal{B}}^r .$$

Of course, the above theorems apply, in particular, to the **semilinear parabolic boundary value problem**

$$\begin{aligned} \partial_t u + \mathcal{A}u &= F(u) && \text{in } \Omega \times (0,\infty) , \\ \mathcal{B}u &= G(u) && \text{on } \partial\Omega \times (0,\infty) , \\ u(\cdot, 0) &= u^0 && \text{on } \Omega , \end{aligned} \tag{13.18}$$

provided $(\mathcal{A}, \mathcal{B}) \in \mathcal{E}^1(\Omega)$. However, in this case we can also apply Remark 12.2(b) to get the following useful complement to the above results.

13.8 Theorem *Let $(\mathcal{A}, \mathcal{B}) \in \mathcal{E}^1(\Omega)$ and suppose that $s, \tau \in [-2, 2] \backslash \Sigma_p$ satisfy $0 \leq s \leq 2$ and $0 \leq s - \tau < 2$. Let $V_{p,\mathcal{B}}^s$ be a nonempty open subset of $W_{p,\mathcal{B}}^s$ and suppose that*

$$(F, G) \in C^{1\text{-}}(V_{p,\mathcal{B}}^s, W_{p,\mathcal{B}}^\tau \times \partial W_p^{\tau+2}) ,$$

where $G = 0$ if $1 + 1/p < s \leq 2$, and $(1 - \delta)G = 0$ if $1/p < s < 1 + 1/p$.

Then the semilinear parabolic boundary value problem (13.18) has for each $u^0 \in V_{p,\mathcal{B}}^s$ a unique maximal W_p^s-solution $u(\cdot, u^0)$, and $(t, u^0) \mapsto u(t, u^0)$ is a semiflow on $V_{p,\mathcal{B}}^s$.

13.9 Remark On the basis of Remark 12.2(a) it is easy to extend the results of this section to nonautonomous parameter-dependent quasilinear parabolic boundary value problems. □

In the given generality, the above results appear here for the first time in print. The case of weak H_p^s-solutions has been treated in [Ama90a], and preliminary results about strong solutions are contained in [Ama86b] and [Ama86c].

By imposing more smoothness assumptions for the maps $(\mathcal{A}, \mathcal{B})$ and (F, G), it is possible to prove differentiability properties of the semiflow $(t, u^0) \mapsto u(t, u^0)$. For this and for the qualitative theory, i.e., stability, bifurcation, etc., we refer to [Ama94].

14 Reaction-Diffusion Equations: Existence and Regularity

We specialize now the results of the preceding section to the case that $(\mathcal{A}, \mathcal{B})$ and (F, G) are local operators. For simplicity we impose more restrictive regularity hypotheses than really needed and leave it to the reader to find weaker assumptions.

We assume that

$$D_0 \text{ is a nonempty open subset of } \mathbb{R}^N ,$$

that

$$a_{jk}, a_j, b_j, a_0 \in C^{2\text{-}}(\overline{\Omega} \times D_0, \mathcal{L}(\mathbb{R}^N)) , \qquad 1 \leq j, k \leq n , \tag{14.1}$$

and that

$$c \in C^{2\text{-}}(\partial\Omega \times D_0, \mathcal{L}(\mathbb{R}^N)) . \tag{14.2}$$

We put

$$\mathcal{A}(\eta)u := -\partial_j \big(a_{jk}(\cdot, \eta)\partial_j u + a_j(\cdot, \eta)u\big) + b_j(\cdot, \eta)\partial_j u + a_0(\cdot, \eta)$$

and

$$\mathcal{B}(\eta)u := \delta\big\{\nu^j \gamma_\partial \big(a_{jk}(\cdot, \eta)\partial_k u + a_j(\cdot, \eta)u\big) + c(\cdot, \eta)\gamma_\partial u\big\} + (1 - \delta)\gamma_\partial u$$

for $\eta \in D_0$. (To be precise: the term $c(\cdot, \eta)\gamma_\partial u$ should be interpreted in the following as $\gamma_\partial\big(\widetilde{c}(\cdot, \eta)u\big)$, where $\widetilde{c} \in C^{2\text{-}}(\overline{\Omega} \times D_0, \mathcal{L}(\mathbb{R}^N))$ is an arbitrary extension of c.) Then

$$\big(\mathcal{A}(\cdot), \mathcal{B}(\cdot)\big) \in C^{2\text{-}}\big(D_0, \mathbb{E}^1(\Omega)\big) , \tag{14.3}$$

and we assume that $\big(\mathcal{A}(\eta), \mathcal{B}(\eta)\big) \in \mathcal{E}(\Omega)$ for at least one $\eta \in \mathbb{R}^N$. Hence

$$D := \big\{ \eta \in D_0 \ ; \ \big(\mathcal{A}(\eta), \mathcal{B}(\eta)\big) \in \mathcal{E}(\Omega) \big\}$$

is a well defined nonempty open subset of D_0, whence of $\mathbb{R}^N$, thanks to (4.8).

Let X be a compact metric space, let E and F be Banach spaces, and let U be a nonempty open subset of E. Then we write

$$\varphi \in C^{1\text{-},2\text{-}}(X \times U, F)$$

provided each point $y_0 \in U$ has a neighborhood $V(y_0)$ in U such that $\varphi(\cdot, y) \in C^{1-}(X, F)$, uniformly with respect to $y \in V(y_0)$, and such that $\partial_2\varphi(x, \cdot) \in C^{1-}(U, \mathcal{L}(E, F))$, uniformly with respect to $x \in X$. If X is a differentiable manifold,

$$\varphi \in C^{1,2}(X \times Y, F)$$

if $\varphi \in C^1(X \times Y, F)$ and $\partial_2^2 \in C(X \times Y, \mathcal{L}^2(E, F))$, where $\mathcal{L}^2(E, F)$ is the Banach space of all continuous bilinear maps from $E \times E$ to F.

We suppose that

$$\begin{aligned} & f \in C^{1-,2-}((\overline{\Omega} \times D) \times \mathbb{R}^{N\times n}, \mathbb{R}^N) \\ & \text{and for each compact subset } K \text{ of } D \text{ there exists a constant } c_K \text{ such that} \\ & |\partial_3 f(x, \eta, \zeta)| \le c_K(1 + |\zeta|)\,, \qquad (x, \eta, \zeta) \in \overline{\Omega} \times K \times \mathbb{R}^{N\times n}\,. \end{aligned} \tag{14.4}$$

Note that, thanks to the mean value theorem, (14.4) implies that

$$|f(x, \eta, \zeta)| \le c_K'(1 + |\zeta|^2)\,, \qquad (x, \eta, \zeta) \in \overline{\Omega} \times K \times \mathbb{R}^{N\times n}\,, \tag{14.5}$$

that is, f 'grows at most quadratically in the gradient'.

Lastly, we assume that

$$g \in C^{1-}(\partial\Omega \times D, \mathbb{R}^N) \quad \text{and} \quad (1 - \delta)g = 0\,. \tag{14.6}$$

Then we consider the 'local' quasilinear parabolic boundary value problem

$$\begin{aligned} \partial_t u + \mathcal{A}(u)u &= f(\cdot, u, \partial u) && \text{in } \Omega \times (0, \infty)\,, \\ \mathcal{B}(u)u &= g(\cdot, u) && \text{on } \partial\Omega \times (0, \infty)\,, \\ u(\cdot, 0) &= u^0 && \text{on } \Omega\,. \end{aligned} \tag{14.7}$$

14.1 Remarks

(a) Of course, by redefining f and g appropriately, we can always assume that $a_0 = 0$ and $c = 0$. This shows, in particular, that it suffices to assume that

$$a \in C^{1-}(\overline{\Omega} \times D, \mathcal{L}(\mathbb{R}^N)) \quad \text{and} \quad c \in C^{1-}(\partial\Omega \times D, \mathcal{L}(\mathbb{R}^N))\,.$$

If we allow f to depend upon ∂u, we can also assume that $b_j = 0$ for $1 \le j \le n$. However, the following results will be slightly different if f depends upon ∂u or if it does not. This may already be seen from the regularity assumptions for f which are weaker if f is independent of ∂u than otherwise.

(b) Suppose that there exists $\widetilde{g} \in C^{2-}(\partial\Omega \times D, \mathcal{L}(\mathbb{R}^N))$ such that

$$g(\cdot, \eta) = \widetilde{g}(\cdot, \eta)\eta\,, \qquad \eta \in D\,. \tag{14.8}$$

Then, by replacing c by $c - \widetilde{g}$, we can assume that $g = 0$. This observation will be important below for deriving regularity properties. Note that (14.8) is satisfied, for example, if $g \in C^{2-,3-}(\partial\Omega \times D_0, \mathbb{R}^N)$ and D_0 is star-shaped with respect to

$0 \in D_0$, and if $g(\cdot,0)=0$. Then

$$\widetilde{g}(\cdot,\eta) := \int_0^1 \partial_2 g(\cdot, t\eta)\, dt\ , \qquad \eta \in D\ ,$$

by the mean value theorem. On the other hand, (14.8) is true if $N=1$ and $0 \notin D$. Then we can put $\widetilde{g}(\cdot,\eta) := g(\cdot,\eta)/\eta$. If $N>1$, similar constructions can be used in many cases. □

In order to apply the results of Section 13 we have to verify that $(\mathcal{A},\mathcal{B})$ and the map (F,G) induced by (f,g) possess suitable continuity properties. For this we need some preparation.

Given $\varphi : X \times U \to F$, we define $\Phi : U^X \to F^X$, the **Nemytskii operator** induced by φ, by

$$\Phi(u)(x) := \varphi(x, u(x))\ , \qquad u \in U^X\ , \quad x \in X\ .$$

In the next lemma we collect some of the mapping properties of Φ which we shall use in the following.

14.2 Lemma

(i) *$C(X,U)$ is open in $C(X,E)$.*
If $\varphi \in C^{1-}(X \times U, F)$ then $\Phi \in C^{1-}(C(X,U), C(X,F))$.

(ii) *If $\varphi \in C^{1-,2-}(X \times U, F)$ then*

$$\Phi \in C^{1-}(C^\rho(X,U), C^\rho(X,F))\ , \qquad \rho \in [0,1) \cup \{1\text{-}\}\ .$$

(iii) *Let μ be a positive Radon measure on X and suppose that $\varphi \in C(X \times E, F)$. Also suppose that $1 \le q \le p < \infty$ and there exists a constant c such that*

$$|\varphi(x,\xi)| \le c(1+|\xi|^{p/q})\ , \qquad (x,\xi) \in X \times E\ .$$

Then

$$\Phi \in C(L_p(X,\mu,E), L_q(X,\mu,F))\ .$$

If, in addition, $\partial_2\varphi \in C(X \times E, \mathcal{L}(E,F))$ and

$$|\partial_2\varphi(x,\xi)| \le c(1+|\xi|^{(p/q)-1})\ , \qquad (x,\xi) \in X \times E\ ,$$

then

$$\Phi \in C^{1-}(L_p(X,\mu,E), L_q(X,\mu,F))\ .$$

In fact, Φ is continuously differentiable provided $p>q$.

Parts (i) and (ii) are easy exercises, thanks to the fact that locally Lipschitz continuous maps are uniformly Lipschitz continuous on compact sets. Part (iii) is a trivial extension of (special cases of) well known results of Krasnosel'skii [Kra64] and Vainberg [Vai64] (cf. [Ama94] for details; also cf. [Goe92], [GoKT92]).

14.3 Remark If $\varphi \in C^{1-}(X \times U, F)$, it is easily seen that $\Phi(C^\rho(X,U)) \subset C^\rho(X,F)$ for $\rho \in (0,1) \cup \{1\text{-}\}$. However, Φ is not continuous, in general, as easy examples show. □

Suppose that $n < p < \infty$ and note that (5.1)–(5.9) and the compactness of $\overline{\Omega}$ imply

$$W_p^r \hookrightarrow C^{r-n/p} \hookrightarrow C \hookrightarrow L_p \tag{14.9}$$

and

$$W_p^{r-1/p}(\partial\Omega) \hookrightarrow C^{r-n/p}(\partial\Omega) \hookrightarrow C(\partial\Omega) \hookrightarrow L_p(\partial\Omega)\ , \tag{14.10}$$

where $W_p^s(\partial\Omega) := W_p^s(\partial\Omega, \mathbb{R}^N)$ etc., provided

$$r > n/p \quad \text{and} \quad r \notin \mathbb{N} + n/p\ . \tag{14.11}$$

Hence Lemma 14.2(i) implies, in particular, that

$$W_p^r(\Omega, D) := \left\{ u \in W_p^r\ ;\ u(\overline{\Omega}) \subset D \right\} \text{ is open in } W_p^r \text{ for } n/p < r \le 2\ . \tag{14.12}$$

Moreover, we deduce from (14.1), (14.2), and Lemma 14.2(ii) that

$$\begin{aligned}&(\mathcal{A}, \mathcal{B}) \in C^{1\text{-}}\big(W_p^r(\Omega, D), \mathcal{E}^{\widehat{\alpha}(r)}(\Omega)\big)\\ &\text{for } n/p < r \le 2, \text{ where } 2\widehat{\alpha}(r) \in [1, 1 + r - n/p) \text{ with } \widehat{\alpha}(r) \le 1\ ,\end{aligned} \tag{14.13}$$

provided we choose σ in (8.5) in such a way that $\rho(\widehat{\alpha}(r)) := \sigma(|2\widehat{\alpha}(r) - 1|) \le r - n/p$, which we can do.

Let now F be the Nemytskii operator induced by f. If f is independent of the gradient, the factorization

$$W_p^r(\Omega, D) \hookrightarrow C(\overline{\Omega}, D) \xrightarrow{\ F\ } C(\overline{\Omega}, \mathbb{R}^N) \hookrightarrow L_p$$

and Lemma 14.2(i) show that

$$F \in C^{1\text{-}}\big(W_p^r(\Omega, D), L_p\big)\ , \qquad n/p < r \le 2\ . \tag{14.14}$$

On the other hand, we deduce from (5.1)–(5.11) that

$$W_p^r(\Omega, D) \xrightarrow{\ (u, \partial u)\ } C(\overline{\Omega}, D) \times (W_p^{r-1})^n \hookrightarrow C(\overline{\Omega}, D) \times (L_q)^n$$

provided

$$\frac{1}{p} \ge \frac{1}{q} \ge \frac{1}{p} - \frac{r-1}{n} > 0\ .$$

Hence it is a consequence of Lemma 14.2(i) and (iii) that

$$F \in C^{1\text{-}}\big(W_p^r(\Omega, D), L_s\big)$$

provided $2 \le q/s$. Suppose that $0 \ge \tau > -2 + 1/p$ with $\tau \notin \Sigma_p$ and note that

$$W_{p', \mathcal{B}^\#}^{-\tau} \hookrightarrow L_{s'} \qquad \text{if } \frac{1}{p'} \ge \frac{1}{s'} \ge \frac{1}{p'} + \frac{\tau}{n} > 0\ .$$

Hence

$$L_s \hookrightarrow W_{p, \mathcal{B}}^{\tau} \qquad \text{if } \frac{1}{p} \le \frac{1}{s} \le \frac{1}{p} - \frac{\tau}{n} < 1\ .$$

Consequently, choosing q and s maximal, we see that

$$F \in C^{1-}\big(W_p^r(\Omega, D), W_{p,\mathcal{B}}^\tau\big) \tag{14.15}$$

provided (14.5) is satisfied and

$$1 \vee \big(1 + (\tau + n/p)/2\big) \leq r < 1 + n/p \quad \text{and} \quad 0 \geq \tau > -2 + 1/p \ , \tag{14.16}$$

where $\tau \notin \Sigma_p$.

Lastly, we denote by $\widehat{g}$ the Nemytskii operator induced by g and put

$$G := \widehat{g} \circ \gamma_\partial \ .$$

Then it is a consequence of (14.8), (14.10), Lemma 14.2(i), and (5.17), (5.18) that

$$G \in C^{1-}\big(W_p^r(\Omega, D), L_p(\partial\Omega)\big) \ , \qquad n/p < r \leq 2 \ .$$

Thus, thanks to $G = \delta G$, we see that

$$G \in C^{1-}\big(W_p^r(\Omega, D), \partial W_p^{\tau+2}\big) \ , \qquad n/p < r \leq 2 \ , \quad \tau < -1 + 1/p \ . \tag{14.17}$$

After these preparations it is easy to apply the theorems of Section 13 to problem (14.7). First assume that

$$0 \in D \qquad \text{if} \quad \delta \neq 1 \ . \tag{14.18}$$

Then it follows from (14.12) that

$$W_{p,\mathcal{B}}^r(\Omega, D) := \big\{\, u \in W_{p,\mathcal{B}}^r \ ; \ u(\overline{\Omega}) \subset D \,\big\}$$

is a nonempty open subset of $W_{p,\mathcal{B}}^r$ if $n/p < r \leq 2$ and either $r < 1 + 1/p$ or $W_{p,\mathcal{B}(v)}^r$ is independent of $v \in \mathrm{dom}\big(\mathcal{A}(\cdot), \mathcal{B}(\cdot)\big)$.

We fix now τ, r, s, and σ such that

$$n/p < r < s < \sigma < \tau + 2 < (1 + 1/p) \wedge (1 + r - n/p) \ , \qquad \sigma > 1 \ , \tag{14.19}$$

if f is independent of the gradient, and such that

$$1 = r < s < \sigma < \tau + 2 < (1 + 1/p) \wedge (2 - n/p) \tag{14.20}$$

otherwise. Then it follows from (14.13) that

$$(\mathcal{A}, \mathcal{B}) \in C^{1-}\big(W_{p,\mathcal{B}}^r(\Omega, D), \mathcal{E}^{\widehat{\alpha}}(\Omega)\big) \ , \tag{14.21}$$

where $\sigma \leq 2\widehat{\alpha} < 2$, and from (14.14)–(14.17) that

$$(F, G) \in C^{1-}\big(W_{p,\mathcal{B}}^r(\Omega, D), W_{p,\mathcal{B}}^\tau \times \partial W_p^{\tau+2}\big) \ . \tag{14.22}$$

Also note that $2 - 2\widehat{\alpha} \leq \sigma \leq 2\widehat{\alpha}$ and that, of course, problem (14.7) has the form (13.1). Hence Theorem 13.1 easily implies

14.4 Theorem *Suppose that $n/p < s < (1+1/p) \wedge (2-n/p)$ and that either*

$$f \text{ is independent of the gradient} \tag{14.23}$$

or

$$s > 1\ . \tag{14.24}$$

Then the boundary value problem (14.7) has for each $u^0 \in W^s_{p,\mathcal{B}}(\Omega, D)$ a unique maximal weak W^s_p-solution $u(\cdot, u^0)$. The map $(t, u^0) \mapsto u(t, u^0)$ is a semiflow on $W^s_{p,\mathcal{B}}(\Omega, D)$, and bounded orbits which are bounded away from ∂D are relatively compact. If there exists $\rho \in (n/p, s)$, where $\rho > 1$ if (14.24) is satisfied, such that $u(\cdot, u^0)|[0,T]$ is bounded in W^ρ_p and bounded away from ∂D for each $T > 0$ then $t^+(u^0) = \infty$.

To see that, indeed, ρ can be chosen arbitrarily close to n/p if (14.23) is satisfied, it suffices to choose $s \leq 1$, so that (14.19) is satisfied with $r < \rho < s$.

Throughout the remainder of this section

> we presuppose the hypotheses of Theorem 14.4, fix $u^0 \in W^s_{p,\mathcal{B}}(\Omega, D)$, and put $u := u(\cdot, u^0)$ and $J := [0, t^+)$ with $t^+ := t^+(u^0)$.

We shall now show that u has better regularity properties on $\dot{J}$ than stated above. First note that

$$u \in C(\dot{J}, W^\sigma_{p,\mathcal{B}}(\Omega, D)) \cap C^1(\dot{J}, W^{\sigma-2}_{p,\mathcal{B}}) \tag{14.25}$$

and that u (whence J) is independent of the particular choice of σ. Thus, by replacing u^0 by $u(t_0)$ for any $t_0 \in \dot{J}$ and t by $t - t_0$, respectively, we can assume that $u^0 \in W^\sigma_{p,\mathcal{B}}(\Omega, D)$. Observe that we can choose r arbitrarily close to 1. Consequently, we can place σ arbitrarily close to $(1+1/p) \wedge (2-n/p)$. Thus we deduce from (14.25) that

$$u \in C(\dot{J}, W^{s_1}_{p_1})\ , \qquad \frac{1}{p} \geq \frac{1}{p_1} > \frac{1}{p} - \frac{(1-n/p) \wedge 1/p}{n} > 0\ , \tag{14.26}$$

provided s_1 is chosen sufficiently close to 1. Similarly as above, we can now assume that $u^0 \in W^{s_1}_{p_1}$ and apply the preceding argument with p replaced by $p_1 > p$. Then we obtain (14.26) with s_1 replaced by some $s_2 > 1$ and p [resp. p_1] replaced by p_1 [resp. p_2]. By repeating this argument a finite number of times we arrive at a $p_k \geq n+1$. Hence, being interested in $u|(0, t^+)$ only, we can assume that $p \geq n+1$ so that $(1-1/p) \wedge 1/p = 1/p$. Then we obtain from (14.26) that $u \in C(\dot{J}, W^{s_1}_{p_1})$, provided $1/p \geq 1/p_1 > (1-1/n)/p$. Thus, given any $k \in \dot{\mathbb{N}}$, we find iteratively that $u \in C(\dot{J}, W^{s_k}_{p_k})$ for some $s_k \in (1, 1+1/p_k)$, provided $1/p \geq 1/p_k > (1-1/n)^k/p$. From this we infer that $u \in C(\dot{J}, W^1_q)$ for each $q \in (1, \infty)$, which implies the following

14.5 Corollary *The function u is a weak W^1_q-solution of (14.7) for each $q \in (1, \infty)$.*

Suppose now that $g = 0$. Then we shall show that u is a classical solution of (14.7). Since this concerns the open interval $\dot{J}$ only, we can assume, as above, that $u^0 \in W^{\sigma}_{p,\mathcal{B}}(\Omega, D)$. Thus

$$u \in C(J, W^{\sigma}_{p,\mathcal{B}}) \cap C^1(J, W^{\sigma-2}_{p,\mathcal{B}}) .$$

Hence it follows from Theorem 7.2 that

$$u \in C^{\rho}(J, W^{\sigma-2\rho}_{p,\mathcal{B}}) , \qquad 0 \leq 2\rho \leq 2 . \tag{14.27}$$

Since we can choose σ arbitrarily close to $1 + 1/p$ (recall that we can assume $p > n + 1$), it follows from (14.13) that

$$\big(\mathcal{A}(u(\cdot)), \mathcal{B}(u(\cdot))\big) \in C^{\rho}\big(J, \mathcal{E}^{\alpha}(\Omega)\big) , \qquad 1 \leq 2\alpha < 2 - 2\rho - (n-1)/p . \tag{14.28}$$

Choose $\alpha, \rho \in (0,1)$ such that

$$2 - 2\alpha - (n-1)/p > 2\rho > 2\alpha - 1 - 1/p > 0 , \tag{14.29}$$

which is easily seen to be possible.

From (14.14)–(14.16) and (14.27) we deduce that

$$F\big(u(\cdot)\big) \in C^{\rho}(J, W^{\tau}_{p,\mathcal{B}}) , \qquad \tau \in (-2 + 1/p, 0] \backslash \Sigma_p , \quad \tau < (2-n)/p - 4\rho .$$

Since, thanks to (14.29), $(2-n)/p - 4\rho > 2\alpha - 2 + 1/p$, it follows that

$$F\big(u(\cdot)\big) \in C^{\rho}(J, W^{2\alpha-2}_{p,\mathcal{B}}) . \tag{14.30}$$

Now we deduce from (14.28)–(14.30) and Theorem 11.3 that the linear parabolic boundary value problem

$$\begin{aligned} \partial_t v + \mathcal{A}\big(u(t)\big)v &= F\big(u(t)\big) && \text{in } \Omega \times \dot{J} , \\ \mathcal{B}\big(u(t)\big)v &= 0 && \text{on } \partial\Omega \times \dot{J} , \\ v(\cdot, 0) &= u^0 && \text{on } \Omega \end{aligned} \tag{14.31}$$

has a unique strong $W^{2\alpha}_p$-solution v. Since a strong $W^{2\alpha}_p$-solution of (14.31) is also a weak W^s_p-solution, since u is a weak W^s_p-solution of (14.31), and since there exists exactly one weak W^s_p-solution of (14.31), it follows that

$$u = v \in C(\dot{J}, W^{2\alpha}_{p,\mathcal{B}}) \cap C^1(\dot{J}, W^{2\alpha-2}_p) . \tag{14.32}$$

Similarly as above, we can assume that (14.32) is true with $\dot{J}$ replaced by J. Thus, using (5.22), we see that

$$u \in C^{\rho}(J, W^{2\alpha-2\rho}_p) , \qquad 0 \leq 2\rho \leq 2 .$$

Choose $\alpha_1 \in (0,1]$ and $\rho_1 \in (0,1)$ such that

$$1 + 2\alpha - 2\alpha_1 - n/p > 2\rho_1 > 2\alpha_1 - 1 - 1/p > 2\alpha - 1 - 1/p ,$$

which is possible. In fact, we can put

$$2\alpha := (1 + 1/p) + (1 - n/p)/4 , \quad 2\alpha_1 := (1 + 1/p) + (1 - n/p)/2 .$$

Then (14.13) and (14.14)–(14.16) imply

$$(\mathcal{A}(u(\cdot)), \mathcal{B}(u(\cdot))) \in C^{\rho_1}(J, \mathcal{E}^{\alpha_1}(\Omega))$$

and

$$F(u(\cdot)) \in C^{\rho_1}(J, W_p^{2\alpha_1 - 2}) ,$$

provided $2\rho_1$ is chosen sufficiently close to $2\alpha_1 - 1 - 1/p = (1 - n/p)/2$. Hence Theorem 13.3 guarantees — similarly as above — that

$$u \in (\dot{J}, W_p^{2\alpha_1}) \cap C^1(\dot{J}, W_p^{2\alpha_1 - 2}) ,$$

where $2\alpha_1 - 2\alpha = (1 - n/p)/2$. By iterating this argument we find after finitely many steps that u is a strong W_p^2-solution of (14.31). Consequently,

$$u \in C^\rho(\dot{J}, C^{1+\mu}) , \qquad 0 \leq 2\rho + \mu \leq 1 - n/p .$$

From this and from Remark 14.3 we deduce that u is a strong W_p^2-solution of a linear parabolic system having Hölder continuous coefficients and a Hölder continuous right hand side. Hence we can invoke the results of Solonnikov [Sol65] to show that u is a classical solution. This proves the following theorem and its corollary.

14.6 Theorem *Suppose that* $g = 0$. *Then* u *is a classical solution of* (14.7), *that is,*

$$u \in C(\overline{\Omega} \times J, D) \cap C^{2,1}(\overline{\Omega} \times \dot{J}, \mathbb{R}^N) ,$$

and u *satisfies* (14.7) *pointwise.*

14.7 Corollary *If* $g = 0$, *and* f *and all coefficients of* $(\mathcal{A}, \mathcal{B})$ *are* C^∞*-smooth,*

$$u \in C^\infty(\overline{\Omega} \times \dot{J}, \mathbb{R}^N) .$$

For simplicity we imposed in Theorem 14.6 and its corollary the assumption $g = 0$, which, in view of Remark 14.1(b), is rather mild. For the general case as well as for better regularity conditions of u at $t = 0$, provided u^0 has more regularity, we refer to [Ama94].

The above results are due to the author [Ama90a]. Earlier results concerning the solvability of quasilinear parabolic systems under nonlinear boundary conditions are contained in [Ama86c]. In the latter paper we have, however, neither been able to prove the uniqeness of the solution in general, nor to show that it generates a semiflow on the natural state space $W^s_{p,\mathcal{B}}(\Omega, D)$. In addition, we had to assume that $s > 1 + n/p$ even if f was affine in ∂u. More recently, Yagi [Yag91] has proven a local existence theorem for problem (14.7) with $g = 0$. He works directly in the pair (L_p, W_p^2), but also uses the fact that the spaces $W^s_{p,\mathcal{B}(u)}$ are independent of u for $0 \leq s < 1 + 1/p$. His result is comparable to the one of [Ama86c] and has the same shortcomings.

Problem (14.7), under homogeneous Dirichlet conditions and with f independent of ∂u, has also been studied in [Ves91]. This author uses extrapolation techniques too, but works with spaces of continuous and Hölder continuous functions (so that even in the case of Dirichlet boundary conditions the domains of the generators $A(u)$ are not constant). In both papers [Yag91],[Ves91] the case of a single equation is considered.

Lastly, there is a local existence result for quasilinear parabolic systems in [GiM87] which is based upon classical Schauder type a priori techniques. In that paper it is assumed that $\mathcal{A}$ satisfies a uniform Legendre condition and the given proofs heavily depend on this assumption (although the authors claim that they can handle strongly uniformly elliptic operators too). Anyhow, the techniques of [GiM87] are pure p.d.e. techniques and cannot be used to handle more general 'abstractly parabolic' problems.

The theory developed in this paper is an L_p-theory. It has the advantage that the basic function spaces, namely the spaces $W^s_{p,\mathcal{B}}$, on which the theory is based, are — from the viewpoint of functional analysis — simple spaces: they are reflexive and separable, and their duals are explicitly known spaces of the same sort. This can be useful in dealing with questions concerning the qualitative behavior of the semiflows generated by our problems, questions which we do not consider here and which involve more sophisticated techniques from nonlinear functional analysis. It also provides us with a simple duality theory in the linear case, of which we have already made extensive use in connection with very weak solutions. Moreover, the concept of a weak W^s_p-solution is a very natural one, which is the straightforward extension of the notion of weak solutions in the 'classical' L_2-case. Lastly, as shown above, there is a good regularity theory, which immediately implies that weak $W^1_{p,\mathcal{B}}$-solutions are in fact classical solutions. Most importantly, this regularity theory applies to general normally elliptic operators $(\mathcal{A}, \mathcal{B})$ without any additional structural conditions.

Of course, there is no unique choice for the functional setting in which a given problem has to be put. There are many other possibilities, like the use of spaces of continuous functions, Hölder spaces, general Besov spaces, etc. However, as explained above, the L_p-setting seems to be a natural one and, perhaps, the simplest one.

14.8 Remark The regularity assumptions (14.1) and (14.2) seem to be overly restrictive. In fact, it is not difficult to see that the ones for the 'lower order coefficients' can be weakened. As for the 'top order coefficients', we have to know that

$$[u \mapsto a_{jk}(u)] \in C^{1\text{-}}(X_\beta, H^{s_1}_p(\Omega, \mathcal{L}(\mathbb{R}^N))) ,$$

where $X_1 \subset W^{s_\beta}_p$. In order to guarantee this we have reduced this continuity problem to one between Hölder spaces and then applied Lemma 14.2(ii), which — thanks to Remark 14.3 — is optimal. Thus, in order to weaken the regularity assumptions for a_{jk}, one has to find rather mild conditions guaranteeing that Nemytskii operators are locally Lipschitz continuous between generalized Sobolev spaces. Although mapping properties of nonlinear operators in generalized Sobolev spaces have been studied intensively during the last few years (cf. [Run86], [Sic89], [Sic93], [Sic92], [Boud91], [BouK92], [BouM91], [Osw92]), the problem of the Lipschitz continuity (or even continuity) of these maps under optimal assumptions is still open. □

15 Positivity and Global Existence

Throughout this section we impose conditions (14.1)–(14.6) and consider the quasilinear parabolic boundary value problem (14.7). In addition,

we presuppose the hypotheses of Theorem 14.4, fix $u^0 \in W^s_{p,\mathcal{B}}(\Omega, D)$, and put $u := u(\cdot, u^0)$ and $J := [0, t^+)$ with $t^+ := t^+(u^0)$.

In many applications — for example, in some of the model problems of Section 1 — it is of importance to know that the solution (or some components of it) of problem (14.7) stays positive for $t \in \dot{J}$, provided the (corresponding component of the) initial value is positive. Although there is no maximum principle for general strongly coupled systems, it is easy to give sufficient conditions for this 'positive invariance of the positive cone'.

Let $r \in \{1, \ldots, N\}$ be fixed and put

$$\widehat{\eta}_r := (\eta_1, \ldots, \eta_{r-1}, \eta_{r+1}, \ldots, \eta_N) , \quad (\widehat{\eta}_r, \eta_r) := \eta = (\eta_1, \ldots, \eta_N) .$$

Suppose that

$$\eta \in D \implies (\widehat{\eta}_r, t\eta_r) \in D_0 , \quad 0 \le t \le 1 , \tag{15.1}$$

and that

$$\begin{gathered} a^{rs}_{jk},\ a^{rs}_j,\ b^{rs}_j,\ a^{rs}_0,\ c^{rs}_r \text{ vanish on} \\ D_r := \{ (\widehat{\eta}_r, 0) \in D_0 \ ; \ \exists\, \tau > 0 : (\widehat{\eta}_r, \tau) \in D \} \\ \text{for } 1 \le j,k \le n \text{ and } 1 \le s \le N \text{ with } s \ne r. \end{gathered} \tag{15.2}$$

Lastly, suppose that

$$\begin{gathered} f_r \in C^1(\overline{\Omega} \times D_0, \mathbb{R}),\ g \in C^1(\partial\Omega \times D_0, \mathbb{R}) \text{ with } (1-\delta)g = 0, \\ \text{and that} \\ f_r(\cdot, (\widehat{\eta}_r, 0)) \ge 0,\ g_r(\cdot, (\widehat{\eta}_r, 0)) \ge 0,\ (\widehat{\eta}_r, 0) \in D_r . \end{gathered} \tag{15.3}$$

Then u_r is a weak W^1_p-solution of the linear parabolic evolution system ($N = 1$)

$$\begin{aligned} \partial_t v + \mathcal{A}_r(t) v &= f^0_r(t) && \text{in } \Omega \times \dot{J} , \\ \mathcal{B}_r(t) v &= g^0_r(t) && \text{on } \partial\Omega \times \dot{J} , \\ v(\cdot, 0) &= u^0_r && \text{on } \Omega , \end{aligned}$$

where

$$((\mathcal{A}_r, \mathcal{B}_r), f^0_r, g^0_r) \in C^\rho(J, \mathcal{E}_p(\Omega) \times W^{-1}_{p,\mathcal{B}} \times \partial W^1_p)$$

for a suitable $\rho \in (0,1)$, provided $u^0 \in W^\sigma_p$. In fact, denoting the coefficients of $(\mathcal{A}, \mathcal{B})$ by $(\alpha_{jk}, \alpha_j, \beta_j, \alpha_0, \gamma)$, it follows — using an obvious simplified notation — that

$$\alpha_{jk}(t) := a^{rr}_{jk}(u(t)) , \quad \beta_j(t) := b^{rr}_j(u(t)) ,$$

and

$$\alpha_j(t) := a^{rr}_j(u(t)) + \sum_{s \ne r} \left[\widetilde{a}^{rs}_{jk}(u(t)) \partial_k u_s(t) + \widetilde{a}^{rs}_j(u(t)) u_s(t) \right] ,$$

and that

$$\alpha_0(t) := a_0^{rr}(u(t)) + \sum_{s\neq r}\big[\widetilde{b}_j^{rs}(u(t))\partial_j u_s(t) + \widetilde{a}_0^{rs}(u(t))u_s(t)\big] - \widetilde{f}_r(u(t))$$

and

$$\gamma(t) := c^{rr}(u(t)) + \sum_{s\neq r}\widetilde{c}^{rs}(\gamma_\partial u(t))\gamma_\partial u_s(t) - \widetilde{g}_r(\gamma_\partial u(t)) ,$$

with

$$\widetilde{d}(\eta) := \int_0^1 \partial_{\eta_r} d(\cdot,(\widehat{\eta}_r, t\eta_r))\, dt , \qquad \eta \in D ,$$

for $d \in \{ a_{jk}^{rs}, a_j^{rs}, b_j^{rs}, a_0^{rs}, c^{rs}, f_r, g_r \ ; \ s \neq r\}$, and where

$$f_r^0(t) := f_r((\widehat{u}_r(t),0)) , \quad g_r^0(t) := g_r(\gamma_\partial(\widehat{u}_r(t),0))$$

for $t \in J$. Hence we deduce from (15.3) and Remark 11.7 that $u_r^0 \geq 0$ implies $u_r(t) \geq 0$ for $t \in J$, provided $u^0 \in W_p^\sigma$. Since, by Theorem 8.7, the positive cone of $W_{p,\mathcal{B}_r}^\sigma$ is dense in the positive cone of $W_{p,\mathcal{B}_r}^s$, given $u^0 \in W_{p,\mathcal{B}}^s$ with $u_r^0 \geq 0$, we can find a sequence $(u_{(j)}^0)$ in $W_{p,\mathcal{B}}^\sigma$ satisfying $u_{(j),r}^0 \geq 0$ such that $u_{(j)}^0 \to u^0$ in $W_{p,\mathcal{B}}^s(\Omega, D)$ as $j \to \infty$. Consequently, given any $t \in J$, we see that $u(t, u_{(j)}^0) \to u(t, u^0)$ in $W_{p,\mathcal{B}}^s(\Omega, D)$ as $j \to \infty$, thanks to the fact that $(t, u^0) \mapsto u(t, u^0)$ is a semiflow on $W_{p,\mathcal{B}}^s(\Omega, D)$. Now the following theorem is obvious.

15.1 Theorem *Let conditions* (15.1)–(15.3) *be satisfied for some* $r \in \{1, \ldots, N\}$ *and let* $u_r^0 \geq 0$. *Then* $u_r(t) \geq 0$ *for* $t \in J$.

15.2 Remarks

(a) For simplicity, we have assumed in the above theorem that f is independent of the gradient. It is easy to extend it to the gradient dependent case too.

(b) Of course, the assumption that f_r be C^1 can be replaced by: there exists $\widetilde{f}_r \in C^{1\text{-}}(\overline{\Omega} \times D, \mathbb{R})$ such that

$$f_r(\cdot, \eta) = \widetilde{f}_r(\cdot, \eta)\eta_r + f_r(\cdot, (\widehat{\eta}_r, 0)) , \qquad \eta \in D .$$

A similar remark applies to g_r. □

The above theorem occurs here for the first time in print although the basic idea for its proof is already contained in [Ama90a] and [Ama89b].

We are now interested in the global existence of u. Since this problem concerns the behavior of u near t^+ only, we can assume — thanks to Corollary 14.5 — that p is arbitrarily large. Using this fact and $C^\varepsilon \hookrightarrow W_p^\rho$ for $\varepsilon > \rho$, the following theorem is an immediate consequence of Theorem 14.4.

Here and in the following, we use c to denote constants, which may differ from occurrence to occurrence. Similar use is made of functions $c(\cdot) : \mathbb{R}^+ \to \mathbb{R}^+$.

15.3 Theorem *Suppose that f is independent of the gradient. Also suppose that $u|[0,T]$ is bounded away from ∂D for each $T > 0$. If there exists $\varepsilon > 0$ such that*

$$\|u(t)\|_{C^\varepsilon} \leq c(T) , \qquad 0 \leq t \leq T < \infty , \quad t < t^+ , \tag{15.4}$$

then $t^+ = \infty$.

Note that (15.4) means that

$$u \in B([0,T] \cap J, C^\varepsilon(\overline{\Omega}, \mathbb{R}^N)) , \qquad T > 0 , \tag{15.5}$$

for some $\varepsilon > 0$. We shall now show that this condition can be replaced by

$$u \in BUC^\varepsilon([0,T] \cap J, C^\rho(\overline{\Omega}, \mathbb{R}^N)) , \qquad T > 0 , \tag{15.6}$$

even if f is gradient dependent, where $\rho = 0$ if f is independent of ∂u, and $\rho > n/(n+1)$ otherwise. For this we suppose first that f is independent of ∂u and that u is bounded away from ∂D on any finite interval. Then, if (15.6) is satisfied and $T > 0$ is fixed, there exists a unique

$$(\overline{u}, \overline{f}, \overline{g}) \in C^\varepsilon([0, T \wedge t^+], C(\overline{\Omega}, \mathbb{R}^N) \times C(\overline{\Omega}, \mathbb{R}^N) \times C(\partial\Omega, \mathbb{R}^N))$$

such that

$$(\overline{u}, \overline{f}, \overline{g})(t) = \big(u(t), f(u(t)), g(\gamma_\partial u(t))\big) , \qquad t \in [0,T] \cap J .$$

Consequently,

$$(\overline{\mathcal{A}}, \overline{\mathcal{B}}) := \big(\mathcal{A}(\overline{u}(\cdot)), \mathcal{B}(\overline{u}(\cdot))\big) \in C^\varepsilon([0, T \wedge t^+], \mathcal{E}^{1/2}(\Omega)) ,$$

and Theorem 11.2 implies that the linear parabolic boundary value problem

$$\begin{aligned} \partial_t v + \overline{\mathcal{A}}(t)v &= \overline{f}(t) && \text{in } \ \Omega \times (0, T \wedge t^+] , \\ \overline{\mathcal{B}}(t)v &= \overline{g}(t) && \text{on } \partial\Omega \times (0, T \wedge t^+] , \\ v(\cdot, 0) &= u^0 && \text{on } \ \Omega \end{aligned} \tag{15.7}$$

has a unique weak W_p^1-solution $\overline{u}$. Since u is such a solution of (15.7) on $[0,T] \cap J$, it follows that $\overline{u} \supset u$. Hence

$$u \in BUC([0,T] \cap J, W^1_{p,\mathcal{B}}) ,$$

which shows that $u|[0,T]$ is bounded in W_p^1 for each $T > 0$. Hence $t^+ = \infty$ by Theorem 14.4. A similar, slightly more sophisticated argument applies if f is gradient dependent. More precisely, we have the following global existence

15.4 Theorem *Suppose that there exists $\varepsilon > 0$ such that (15.6) is true, where*

$$\rho \begin{cases} = 0 & \text{if } f \text{ is independent of } \partial u , \\ > 0 & \text{if (13.12) is satisfied and } g = 0 , \\ > n/(n+1) & \text{otherwise} . \end{cases}$$

Then $t^+ = \infty$, provided $u|[0,T]$ is bounded away from ∂D for each $T > 0$. If, in addition,

$$u \in BUC^\varepsilon(J, C^\rho(\overline{\Omega}, \mathbb{R}^N)) \tag{15.8}$$

and if the orbit $\gamma^+(u^0)$ is relatively compact in $C(\overline{\Omega}, D)$ then $\gamma^+(u^0)$ is relatively compact in $W^s_{p,\mathcal{B}}(\Omega, D)$.

These global existence theorems are due to the author. Theorem 15.3 is given here explicitly for the first time, and Theorem 15.4 is contained in [Ama89a] (where, in fact, it is shown that condition (15.8) implies the boundedness of u in 'higher norms').

The above theorems reduce the problem of global existence to the problem of establishing suitable a priori bounds. Note that (15.6) is satisfied, if

$$\|u\|_{C^{\rho,\varepsilon}(\overline{\Omega}\times[0,t],\mathbb{R}^N)} \leq c(T) \ , \qquad 0 \leq t \leq T < \infty \ , \quad t < t^+ \ . \tag{15.9}$$

Thus (15.9) and — of course — (15.4) are trivially satisfied, if there exists $\varepsilon \in (0,1)$ such that

$$\|u\|_{C^{\varepsilon,\varepsilon/2}(\overline{\Omega}\times[0,t],\mathbb{R}^N)} \leq c(T) \ , \qquad 0 \leq t \leq T < \infty \ , \quad t < t^+ \ , \tag{15.10}$$

where $\varepsilon > n/(n+1)$ if f is gradient dependent. Recall that $C^{\varepsilon,\varepsilon/2}(Q_T)$ is the space of ε-Hölder continuous functions on $Q_T := \overline{\Omega} \times [0,T]$, provided Q_T is given the 'parabolic metric'

$$\big((x,s),(y,t)\big) \mapsto \sqrt{|x-y|^2 + |s-t|} \ .$$

These spaces are usually employed in connection with parabolic equations (e.g., [LadSU68]).

It is an important open question whether the Hölder norms occurring in (15.4) and (15.8), respectively, can be replaced by a weaker norm, in particular, by the supremum norm. More precisely, we can formulate the following **a priori bound problem**:

> Suppose that
>
> $\|u(t)\|_\infty \leq c_0(T) \ , \qquad 0 \leq t \leq T \ , \quad t < t^+ \ .$
>
> Do there exist $\varepsilon, \rho \in (0,1)$ such that (15.4) or (15.9) are true, where $\rho > n/(n+1)$ if f is gradient dependent?

Of course, a positive answer to this problem guarantees that $t^+ = \infty$, provided $u|[0,T]$ is bounded away from ∂D for each $T > 0$.

So far there are only a few cases for which a positive answer to the a priori bound problem has been found. First, if $N = 1$, this is well known to be the case (e.g., [LadSU68]). For systems the situation is much more complicated. The only result known to the author, which applies to strongly coupled systems with 'fully occupied diffusion matrices', is due to Wiegner [Wie92]. He has to assume that $\mathcal{A}$ satisfies a uniform Legendre condition in addition to further structural hypotheses. Then he proves that (15.9) is true for any $\varepsilon \in (0,1)$.

The situation is much better if we restrict ourselves to triangular systems. To be precise, $(\mathcal{A},\mathcal{B})$ is said to be a (lower) **triangular system** if each one of the matrices a_{jk}, a_j, b_j, a_0, and c is lower triangular, that is, all entries above the diagonal are zero. An obvious modification of the proof of [Ama89a, Theorems 5.2 and 6.2] gives a positive answer to the a priori bound problem, provided assumptions (15.11) or (15.12) below are satisfied. Thus we obtain the following theorem which, in fact, is true for more general 'block triangular systems' (cf. [Ama94]).

15.5 Theorem *Suppose that $(\mathcal{A},\mathcal{B})$ is lower triangular and that either*

$$\begin{aligned} &\textit{each } a_{jk} \textit{ is a diagonal matrix and} \\ &|f_r(u,\partial u)| \leq c(|u|)(1+|\nabla u_r|^\lambda) \ , \qquad 1 \leq r \leq N \ , \end{aligned} \tag{15.11}$$

or

$$\begin{aligned}&\text{the } r\text{-th row of each } a_{jk} \text{ is independent of } u_s \text{ for } s > r, \text{ and}\\ &|f_r(u,\partial u)| \le c(|u|)\big(1 + |\nabla u_r|^\lambda + \textstyle\sum_{s<r}|\nabla u_s|^2\big)\ , \qquad 1 \le r \le N\ ,\end{aligned} \tag{15.12}$$

where $\lambda = 2$ if condition (13.12) is satisfied and $g = 0$, and $\lambda = 1$ otherwise. Also suppose that

$$\|u(t)\|_\infty \le c(T)\ , \qquad 0 \le t \le T < \infty\ , \quad t < t^+\ . \tag{15.13}$$

Then $t^+ = \infty$ if $u|[0,T]$ is bounded away from ∂D for each $T > 0$. If c in (15.13) can be chosen to be independent of $T \in \dot{\mathbb{R}}^+$, and if the orbit $\gamma^+(u^0)$ is bounded away from ∂D then $\gamma^+(u^0)$ is relatively compact in $W^s_{p,\mathcal{B}}(\Omega, D)$.

The above theorem has been proven in [Ama89a]. It generalizes, to a certain extent, considerably earlier results of Ladyženskaja and Ural'ceva [LadU65], [LadU66], [LadU67] (also cf. [LadSU68, Theorem VII.7.1] and of Giaquinta and Struwe [GiS81]. These authors considered Dirichlet boundary conditions ($\delta = 0$) and diagonal systems with one and the same strongly uniformly elliptic operator in each row, that is, they assume that $a^{rs}_{jk} = 0$ for $s \ne r$ and $a^{rr}_{jk} = a^{11}_{jk}$ for $1 \le r \le N$. They impose a growth restriction of the form

$$|f(u,\partial u)| \le \alpha\,|\partial u|^2 + \beta\ ,$$

involving the gradient of the full vector u, and not only of u_r as we do, but they have to assume that α is sufficiently small. In fact, the size of α is not specified in the work of Ladyženskaja and Ural'ceva, whereas in [GiS81] it is assumed that

$$\alpha \sup_{Q_T} |u| < \underline{\alpha}\ ,$$

where $\underline{\alpha}$ is the constant of ellipticity. Given these conditions, those authors then show that bounded weak W^1_2-solutions are uniformly Hölder continuous (in x and t).

Of course, if $(\mathcal{A}, \mathcal{B})$ is independent of u, that is, in the semilinear case, a positive answer to the a priori bound problem can easily be obtained from the variation of constants formula. In this case the L_∞-norm can often be replaced by weaker L_q-norms (cf. [Ama85], [Red87], [Red89]. We also refer to [Rhe84] as a protoype for the vast literature dealing with 'weakly coupled' semilinear systems, that is, semilinear systems with diagonal operators $(\mathcal{A}, \mathcal{B})$).

It should be observed that the a priori bound problem is *not* a regularity problem. In fact, it follows from (14.27) that

$$u \in C(\dot{J}, C^\rho) \cap C^{\rho/2}(\dot{J}, C)$$

with $\rho := 1 - n/p$. Since we can choose p arbitrarily large, we see that

$$u \in C^{\rho,\rho/2}(\overline{\Omega} \times \dot{J}, \mathbb{R}^N)\ , \qquad 0 < \rho < 1\ . \tag{15.14}$$

This regularity result shows the difference between our method and the 'classical' approach to quasilinear parabolic systems. The classical method proceeds by establishing — usually by a Galerkin method or by means of the theory of monotone operators — a weak W^1_2-solution.

Then one tries to prove 'regularity results' to show that theses weak solutions are Hölder continuous and — furthermore — classical . However, it is known that weak W_2^1-solutions of *systems* are, in general, not Hölder continuous (e.g., [Gia83] and note that weak W_2^1-solutions to elliptic systems are stationary weak W_2^1-solutions to the corresponding parabolic systems). Our approach avoids this difficulty since we work from the very beginning with weak W_p^1-solutions with $p > n$, which implies the Hölder regularity result (15.14), as we have seen.

As already mentioned earlier, the Galerkin approach cannot work in our case since — even in the W_2^1-case — the Dirichlet form associated with $(\mathcal{A}, \mathcal{B})$ is not coercive, in general. Thus Theorem 14.4 cannot be proven by the 'classical' method, except rather restrictive hypotheses are imposed. The theory of monotone operators also applies to a rather limited class of problems only.

There are examples of parabolic systems possessing weak W_2^1-solutions which are smooth for small t and develop a discontinuity for some $\tau \in (0, t^+)$ (cf. [Str84], [StJM86]). However, in these examples the operators and the nonlinearities do not satisfy our regularity assumptions. Hence they do not provide negative answers to the a priori bound problem.

Lastly, it should be mentioned that it is by no means trivial to find L_∞-a priori bounds for a given solution in the case of systems. This is mainly due to the fact that there is no good maximum principle for systems. The only result in this direction, which applies to a whole (though rather restricted) class of strongly coupled quasilinear parabolic problems with 'fully occupied diffusion matrices', seems to be [Wie92, Theorem 2]. For triangular systems it is in certain cases possible to get L_∞-a priori bounds, provided suitable growth restrictions are satisfied. We refrain from giving details and refer to [Ama89c] and [Ama94].

16 Strong and Very Weak Solutions

We turn now briefly to the case of 'constant boundary operators' and strong solutions. In this case we do not assume that $(\mathcal{A}, \mathcal{B})$ is of divergence form. More precisely, we assume that

$$D \text{ is open in } \mathbb{R}^N \times \mathbb{R}^{N \times n} \tag{16.1}$$

and

$$a_{jk} \in C^{2-}(\overline{\Omega} \times D, \mathcal{L}(\mathbb{R}^N)) , \qquad 1 \leq j,k \leq n . \tag{16.2}$$

We put

$$\mathcal{A}(\eta)u := -a_{jk}(\cdot, \eta)\partial_j \partial_k u , \qquad \eta \in D .$$

We also assume that

$$b_j \in C^1(\partial\Omega, \mathcal{L}(\mathbb{R}^N)) , \quad b_0 \in C^2(\partial\Omega, \mathcal{L}(\mathbb{R}^N)) \tag{16.3}$$

and put

$$\mathcal{B}u := b_j \gamma_\partial \partial_j u + b_0 \gamma_\partial u . \tag{16.4}$$

Then we suppose that

$$(\mathcal{A}(\eta), \mathcal{B}) \text{ is normally elliptic for } \eta \in D . \tag{16.5}$$

Note that, thanks to Example 4.3(c), this assumption is satisfied, for example, if $\mathcal{A}(\eta)$ is uniformly strongly elliptic and $\mathcal{B} = \gamma_\partial$, the Dirichlet boundary operator.

Lastly, we suppose that

$$f \in C^{2\text{-}}(\overline{\Omega} \times D, \mathbb{R}^N) \tag{16.6}$$

and consider the quasilinear parabolic boundary value problem

$$\begin{aligned} \partial_t u + \mathcal{A}(u, \partial u)u &= f(\cdot, u, \partial u) && \text{in} \quad \Omega \times (0,\infty) , \\ \mathcal{B}u &= 0 && \text{on} \quad \partial\Omega \times (0,\infty) , \\ u(\cdot, 0) &= u^0 && \text{on} \quad \Omega , \end{aligned} \tag{16.7}$$

where we use an obvious notation.

We fix $p \in (n, \infty)$ and numbers

$$0 < \tau < 1/p < 1 + n/p < r < s < 1 , \quad 0 < \tau < r - n/p ,$$

and put

$$\alpha := s/2 , \quad \beta := r/2 , \quad \gamma := \tau/2 ,$$

and

$$E_\xi := W^{2\xi}_{p,\mathcal{B}} , \qquad \xi \in \{1, \alpha, \beta, \gamma, 0\} .$$

We also put

$$X_\beta := W^r_{p,\mathcal{B}}(\Omega, D) := \left\{ u \in W^r_{p,\mathcal{B}} \ ; \ (u, \partial u)(\overline{\Omega}) \subset D \right\} .$$

Then X_β is an open subset of E_β, which we assume to be nonempty. Given $u \in X_\beta$, we denote by $A(u)$ the L_p-realization of $\big(\mathcal{A}(u, \partial u), \mathcal{B}\big)$ and put $F(u) := f(\cdot, u, \partial u)$. It is not difficult to see that

$$(A, F) \in C^{1\text{-}}\big(X_\beta, \mathcal{H}(E_1, E_0) \times E_\gamma\big)$$

and that the Cauchy problem

$$\dot{u} + A(u)u = F(u) , \quad 0 < t < \infty , \qquad u(0) = u^0$$

is the strong W^2_p-formulation of problem (16.7).

The following theorem is now an easy consequence of Theorems 7.1, 12.1, and 12.3 combined with the fact that we can apply Solonnikov's 'classical theory' [Sol65] to the linear boundary value problem which is obtained from (16.7) by substituting the strong solution $u(\cdot, u^0)$ into $\mathcal{A}(\cdot)$ and $F(\cdot)$.

16.1 Theorem *Let assumptions* (16.1)–(16.6) *be satisfied and suppose that*

$$1 + n/p < r < s < 2 .$$

Then, given $u^0 \in W^s_{p,\mathcal{B}}(\Omega, D)$, the quasilinear parabolic boundary value problem (16.7) *has a unique maximal classical solution*

$$u(\cdot, u^0) \in C\big(\overline{\Omega} \times [0, t^+(u^0)), D\big) \cap C^{2,1}\big(\overline{\Omega} \times (0, t^+(u^0)), \mathbb{R}^N\big) .$$

The map $(t, u^0) \mapsto u(t, u^0)$ is a semiflow on $W^s_{p,\mathcal{B}}(\Omega, D)$. Bounded orbits, which are bounded away from ∂D, are relatively compact. If there exists $\rho \in (r, s)$ such that $u(\cdot, u^0)|[0, T]$ is bounded in W^ρ_p and bounded away from ∂D for each $T > 0$ then $u(\cdot, u^0)$ is a global solution.

16.2 Remarks

(a) Suppose that a_{jk} is independent of ∂u and f is affine in ∂u, that is, f is of the form

$$f(\cdot, u, \partial u) = f^0(\cdot, u) + a_j(\cdot, u)\partial_j u \ ,$$

where now D is open in $\mathbb{R}^N$ and

$$f^0 \in C^{1\text{-}}(\overline{\Omega} \times D, \mathbb{R}^N) \ , \quad a_j \in C^{2\text{-}}(\overline{\Omega} \times D, \mathcal{L}(\mathbb{R}^N)) \ .$$

Then, by replacing $\mathcal{A}$ in the obvious way, it is clear that it suffices to assume that $r > n/p$ with $r, s \neq 1 + 1/p$.

(b) It is clear how to apply the abstract results of Section 12 to the above case in order to deduce further information.

(c) Given suitable growth restrictions for f, the global existence of $u(\cdot, u^0)$ can be guaranteed provided an a priori bound in some weaker norm than the one of W_p^ρ is known. This follows by arguments which are similar to the ones of Section 15. □

Theorem 16.1 generalizes and simplifies the main result of [Ama86b], except that we restrict ourselves here to the second order case. It is, however, clear from Remark 4.2(b) that Theorem 16.1 extends to the higher order case too.

Lastly, we consider the semilinear parabolic boundary value problem (13.18) with local nonlinearities. More precisely, we restrict ourselves to the simplest case, namely to the problem

$$\begin{aligned} \partial_t u + \mathcal{A}u &= f(\cdot, u) && \text{in } \ \Omega \times (0, \infty) \ , \\ \mathcal{B}u &= 0 && \text{on } \partial\Omega \times (0, \infty) \ , \\ u(\cdot, 0) &= u^0 && \text{on } \ \Omega \ , \end{aligned} \tag{16.8}$$

where we assume that $(\mathcal{A}, \mathcal{B}) \in \mathcal{E}^1(\Omega)$. This problem has been studied by many authors in great detail under various aspects (e.g., [Hen81], [DaK92], [Rhe84], [vW85a] to mention just a few general references which, however, almost exclusively consider the case of a single equation (i.e., $N = 1$) or of weakly coupled systems. For the case of strongly coupled systems we refer to [Ama84], [Ama85]). For this reason we discuss here one point only, namely the solvability of (16.8) on L_p. Of course, it is well known that (16.8) generates a semiflow on L_p provided F, the Nemytskii operator induced by f, maps L_p locally Lipschitz continuously into itself. It follows from Lemma 14.2 that this is the case if $\partial_2 f$ is uniformly bounded on $\overline{\Omega} \times \mathbb{R}^N$, which implies that f is linearly bounded. In fact, it is known (e.g., [Kra64], [Vai64]) that the linear boundedness of f is necessary for F to map L_p into itself. However, we can get a much better result by studying very weak solutions of (16.8).

Suppose that

$$f \in C^1(\overline{\Omega} \times \mathbb{R}^N, \mathbb{R}^N) \ . \tag{16.9}$$

Also suppose that

$$1 < p < \infty \quad \text{and} \quad 1 < \nu < 1 + 2p/n \ , \tag{16.10}$$

and there exists a constsnt c such that

$$|\partial_2 f(x, \xi)| \le c(1 + |\xi|^{\nu - 1}) \ , \qquad (x, \xi) \in \overline{\Omega} \times \mathbb{R}^N \ . \tag{16.11}$$

Fix $\tau \in (0,2)$ such that $\nu \le 1 + \tau p/n$ and note that $W^{\tau}_{p',\mathcal{B}\#} \hookrightarrow L_{q'}$ so that $L_q \hookrightarrow W^{-\tau}_{p',\mathcal{B}\#}$, provided $1/q = (1/p) + (\tau/n)$. Thus $\nu - 1 \le \tau p/n = p/q$ and, thanks to Lemma 14.2(ii),

$$F \in C^{1\text{-}}(L_p, L_q) \hookrightarrow C^{1\text{-}}(L_p, W^{-\tau}_{p',\mathcal{B}\#}) \ .$$

Now we deduce from Theorem 13.8 and a standard boot strapping argument based upon the results of Section 11 and of [Sol65] the following

16.3 Theorem *Suppose that $(\mathcal{A}, \mathcal{B}) \in \mathcal{E}^1(\Omega)$ and let conditions (16.9)–(16.11) be satisfied. Then the semilinear parabolic boundary value problem (16.8) possesses for each $u^0 \in L_p$ a unique maximal solution*

$$u(\cdot, u^0) \in C\big([0,t^+), L_p\big) \cap C^{2,1}\big(\overline{\Omega} \times (0,t^+), \mathbb{R}^N\big) \ .$$

The map $(t, u^0) \mapsto u(t, u^0)$ is a semiflow on L_p. If $u|[0,T]$ is bounded in L_p for each $T > 0$ then $t^+ = \infty$.

The last part of the assertion follows easily from [Ama84, Corollary 3.2], for example.

Theorem 16.3 is a vast generalization of earlier results of Weissler [Wei80], who studies the case $N = 1$ with $(\mathcal{A}, \mathcal{B}) = (-\Delta, \gamma_\partial)$ and with $f(x,\xi) := |\xi|^{\nu-1}\xi$. However, that author admits the case $\nu = 1 + 2p/n$ too and also shows that this exponent is optimal.

Theorem 16.3 is included here mainly to give another illustration of the usefulness and simplicity of our extrapolation technique. Of course, it applies to a great many other situations too.

16.4 Remarks

(a) It is clear that the solution of Theorems 16.1 and 16.3 are smooth, that is, C^∞, on $\overline{\Omega} \times (0, t^+)$ if all data are smooth.

(b) By employing Remark 12.2(a) it is not difficult to see that all results of Sections 13–16 are valid — modulo obvious modifications — for nonautonomous quasilinear parabolic systems. Moreover, the solutions depend continuously (or smoothly, if all data are smooth) in the W^s_p-topology on parameters (e.g., the coefficients etc.). □

17 The Model Problems Revisited

As already mentioned in Section 1, we indicate now briefly how the general theory described in this paper applies to our model problems. Unless explicitly stated otherwise, all data are supposed to be smooth.

17.1 The Shigesada-Kawasaki-Teramoto Model In order to guarantee the applicability of our theory to problem (1.11), and since we are interested in positive solutions only, we have to verify that there exists a neigborhood D of $(\mathbb{R}^2)^+$ in $\mathbb{R}^2$ such that

$$\sigma\big(a(\eta)\big) \subset [\operatorname{Re} z > 0] \ , \qquad \eta \in D \ , \tag{17.1}$$

where a is given by (1.12) (cf. Example 4.3(e)). Note that (17.1) is equivalent to

$$\operatorname{trace}(a(\eta)) > 0 \quad \text{and} \quad \det(a(\eta)) > 0 \ , \qquad \eta \in D \ .$$

Since $\alpha_r > 0$ and $\beta_{rs} \geq 0$, it is obvious that $\operatorname{trace}(a(\eta)) > 0$ for η in some neighborhood D of $(\mathbb{R}^2)^+$. It is easily verified that

$$\det(a(\eta)) = \alpha_1\alpha_2 + p_2(\eta) \ , \qquad \eta \in \mathbb{R}^2 \ ,$$

where p_2 is a polynomial of degree 2 in $\eta = (\eta_1, \eta_2)$ with nonnegative coefficients. Hence we can find an open neighborhood D of $(\mathbb{R}^2)^+$ such that (17.1) is true and such that ∂D is bounded away from $(\mathbb{R}^2)^+$.

Now the following theorem is an immediate consequence of Theorems 14.4, 14.6, 15.1, 15.3, and 15.5, where Theorem 15.1 guarantees the positive invariance of the positive cone $(W_p^1)^+$ of W_p^1.

17.1 Theorem *Let $p > n$ $(= 2)$ be fixed. The population dynamical model (1.11) has for each $u^0 \in (W_p^1)^+$ a unique maximal classical solution $u(\cdot, u^0)$, which is smooth on $\overline{\Omega} \times (0, t^+(u^0))$. The map $(t, u^0) \mapsto u(t, u^0)$ is a semiflow on $(W_p^1)^+$ such that bounded orbits are relatively compact. If there exist $\varepsilon > 0$ and $c \colon \mathbb{R}^+ \to \mathbb{R}^+$ such that*

$$\|u(t, u^0)\|_{C^\varepsilon} \leq c(T) \ , \qquad 0 \leq t \leq T < \infty \ , \quad t < t^+(u^0) \ , \tag{17.2}$$

or if $\beta_{12}\beta_{21} = 0$ and (17.2) is true with $\varepsilon = 0$, then $t^+(u^0) = \infty$.

Note that (1.11) generates for each s with $n/p < s < (1 + 1/p) \wedge (2 - n/p)$ a semiflow on $W_p^s(\Omega, D)$, and that Theorem 15.1 implies that $(W_p^s)^+$ is positively invariant. Moreover, Theorem 17.1 remains true if the coefficients of γ_r depend smoothly on $x \in \overline{\Omega}$, provided the positivity assumptions are preserved.

17.2 Remark Problem (1.11) can be used to illustrate the importance of the fact that we can handle normally elliptic boundary value problems $(\mathcal{A}, \mathcal{B})$. Indeed, suppose that we were able to treat strongly uniformly elliptic systems only. Then we would have to define the set D by

$$D := \left\{ \eta \in \mathbb{R}^2 \ ; \ a(\eta) + {}^t a(\eta) \text{ is positive definite} \right\} \ .$$

It is easily seen that in this case $D \not\supset (\mathbb{R}^2)^+$ if either $\beta_{11} = 0$ and $\beta_{12} \neq 0$, or $\beta_{22} = 0$ and $\beta_{21} \neq 0$. Thus in this case we could not solve (1.11) for arbitrary positive initial values $u^0 \in W_p^1$, which would be unreasonable from the point of view of the model. □

The classical solvability of problem (1.11) and the invariance of the positive cone $(W_p^1)^+$ in the above generality has first been indicated by the author in [Ama90a] as an application of the general results proven there. Recently, Yagi [Yag92] has proven a weaker result. Moreover, he assumes that

$$g_r(u) := \rho_r - \sigma_{r1}u_1 - \sigma_{r2}u_2 \ , \qquad \rho_r, \sigma_{rs} \geq 0 \ , \quad 1 \leq r, s \leq 2 \ . \tag{17.3}$$

The question of global existence of positive solutions to (1.11) has already attracted a fair amount of interest. The first global existence theorem seems to be due to Kim [Kim84] for the very special case $\alpha_1 = \alpha_2$, $\beta_{11} = \beta_{22} = \gamma_1 = \gamma_2 = 0$, and $n = 1$, and with g satisfying (17.3) with positive ρ_r and σ_{rs}. Later Deuring [Deu87] proved global existence under the same assumptions as Kim, but for arbitrary n and allowing $\alpha_1 \neq \alpha_2$, provided the coefficients of $(\mathcal{A}, \mathcal{B})$ are sufficiently small, their size depending on u^0. In [PoT90] there is a global existence result under the assumption that $\beta_{11} = \beta_{21} = \beta_{22} = \gamma_1 = \gamma_2 = 0$ and that g is sufficiently nonlinear, but for Dirichlet boundary conditions. In [Yag92] global existence is proven under the assumption (17.3), provided either $0 < \beta_{21} < 8\beta_{11}$ and $0 < \beta_{12} < 8\beta_{22}$, or $\beta_{11} > 0$, $\beta_{21} = \beta_{22} = 0$, and $\sigma_{11} > 0$. Lastly, Redlinger [Red92] considers the case $\beta_{11} = \beta_{21} = \beta_{22} = 0$ and $\gamma_1 = \gamma_2 = 0$ and assumes that g is of the form (17.3) with positive constants, except that the term $\sigma_{11}u_1$ is replaced by $h(u_1)$, where $\liminf_{t\to\infty} h(t)t^{-\nu} > 0$ for some $\nu > 0$ (with $\nu > 1$ if $n \geq 2$) and h is polynominally bounded. Then he proves the existence of a global attractor, that is, of a compact invariant set for the semiflow on $(W_p^1)^+$, which attracts each bounded set.

Note that $u = 0$ is a stationary point of the semiflow $(t, u^0) \mapsto u(t, u^0)$ and that the linearization at $u = 0$ of the stationary part of (1.11) is given by

$$\begin{aligned} -\partial_j\bigl(a(0)\partial_j v + a_j v\bigr) &= g(0)v \qquad \text{in } \Omega\ , \\ a(0)\partial_\nu v + a_j \nu^j v &= 0 \qquad \text{on } \partial\Omega\ , \end{aligned} \tag{17.4}$$

where $a(0) = \operatorname{diag}[\alpha_1, \alpha_2]$, the a_j are given by (1.13), and $g(0) := \operatorname{diag}\bigl[g_1(0), g_2(0)\bigr]$. Hence, if we consider, for simplicity, the special case $\gamma_1 = \gamma_2 = 0$, we see that all eigenvalues of the linear boundary value problem (17.4) are strictly positive, provided $g_r(0) > 0$ for $r = 1, 2$. This is true, in particular, if g_r is given by (17.3) with $\sigma_r > 0$ for $r = 1, 2$. Then the principle of linearized stability implies that 0 is an asymptotically stable critical point of the semiflow on $(W_p^1)^+$ (cf. [Dra88], [Dra89], where it has even been shown that 0 is asymptotically stable with respect to the W_p^2-topology). This gives a certain generalization of the result in [Deu87].

17.2 The Penrose-Fife Model In order to study the well-posedness of (1.14) under the general boundary conditions (1.23) we put $D := \mathbb{R} \times (0, \infty)$. Since

$$\sigma\bigl(a(\eta)\bigr) = \bigl\{\kappa(\eta_2), M(\eta)/\mu(\eta_1)\eta_2^2\bigr\} \subset \dot{\mathbb{R}}^+\ , \qquad \eta \in D\ ,$$

we can directly apply Theorem 14.4. By taking into consideration Remark 14.1(b) too, we obtain the following

17.3 Theorem *Suppose that $p > n$ and $1 < s < (1 + 1/p) \wedge (2 - n/p)$. Then the phase field model (1.14) under the general boundary condition (1.23) generates a semiflow on $W_p^s(\Omega, \dot{\mathbb{R}}^+)$ such that bounded orbits, satisfying*

$$T(t) \geq \varepsilon > 0 \qquad \textit{for some} \quad \varepsilon > 0 \quad \textit{and all} \quad t \in [0, t^+)\ , \tag{17.5}$$

are relatively compact. If $g_1 = 0$,

$$(\varphi, T) := \bigl(\varphi(\cdot, \varphi^0, T^0), T(\cdot, \varphi^0, T^0)\bigr) \tag{17.6}$$

is a classical solution which is smooth on $\overline{\Omega} \times (0, t^+)$.

As for global existence, we can, of course, apply Theorm 15.4. However, if $g_1 = 0$, it follows from (1.23) and Example 13.4(b) that (13.12) is satisfied. Hence we deduce from (1.15)–(1.18) and from Theorem 15.5 the following global existence

17.4 Theorem *Suppose that $g_1 = 0$. Then each solution (17.6) satisfying (17.5) exists globally, provided*

$$\|\varphi(t)\|_\infty + \|T(t)\|_\infty \leq c(\tau) , \qquad 0 \leq t \leq \tau < \infty , \quad t < t^+ . \tag{17.7}$$

If c is bounded, the orbit through (φ^0, T^0) is relatively compact in $W_p^s(\Omega, \mathbb{R} \times \dot{\mathbb{R}}^+)$.

Of course, the above theorem remains valid if all functions depend smoothly upon $x \in \overline{\Omega}$ (and on t).

The above results are new. The well-posedness of the phase field model (1.14), (1.23) has — so far — only been proven in special cases. In [SpZ92] the authors derive an existence theorem guaranteeing a unique classical global solution, provided assumptions (1.19)–(1.21) hold and, in addition, the function s (the double well potential) satisfies suitable boundedness restrictions. Much of the work in [SpZ92] consists of the derivation of the a priori bounds (17.5) and (17.7) which enable the authors to apply our Theorem 15.5 to get global solutions.

In [Zhe92] a global existence theorem for problem (1.14), (1.19), (1.20) is obtained in space dimension one and for homogeneous Neumann boundary conditions (that is, conditions (1.21) with $\alpha = 0$). Additional results in even simpler situations are contained in [BaZ90] and [ElZ89].

It should be of interest to derive a priori bounds of the form (17.5) and (17.7) for getting global smooth solutions of (1.14), (1.23) in more general cases.

17.3 The Groundwater Flow Problem In order to treat the parabolic-elliptic problem (1.24)–(1.29) we use the fact that in (1.31) it has been reformulated as a nonlocal parabolic problem of the form (13.1) with $F = 0$ and $G = 0$. Thus it remains to verify (13.3) for appropriate choices of the parameters (13.2).

It is obvious that

$$M \in C^{1-}(\mathbb{R}^2, \mathcal{L}(\mathbb{R}^n)) \tag{17.8}$$

with $n = 2$, and that M is symmetric and uniformly positive definite. Note that Theorem 9.2, Remark 9.3(a), and the commutativity of the diagram

$$\begin{array}{ccc} W_p^s & \xrightarrow{\quad \partial_1 \quad} & W_{p,\gamma_\partial}^{s-1} \\ {\scriptstyle Q_j} \big\downarrow & & \big\downarrow {\scriptstyle (-\Delta_D)^{-1}} \\ W_{p,\gamma_\partial}^s & \xleftarrow{\quad (-1)^{j-1}\partial_j \quad} & W_{p,\gamma_\partial}^{s+1} \end{array} \tag{17.9}$$

where $W_p^s := W_p^s(\Omega, \mathbb{R})$ etc., imply

$$Q_j \in \mathcal{L}(W_p^s, W_{p,\gamma_\partial}^s) , \qquad s \in [0, 1 + 1/p) \backslash \{1/p\} , \quad j = 1, 2 . \tag{17.10}$$

Fix $p > n$ and $n/p < r < s < \sigma := 1$. Then we deduce from (17.8), (17.10), (1.30), the imbedding $W_p^r \hookrightarrow C$, and from Lemma 14.2(i) that

$$(\mathcal{A}, \mathcal{B}) \in C^{1\text{-}}(W_p^r, \mathcal{E}^{1/2}(\Omega)) .$$

Consequently, Theorem 13.1 implies the following well-posedness

17.5 Theorem *Let $0 < 2/p < s < 1$. Then the porous media problem (1.24)–(1.29) generates a semiflow of weak W_p^1-solutions on W_p^s. The positive cone $(W_p^s)^+$ is positively invariant, and bounded orbits are relatively compact.*

This theorem is essentially contained in [CDL92]. In that paper the authors also consider the much simpler semilinear problem with $a = b = 0$. Then (1.31) takes the form

$$\begin{aligned} \partial_t u - m\Delta u &= -\partial_j\big(a_j(u)u\big) && \text{in } \ \Omega \times (0,\infty) , \\ \partial_\nu u &= 0 && \text{on } \partial\Omega \times (0,\infty) . \end{aligned} \tag{17.11}$$

Since $W_{p'}^{n/p} \hookrightarrow L_{(p/2)'}$ for $p > n$, we deduce from (17.9), Hölder's inequality, and the factorization

$$L_p \xrightarrow{\ u \mapsto a_j(u)u\ } L_{p/2} \hookrightarrow W_{p,\mathcal{B}}^{-n/p} , \tag{17.12}$$

where $\mathcal{B} := \partial_\nu$, that

$$\big[u \mapsto \partial_j\big(a_j(u)u\big)\big] \in C^{1\text{-}}(L_p, W_{p,\mathcal{B}}^{-\tau}) \tag{17.13}$$

with $\tau := 1 + n/p < 2$. Thus Theorem 13.8 implies that problem (17.11) generates a semiflow of very weak L_p-solutions on L_p. Then an obvious boot strapping argument based upon the results of Section 11 guarantees that, given any $u^0 \in L_p$, the unique maximal very weak L_p-solution is, in fact, a classical solution on $\big(0, t^+(u^0)\big)$ of (17.11).

Put $\varphi(\xi) := |\xi|^p$ for $\xi \in \mathbb{R}$ and note that φ is twice continuously differentiable and convex. By multiplying the first equation in (17.11) by $\varphi'(u)$, where $u := u(\cdot, u^0)$, and integrating over Ω we obtain

$$\langle \varphi'(u), \dot u\rangle + m\langle \partial_j \varphi'(u), \partial_j u\rangle = -\big\langle \varphi'(u), \partial_j\big(a_j(u)u\big)\big\rangle . \tag{17.14}$$

Observe that

$$m\langle \partial_j \varphi'(u), \partial_j u\rangle = m\langle \varphi''(u)\partial_j u, \partial_j u\rangle \geq 0 . \tag{17.15}$$

Moreover, since $\partial_j a_j(u) = 0$ and $\gamma_\partial a_j(u)\nu^j = 0$,

$$\begin{aligned} \big\langle \varphi'(u), \partial_j\big(a_j(u)u\big)\big\rangle &= \langle \varphi'(u), a_j(u)\partial_j u\rangle = \langle \partial_j\varphi(u), a_j(u)\rangle \\ &= \big\langle \partial_j\big(\varphi(u)a_j(u)\big), \mathbf{1}\big\rangle = \big\langle \gamma_\partial\varphi(u), \gamma_\partial a_j(u)\nu^j\big\rangle_\partial = 0 . \end{aligned} \tag{17.16}$$

Thus we deduce from (17.13)–(17.16) that

$$\frac{d}{dt}\|\varphi(u)\|_1 = \langle \varphi'(u), \dot u\rangle \leq 0 \qquad \text{on } \dot J , \tag{17.17}$$

which implies

$$\|u(t)\|_p \le \|u^0\|_p \, , \qquad t \in J \, . \tag{17.18}$$

From this estimate and from (17.9) and (17.11) we infer that

$$\sup_{t\in J} \left\| \partial_j \left(a_j(u(t))u(t) \right) \right\|_{W^{-\tau}_{p,\mathcal{B}}} < \infty$$

which, by an easy Gronwall inequality type argument, implies $t^+(u^0) = \infty$ (e.g., [Ama84, Corollary 3.2]). This proves

17.6 Theorem *Suppose that $a = b = 0$ and $p > 2$. Then the porous media problem (1.24)–(1.29) generates a global semiflow on L_p. The positive cone L_p^+ is positively invariant and $u(\cdot, u^0)$ is a classical solution of (17.11) for $t > 0$.*

In [CDL92] the above theorem has been proven by a completely different method which is based on an abstract theorem for semilinear parabolic equations due to v. Wahl [vW85b]. Our simple proof gives another illustration for the usefulness of very weak solutions even in the case of nonlocal operators.

17.4 The Dissolution-Growth Model We discuss now the well-posedness of problem (1.32). For this we fix numbers

$$0 < 1/p < r < s < \sigma_1 < \tau + 2 < 1 + 1/p < 2 \tag{17.19}$$

and put

$$E_j := H^{\sigma_1 - 2(1-j)}_{p,\mathcal{B}_{11}} \, , \qquad j = 0, 1 \, , \tag{17.20}$$

where $H^\sigma_p := H^\sigma_p(\Omega, \mathbb{R})$ etc. We also put

$$E := \mathbb{R} \, , \quad \boldsymbol{E}_j := E_j \times E \, , \qquad j = 0, 1 \, ,$$

and define $\mathfrak{F}$ by (13.8) and $0 < \gamma < \beta < \alpha < 1$ by the line subsequent to (13.9). Then it follows that

$$\boldsymbol{E}_\theta := (\boldsymbol{E}_0, \boldsymbol{E}_1)_\theta \doteq E_\theta \times \mathbb{R} \, , \quad E_\theta := (E_0, E_1)_\theta \doteq W^{\sigma_1 - 2 + 2\theta}_{p,\mathcal{B}_{11}} \, , \qquad 0 < \theta < 1 \, ,$$

(cf. [DaK92, Proposition 3.4]). Note that

$$X_\beta := W^r_{p,\mathcal{B}_{11}} \times (0, 1)$$

is an open subset of $\boldsymbol{E}_\beta$. It is easily verified that

$$(\mathcal{A}_{11}, \mathcal{B}_{11}) \in C^{1\text{-}}\left(X_\beta, \mathcal{E}^1(\Omega)\right) \, .$$

Thus, denoting by A_{11} the E_0-realization of $(\mathcal{A}_{11}, \mathcal{B}_{11})$,

$$A_{11} \in C^{1\text{-}}\left(X_\beta, \mathcal{H}(E_1, E_0)\right) \, ,$$

thanks to Theorem 8.5. From this it is obvious that

$$A := \begin{bmatrix} A_{11} & 0 \\ 0 & 0 \end{bmatrix} \in C^{1-}(\mathbb{R}^+ \times X_\beta, \mathcal{H}(\boldsymbol{E}_1, \boldsymbol{E}_0)) \ . \tag{17.21}$$

Observe that $\partial W_p^\sigma \doteq \mathbb{R} \oplus \mathbb{R}$ for $0 \leq \sigma \leq 2$. From this we easily deduce that

$$F := \begin{bmatrix} f_1 + \mathcal{R} g_1 \\ f_2 \end{bmatrix} \in C^{1-}(\mathbb{R}^+ \times X_\beta, \boldsymbol{E}_\gamma) \ ,$$

where $\mathcal{R} \in \mathcal{L}(\mathbb{R}, E_\gamma)$ is the dual of the trace operator $\gamma_1 \in \mathcal{L}(W_{p',\mathcal{B}_{11}}^{-\tau}, \mathbb{R})$ (on $x = 1$). Then the quasilinear evolution equation

$$\dot{u} + A(t,u)u = F(t,u) \ , \qquad t > 0 \ ,$$

is equivalent to the weak formulation — in the obvious sense — of problem (1.33)–(1.40). Thus, by applying Theorem 12.1, Remark 12.2(a), and an obvious boot strapping argument, we obtain

17.7 Theorem *The moving boundary value problem* (1.32) *has for each* $R^0 \in (0,1)$ *and* $c^0 \in W_p^s((R^0,1),\mathbb{R})$ *with* $c^0(1) = \varphi(0)$ *a unique maximal classical solution* (c, R) *depending continuously in the* W_p^s*-topology on* (c^0, R^0).

Of course, the precise formulation of the continuous dependence on the initial data is given via the transformed quantities (u_1, u_2) of the 'fixed boundary value problem'.

It is an easy consequence of Theorem 12.5 that $t^+ := t^+(c^0, R^0) < \infty$ implies $R(t) \to 0$ or $R(t) \to 1$ as $t \to t^+$, provided

$$\sup_{0<t<t^+} \|c(\cdot,t)\|_\infty < \infty \ . \tag{17.22}$$

This follows from the fact that, by standard easy a priori estimation techniques (use u_1 as a test function in the weak W_2^1-formulation of the first two equations in (1.33)), the estimate (17.22) implies a bound for the W_2^1-norm of $u_1(\cdot,t)$, uniformly with respect to $t \in (t_0, t^+)$, where $t_0 \in (0, t^+)$ is arbitrary. Details are left to the reader.

Theorem 17.7 generalizes the results in [CoHS89] and [CoHS90], respectively, where by a rather lengthy argument the well-posedness of problem (1.32) in the framework of weak solutions has been shown. In contrast to this, our proof is nothing but an almost trivial application of our general theorems. It serves as another illustration for the power of the techniques presented in this paper.

17.5 Highly Degenerate Problems In order to prove the well-posedness of (1.41) we study its equivalent form (1.43). For this we fix p, r, s, σ_1, and τ as in (17.19) and define E_j, $j = 0, 1$, by (17.20), where now $\mathcal{B}_{11} = \gamma_\partial$, of course. We also put

$$E := H_p^{\sigma_1} \ , \quad \boldsymbol{E}_j := E_j \times E \ , \qquad j = 0, 1 \ ,$$

and define $\mathfrak{F}$ and $0<\gamma<\beta<\alpha<1$ as in Subsection 17.4. Then, similarly as above,

$$\boldsymbol{E}_\theta := (\boldsymbol{E}_0, \boldsymbol{E}_1)_\theta \doteq E_\theta \times E\ , \quad E_\theta := (E_0, E_1)_\theta \doteq W^{\sigma_1-2+2\theta}_{p,\mathcal{B}_{11}}\ , \qquad 0<\theta<1\ .$$

We denote by $A_{11}(u)$ the E_0-realization of $\big(\mathcal{A}_{11}(u), \gamma_\partial\big)$ and define $A_{12}(u)$ by

$$\big\langle w_1, A_{12}(u)v_2\big\rangle := \big\langle \partial_x w_1, H(u_1+\Psi)\partial_x v_2\big\rangle_\Omega\ , \qquad w_1 \in W^2_{p',\gamma_\partial}\ , \quad v_2 \in E\ .$$

Thanks to (1.44) it is not difficult to verify that

$$(A_{11}, A_{12}) \in C^2\big(X_\beta, \mathcal{H}(E_1, E_0) \times \mathcal{L}(E_1, E_0)\big)\ , \tag{17.23}$$

where $X_\beta := \boldsymbol{E}_\beta$. (For simplicity, we assume that Ψ is independent of t. The more general case can easily be handled by means of Remark 12.2.) This implies

$$A := \begin{bmatrix} A_{11} & A_{12} \\ 0 & 0 \end{bmatrix} \in C^2\big(\mathbb{R}^+ \times X_\beta, \mathcal{H}(\boldsymbol{E}_1, \boldsymbol{E}_0)\big) \tag{17.24}$$

thanks to [Ama91b, Theorem 2.1], [Ang90, Lemma 2.6], or [Ama94, Corollary 1.6.3]. On the other hand, it is easily seen that

$$F := \begin{bmatrix} f_1 \\ f_2 \end{bmatrix} \in C^2(X_1, \boldsymbol{E}_\gamma)\ , \tag{17.25}$$

where $X_1 := \boldsymbol{E}_1$. Of course, the evolution equation

$$\dot{u} + A(u)u = f(u)\ , \qquad t>0\ , \tag{17.26}$$

is the weak $H^{\sigma_1}_p$-formulation — in the obvious sense — of problem (1.43).

Since f_2 is a (smooth) local function depending upon the component u_1 of u, in particular, it is clear that its domain $X_1 = \boldsymbol{E}_1$ in (17.25) cannot be replaced by X_β, no matter how $\beta \in (0,1)$ has been chosen, if one wants to be sure that the image of F is contained in $\boldsymbol{E}_\gamma$. Thus Theorem 12.1 is not applicable to equation (17.26). Therefore we put

$$\varphi(u) := A(u)u - F(u)$$

and note that $\varphi \in C^2(X_1, \boldsymbol{E}_0)$ with

$$\partial\varphi(u) = A(u) + \partial A(u)[\cdot, u] - \partial F(u) =: A(u) + B(u) \in \mathcal{L}(\boldsymbol{E}_1, \boldsymbol{E}_0)\ , \qquad x \in X_1\ .$$

Observe that

$$\partial\varphi(u) = \begin{bmatrix} A_{11}(u)+B_{11}(u) & A_{12}(u)+B_{12}(u) \\ B_{21}(u) & B_{22}(u) \end{bmatrix}$$

where

$$B_{11}(u) = \partial_1 A_{11}(u)[\cdot, u_1] - \partial_1 f_1(u)\ .$$

Thus, thanks to (1.45) and (17.23), it follows that

$$B_{11}(u) \in \mathcal{L}(E_\beta, E_0)\ , \qquad u \in X_1\ .$$

Hence (17.23) and a well-known perturbation theorem (e.g., [Ama91b, Proposition 1.1(iii)]) imply

$$A_{11}(u) + B_{11}(u) \in \mathcal{H}(E_1, E_0) .$$

Now [Ama91b, Theorem 2.1] guarantees that

$$\partial\varphi(u) \in \mathcal{H}(\boldsymbol{E}_1, \boldsymbol{E}_0) , \qquad u \in X_1 ,$$

and we can apply Theorem 12.7 to deduce the following well-posedness result in which we specify σ_1 to be 1.

17.8 Theorem *The highly degenerate problem* (1.43) *generates a semiflow of weak* W_p^1*-solutions on* $W^1_{p,\gamma_\theta} \times W^1_p$.

It is clear that this result guarantees the well-posedness of the original system (1.41) in the sense that the first equation of (1.41) is satisfied in the weak W_p^1-sense and the second one pointwise. If $H(0) = 0$, we can apply Theorem 11.6 in the obvious way to the first equation of (1.41) to deduce that $c(t, c^0, \sigma^0) \geq 0$ for $0 < t < t^+(c^0, \sigma^0)$, provided $c^0 \geq 0$ and $\psi \geq 0$.

Theorem 17.8 is a special case of the results in [Ama91b], where more general situations have been studied too. In the particular semilinear situation where $D = H =$ const, φ and M vanish identically, and $\rho = 1$, it has been shown in [Ama91a] that the corresponding system generates a global semiflow, provided β is uniformly bounded.

Although the operator matrices (17.21) and (17.24) have, formally, the same structure, in as much as their second rows consist of zeros, there is a big difference between the two evolution equations of Subsections 17.4 and 17.5, respectively. This is due to the fact that the second row, that is, the second factor E of the products $\boldsymbol{E}_j = E_j \times E$, is finite dimensional in the first case, but infinite dimensional in the second one. For this reason systems of the latter form are said to be 'highly degenerate'. Although the corresponding evolution equations are 'abstractly parabolic' in the sense that $A(u) \in \mathcal{H}(\boldsymbol{E}_1, \boldsymbol{E}_0)$, highly degenerate problems lack the regularizing property of 'standard' quasilinear parabolic systems which can be put in a framework to which Theorem 12.1 applies.

17.6 The Electrolysis Problem By using the notations introduced in Subsection 1.6 we can write the system under consideration in the form

$$\begin{aligned} \partial_t u - \partial_j\big(a(u)\partial_j u + a_j(u,\varphi)u\big) &= 0 && \text{in } \Omega \times (0,\infty) , \\ a(u)\partial_\nu u + a_j(u,\varphi)\nu^j u &= g(u,\varphi) && \text{on } \partial\Omega \times (0,\infty) . \end{aligned} \tag{17.27}$$

It has to be complemented by the initial condition

$$u(\cdot, 0) = u^0 \qquad \text{on } \Omega \tag{17.28}$$

and the electroneutrality condition

$$z(u) := \textstyle\sum_r z_r u_r = 0 \qquad \text{on } \Omega \times [0,\infty) . \tag{17.29}$$

Note that (17.29) is an invariance condition for the semiflow hopefully generated by (17.27). Thus, if (u, φ) is a solution to (17.27)–(17.29), by applying $z(\cdot)$ to (17.27) it is easily seen that (u,φ) necessarily satisfies the stationary problem

$$\begin{aligned} -\partial_j z\big(a(u)\partial_j u\big) - \nabla\cdot\big(\kappa(u)\nabla\varphi\big) &= 0 && \text{in } \Omega\,, \\ z\big(a(u)\partial_\nu u\big) + \kappa(u)\partial_\nu\varphi &= z\big(g(u,\varphi)\big) && \text{on } \partial\Omega\,, \end{aligned} \tag{17.30}$$

where

$$\kappa(u) := \textstyle\sum_r z_r m_r u_r\,.$$

In fact, it can be shown that (17.30) is necessary too. More precisely, (u,φ) is a solution to (17.27)–(17.29) iff it is a solution to (17.27), (17.28) with $z(u^0)=0$, and (17.30).

Observe that, given u, the system (17.30) is an elliptic boundary value problem with nonlinear boundary conditions for φ, provided

$$\kappa(u)(x) > 0\,, \qquad x \in \overline{\Omega}\,. \tag{17.31}$$

Thanks to (1.49), this is certainly the case if u is in the interior of the positive cone of $C(\overline{\Omega},\mathbb{R}^N)$ and if $z_r m_r > 0$ for at least one $r \in \{1,\dots,N\}$.

Suppose that, given u^0 satisfying (17.31), there exists φ^0 such that (u^0,φ^0) is a solution of (17.30). Also assume that in a suitable neighborhood U of u^0 every solution of (17.30) near φ^0 can be represented in the form $\big(u,\varphi(u)\big)$. By substituting $\varphi(u)$ in (17.27) we obtain a system of the form

$$\begin{aligned} \partial_t u_r + \mathcal{A}(u)u &= F(u) && \text{in } \Omega\,, \\ \mathcal{B}(u)u &= G(u) && \text{on } \partial\Omega\,, \end{aligned} \tag{17.32}$$

where F and G are suitable nonlocal operators and

$$\mathcal{A}(u)v := -\nabla\cdot\big(\textstyle\sum_s a_{rs}(u)\nabla v_s\big)\,,\quad \mathcal{B}(u)v := \textstyle\sum_s a_{rs}(u)\partial_\nu v_s \tag{17.33}$$

with

$$a_{rs}(u) := a_r(\cdot,u)\delta_{rs} - \frac{m_r u_r}{\kappa(u)} z_s a_s(\cdot,u)\,, \qquad 1 \le r,s \le N\,.$$

Note that (17.32) is a strongly coupled quasilinear system even if the diffusion coefficients a_r of the original system are constant.

Observe that $\sum_r z_r a_{rs}(u) = 0$. Hence it follows that the matrix $\big[a_{rs}(u)\big]_{1\le r,s\le N}$ defines at each $x \in \overline{\Omega}$ a linear operator $\alpha(u)(x) \in \mathcal{L}(Z)$, where

$$Z := \big\{\,\eta \in \mathbb{R}^N\ ;\ \textstyle\sum_r z_r \eta_r = 0\,\big\}\,.$$

It can be shown that $\sigma\big(\alpha(u)(x)\big) \subset [\operatorname{Re} z > 0]$ for $x \in \overline{\Omega}$ and each u satisfying (17.31). Consequently, it follows from Example 4.3(e) that (17.33) defines a normally elliptic boundary value problem in spaces of Z-valued functions. Thus we can apply the results of Section 13 to (17.32), provided the latter system is well defined.

By means of the above arguments it can be shown that the reaction-diffusion problem of electrolysis is well-posed, provided suitable hypotheses for the nonlinearity g are met. Details can be found in [AmR93]. Note that this problem is rather more complex than the other model cases since it involves a solvability theory for the nonlinear elliptic boundary value problem (17.30).

17.7 Stefan Problems In order to handle problem (1.50)–(1.53) we study the system (1.57)–(1.59). It is an important observation that the boundary condition $\mathcal{B}_{11}(u,s)u = 0$ is equivalent to the 'constant boundary condition'

$$\mathcal{B}u := \delta\partial_\nu u + (1-\delta)u = 0 \ ,$$

where $\delta(1) = 1_2$. Thus we fix numbers p, r, s, and τ such that

$$0 < \tau < 1/p < 1 + 1/p < r < s < 2 \ .$$

Let

$$E_j := W^{2j}_{p,\mathcal{B}} \ , \qquad j = 0,1 \ ,$$

where now $N = 2$, and define $\mathfrak{F}$, $0 < \gamma < \beta < \alpha < 1$, and the spaces E and $\boldsymbol{E}_\theta$, $0 \le \theta \le 1$, as in Subsection 17.4 (but with $N = 2$). Also put $X_\beta := W^r_{p,\mathcal{B}} \times (0,1)$ and note that it is an open subset of $\boldsymbol{E}_\beta$. Let $\boldsymbol{u} := (u,s)$ and $\big(\mathcal{A}_{11}(\boldsymbol{u}), \mathcal{B}_{11}\big) := \big(\mathcal{A}(u,s), \mathcal{B}\big)$. Then

$$(\mathcal{A}_{11}, \mathcal{B}_{11}) \in C^{1\text{-}}(X_\beta, \mathcal{E}^1(\Omega)) \ .$$

Thus, denoting by A_{11} the E_0-realization of $(\mathcal{A}_{11}, \mathcal{B}_{11})$, it follows that

$$A := \begin{bmatrix} A_{11} & 0 \\ 0 & 0 \end{bmatrix} \in C^{1\text{-}}(X_\beta, \mathcal{H}(\boldsymbol{E}_1, \boldsymbol{E}_0)) \ .$$

It is also not difficult to see that the map

$$\boldsymbol{u} \mapsto F(\boldsymbol{u}) := \begin{bmatrix} f(u) \\ h(s,u) \end{bmatrix}$$

belongs to $C^{1\text{-}}(X_\beta, \boldsymbol{E}_\gamma)$. Of course, the evolution equation

$$\dot{\boldsymbol{u}} + A(\boldsymbol{u})\boldsymbol{u} = F(\boldsymbol{u}) \ , \qquad t > 0 \ ,$$

is a reformulation of (1.59). Hence we can apply Theorems 12.1 and 12.5, an obvious boot strapping argument based upon the classical regularity theory for linear parabolic equations in one space dimension, and standard a priori techniques to deduce

17.9 Theorem *The Stefan problem possesses for each $s^0 \in (0,1)$ and $T^0 \in C([0,1],\mathbb{R})$, satisfying $T^0|\Omega_i(0) \in W^s_p(\Omega_i(0),\mathbb{R})$ and the boundary conditions (1.51), a unique maximal classical solution $t \mapsto \big(T(t), s(t)\big)$ such that $\big(T(0), s(0)\big) = (T^0, s^0)$. If t^+, the maximal existence time of $t \mapsto \big(T(t), s(t)\big)$, is finite and if*

$$\sup_{0<t<t^+} \|T(t)\|_\infty < \infty$$

then $s(t) \to 0$ or $s(t) \to 1$ as $t \to t^+$.

There is an enormous literature on Stefan and related free boundary value problems for parabolic equations. We simply refer to [Tar88] for an exhaustive bibliography covering the years up to 1987. Nevertheless, our approach seems to be new. Of course, (1.50)–(1.53) is just a simple model problem, chosen to demonstrate our techniques. It is clear that much more general situations can be treated this way. But even in the simple case of our model problem, Theorem 17.9 does not seem to be in the literature. Closely related, though different, results are contained in [FasP77], [FasPK77], [FasP83].

17.8 Dynamic Boundary Conditions In this subsection we briefly indicate how quasilinear parabolic systems with dynamic boundary conditions of the type (1.60) can be handled within the framework of abstract quasilinear parabolic evolution equations.

First we assume that $(\mathcal{A},\mathcal{B})\in\mathcal{E}^1(\Omega)$ with $\delta=1_N$, and consider the linear system

$$\begin{aligned} \partial_t u+\mathcal{A}u &= f(t) && \text{in } \Omega\times(0,\infty)\ ,\\ \partial_t u+\mathcal{B}u &= g(t) && \text{on } \partial\Omega\times(0,\infty)\ ,\\ u(\cdot,0) &= u^0 && \text{on } \Omega\ , \end{aligned} \tag{17.34}$$

where, of course, $\partial_t u$ in the second equation means $\gamma_\partial(\partial_t u)=\partial_t(\gamma_\partial u)$. Suppose that $(\mathcal{A},\gamma_\partial)$ is normally elliptic, fix $p\in(1,\infty)$, and assume, for simplicity, that the spectrum of the L_p-realization of $(\mathcal{A},\gamma_\partial)$ is contained in $[\operatorname{Re} z>0]$. Then, cf. (11.17),

$$W_p^2=W_{p,\gamma_\partial}^2\oplus W_{p,\mathcal{A}}^2$$

and

$$\gamma_\partial^c:=(\gamma_\partial|W_{p,\mathcal{A}}^2)^{-1}\in\mathcal{L}\mathrm{is}(\partial_D W_p^2,W_{p,\mathcal{A}}^2)\hookrightarrow\mathcal{L}(\partial_D W_p^2,W_p^2)\ ,$$

where $\partial_D W_p^s:=W_p^{s-1/p}(\partial\Omega,\mathbb{R}^N)$ for $s\in\mathbb{R}$. It follows that (17.34) is equivalent to the system

$$\begin{aligned} \partial_t u+\mathcal{A}(u-\gamma_\partial^c z) &= f(t) && \text{in } \Omega\times(0,\infty)\ ,\\ \partial_t z+\mathcal{B}u &= g(t) && \text{on } \partial\Omega\times(0,\infty)\ , \end{aligned}$$

where $z=\gamma_\partial u$. This suggests that we introduce the spaces

$$\mathbb{E}_0:=L_p\times\partial W_p^1\ ,\quad \mathbb{E}_1:=\big\{\,(u,z)\in W_p^2\times\partial_D W_p^2\ ;\ z=\gamma_\partial u\,\big\}=\operatorname{graph}(\gamma_\partial|W_p^2)$$

and the operator

$$\mathbb{A}:\mathbb{E}_1\to\mathbb{E}_0\ ,\quad (u,z)\mapsto\big(\mathcal{A}(u-\gamma_\partial^c z),\mathcal{B}u\big)\ .$$

It has been shown in [Hin88] (also see [Hin89]) that

$$\mathbb{A}\in\mathcal{H}(\mathbb{E}_1,\mathbb{E}_0)\ .$$

Let

$$E_0:=\big\{\,(u,z)\in W_{p'}^1\times\partial_D W_{p'}^1\ ;\ \gamma_\partial u=z\,\big\}'$$

with respect to the duality pairing naturally induced by the $L_p\times L_p(\partial\Omega,\mathbb{R}^N)$ duality pairing $\langle\cdot,\cdot\rangle+\langle\cdot,\cdot\rangle_\partial$, and put

$$E_1:=\big\{\,(u,z)\in W_p^1\times\partial_D W_p^1\ ;\ \gamma_\partial u=z\,\big\}\ .$$

Then (E_0, E_1) is a densely injected Banach couple, and Escher [Esc91] (also cf. [Esc93b]) has proven that there exists a unique $A \supset \mathbb{A}$ such that

$$A \in \mathcal{H}(E_1, E_0) .$$

Moreover, the abstract evolution equation

$$\dot{v} + Av = F(t) , \quad 0 < t < \infty , \qquad v(0) = v^0 := (u^0, \gamma_\partial u^0)$$

in E_0, where $F := (f, g)$, is equivalent to the weak formulation

$$\begin{aligned} &\int_0^T \{-(\langle\dot{\varphi}, u\rangle + \langle\dot{\psi}, \gamma_\partial u\rangle_\partial) + a(\varphi, u)\}\, dt \\ &\qquad = \int_0^T (\langle\varphi, f\rangle + \langle\psi, g\rangle_\partial)\, dt + \langle\varphi(0), u^0\rangle + \langle\psi(0), \gamma_\partial u^0\rangle_\partial \end{aligned} \tag{17.35}$$

for all $T > 0$, all $(\varphi, \psi) \in C^1([0,T], E_1') \cap C([0,T], E_0')$ vanishing at $t = T$. Observe that (17.35) is precisely the relation one formally obtains from (17.34) by 'multiplying with a test function' $\varphi \in C^1([0,T], W_{p'}^1)$ and integrating by parts.

Thanks to these results it is now not surprising that one can handle quasilinear parabolic systems of the form (17.34) by reducing them to quasilinear parabolic evolution equations in the interpolation-extrapolation scales generated by A and suitable interpolation functors $\mathfrak{F}$. For details, generalizations, regularity theory, related results, and qualitative studies we refer to [Esc92], [Esc93a], [Esc93c].

It should be mentioned that in certain cases the theory of monotone operators can be applied to problems with dynamic boundary conditions (cf. [Kač80], [Lio69], [Sho74]). In [Gro87] the concept of a 'closed pair of operators' has been applied to problems of this type. All these techniques require coerciveness assumptions, which is not the case for the semigroup technique sketched above.

References

[AcT85] P. Acquistapace, B. Terreni. On the abstract non-autonomous Cauchy problem in the case of constant domains. *Ann. Mat. Pura Appl.* (4), **140** (1985), 1–55.

[AcT86] P. Acquistapace, B. Terreni. Linear parabolic equations in Banach spaces with variable domain but constant interpolation spaces. *Ann. Scuola Norm. Sup. Pisa, Ser.* IV, **13** (1986), 75–107.

[AcT87] P. Acquistapace, B. Terreni. A unified approach to abstract linear non-autonomous parabolic equations. *Rend. Sem. Mat. Univ. Padova*, **78** (1987), 47–107.

[Ada75] R.A. Adams. *Sobolev Spaces.* Academic Press, New York, 1975.

[AgV64] M.S. Agranovich, M.I. Vishik. Elliptic problems with a parameter and parabolic problems of general type. *Russ. Math. Surveys*, **19** (1964), 53–157.

[Ama83] H. Amann. Dual semigroups and second order linear elliptic boundary value problems. *Israel J. Math.*, **45** (1983), 225–254.

[Ama84] H. Amann. Existence and regularity for semilinear parabolic evolution equations. *Ann. Scuola Norm. Sup. Pisa, Ser.* IV, **11** (1984), 593–676.

[Ama85] H. Amann. Global existence for semilinear parabolic systems. *J. reine angew. Math.*, **360** (1985), 47–83.

[Ama86a] H. Amann. Parabolic evolution equations with nonlinear boundary conditions. In F.E. Browder, editor, *Nonlinear Functional Analysis and its Applications*, pages 17–27, Proc. Symp. Pure Math. 45, Amer. Math. Soc., Providence, R.I., 1986.

[Ama86b] H. Amann. Quasilinear evolution equations and parabolic systems. *Trans. Amer. Math. Soc*, **293** (1986), 191–227.

[Ama86c] H. Amann. Quasilinear parabolic systems under nonlinear boundary conditions. *Arch. Rat. Mech. Anal.*, **92** (1986), 153–192.

[Ama86d] H. Amann. Semigroups and nonlinear evolution equations. *Lin. Alg. & Appl.*, **84** (1986), 3–32.

[Ama87] H. Amann. On abstract parabolic fundamental solutions. *J. Math. Soc. Japan*, **39** (1987), 93–116.

[Ama88a] H. Amann. Dynamic theory of quasilinear parabolic equations – I. Abstract evolution equations. *Nonlin. Anal., T.M.&A.*, **12** (1988), 895–919.

[Ama88b] H. Amann. Parabolic evolution equations and nonlinear boundary conditions. *J. Diff. Equ.*, **72** (1988), 201–269.

[Ama88c] H. Amann. Parabolic evolution equations in interpolation and extrapolation spaces. *J. Funct. Anal.*, **78** (1988), 233–270.

[Ama89a] H. Amann. Dynamic theory of quasilinear parabolic systems-III. Global existence. *Math. Z.*, **202** (1989), 219–250; Erratum **205** (1990), 331.

[Ama89b] H. Amann. Feedback stabilization of linear and semilinear parabolic systems. In Clément et al., editors, *Semigroup Theory and Applications*, pages 21–57, Lecture Notes Pure Appl. Math. 116, M. Dekker, New York, 1989.

[Ama89c] H. Amann. Global existence and positivity for triangular quasilinear reaction-diffusion systems. 1989. Unpublished manuscript.

[Ama90a] H. Amann. Dynamic theory of quasilinear parabolic equations – II. Reaction-diffusion systems. *Diff. Int. Equ.*, **3** (1990), 13–75.

[Ama90b] H. Amann. *Ordinary Differential Equations. An Introduction to Nonlinear Analysis.* W. de Gruyter, Berlin, 1990.

[Ama91a] H. Amann. Global existence for a class of highly degenerate parabolic systems. *Japan J. Indust. Appl. Math.*, **8** (1991), 143–151.

[Ama91b] H. Amann. Highly degenerate quasilinear parabolic systems. *Ann. Scuola Norm. Sup. Pisa, Ser.* IV, **18** (1991), 135–166.

[Ama91c] H. Amann. Hopf bifurcation in quasilinear reaction-diffusion systems. In S. Busenberg, M. Martelli, editors, *Delay Differential Equations and Dynamical Systems, Proceedings, Claremont 1990*, pages 53–63, Lecture Notes in Math. #1475, Springer Verlag, New York, 1991.

[Ama91d] H. Amann. Multiplication in Sobolev and Besov Spaces. In *Nonlinear Analysis, A Tribute in Honour of G. Prodi*, pages 27–50, Quaderni, Scuola Norm. Sup. Pisa, 1991.

[Ama94] H. Amann. *Linear and Quasilinear Parabolic Problems.* 1994. Book in preparation.

[AmR93] H. Amann, M. Renardy. Reaction-diffusion problems in electrolysis. *Nonlin. Diff. Equ. and Appl.*, (1993). To appear.

[Ang90] S.B. Angenent. Nonlinear analytic semiflows. *Proc. Royal Soc. Edinburgh*, **115A** (1990), 91–107.

[Bai80] J.B. Baillon. Caractère borné de certains genérateurs de semigroupes linéaires dans les espaces de Banach. *C.R. Acad. Sc. Paris*, **290** (1980), 757–760.

[Bal76] A.V. Balakrishnan. *Applied Functional Analysis.* Springer Verlag, New York, 1976.

[BaZ90] P.W. Bates, S. Zheng. Inertial manifolds and inertial sets for the phase-field equations. 1990. Preprint.

[BeL76] J. Bergh, J. Löfström. *Interpolation Spaces. An Introduction.* Springer Verlag, Berlin, 1976.

[BoS86] A. Boukricha, M. Sieveking. Why second order parabolic systems? *Rocky Mountain J. Math.*, **16** (1986), 33–54.

[Boud87] G. Bourdaud. *Analyse Fonctionelle dans l'Espace Euclidien.* Pub. Math. Paris VII, Vol. 23, 1987.

[Boud91] G. Bourdaud. Le calcul fonctionnel dans les espaces de Sobolev. *Invent. Math.*, **104** (1991), 435–441.

[BouK92] G. Bourdaud, M.E.D. Kateb. Calcul fonctionnel dans l'espace de Sobolev fractionnaire. *Math. Z.*, **210** (1992), 607–613.

[BouM91] G. Bourdaud, Y. Meyer. Fonctions qui opèrent sur les espaces de Sobolev. *J. Funct. Anal.*, **97** (1991), 351–360.

[CDL92] Ph. Clément, C.J. van Duijn, S. Li. On a nonlinear elliptic-parabolic pde system in a two dimensional ground water flow problem. *SIAM J. Appl. Math.*, **23** (1992), 836–851.

[CH*87] Ph. Clément, H.J.A.M. Heijmans et al. *One-Parameter Semigroups.* North-Holland CWI Monograph, Amsterdam, 1987.

[ChC89] D.S. Cohen, R.W. Cox. A mathematical model for stress-driven diffusion in polymers. *J. Polymer Sci., Part B: Polymer Physics*, **27** (1989), 589–602.

[ChW89] D.S. Cohen, A.B. White Jr. Sharp fronts due to diffusion and stress at the glass transition in polymers. *J. Polymer Sci., Part B: Polymer Physics*, **27** (1989), 1731–1747.

[ChW91] D.S. Cohen, A.B. White Jr. Sharp fronts due to diffusion and viscoelastic relaxation in polymers. *SIAM J. Appl. Math.*, **51** (1991), 472–483.

[CoHS89] F. Conrad, D. Hilhorst, T. Seidman. On a reaction-diffusion equation with a moving boundary. In P. Bénilan, M. Chipot, L.C. Evans, M. Pierre, editors, *Recent advances in nonlinear elliptic and parabolic problems*, pages 50–58, Pitman Research Notes in Math. #208, 1989.

[CoHS90] F. Conrad, D. Hilhorst, T. Seidman. Well-posedness of a moving boundary problem arising in a dissolution-growth process. *Nonlin. Anal., T.M.&A.*, **15** (1990), 445–465.

[Cra56] J. Crank. *The Mathematics of Diffusion.* Clarendon Press, Oxford, 1956.

[Cra84] J. Crank. *Free and Moving Boundary Problems.* Clarendon Press, Oxford, 1984.

[DaK92] D. Daners, P. Koch Medina. *Abstract Evolution Equations, Periodic Problems and Applications.* Pitman Research Notes in Math. #279, Longman Sci. & Tech., Essex, 1992.

[DaP84] G. Da Prato. Abstract differential equations and extrapolation spaces. In *Infinite-Dimensional Systems, Retzhof 1983*, pages 53–61, Lecture Notes in Math. #1076, Springer Verlag, Berlin, 1984.

[DaPG79] G. Da Prato, P. Grisvard. Equations d'évolutions abstraites non linéaires de type parabolique. *Ann. Mat. Pura Appl.* (4), **120** (1979), 329–396.

[DaPG84] G. Da Prato, P. Grisvard. Maximal regularity for evolution equations by interpolation and extrapolation. *J. Funct. Anal.*, **58** (1984), 107–124.

[DauL90] R. Dautray, J.-L. Lions. *Mathematical Analysis and Numerical Methods for Science and Technology.* Springer Verlag, Berlin, 1990.

[Dav80] E.B. Davies. *One-Parameter Semigroups.* Academic Press, London, 1980.

[deF63] D.G. de Figueiredo. The coerciveness for forms over vector valued functions. *Comm. Pure Appl. Math.*, **16** (1963), 63–94.

[deGM84] S.R. de Groot, P. Mazur. *Non-Equilibrium Thermodynamics.* Dover Publications Inc., New York, 1984.

[Deu87] P. Deuring. An initial-boundary value problem for a certain density-dependent diffusion system. *Math. Z.*, **194** (1987), 375–396.

[DoF87] G. Dore, A. Favini. On the equivalence of certain interpolation methods. *Boll. UMI* (7), **1-B** (1987), 1227–1238.

[DoV87] G. Dore, A. Venni. On the closedness of the sum of two closed operators. *Math. Z.*, **196** (1987), 189–201.

[Dra88] A.-K. Drangeid. *Das Prinzip der linearisierten Stabilität bei quasilinearen Evolutionsgleichungen.* PhD thesis, Univ. Zürich, Switzerland, 1988.

[Dra89] A.-K. Drangeid. The principle of linearized stability for quasilinear parabolic evolution equations. *Nonlin. Anal., T.M.&A.*, **13** (1989), 1091–1113.

[ElF87] A. El-Fiky. On the well-posedness of the Cauchy problem for p-parabolic systems. *J. Math. Kyoto Univ.*, (I) **27** (1987), 569–586; (II) **28** (1988), 165–172.

[ElZ89] C.M. Elliott, S. Zheng. Global existence and stability of solutions to the phase-field equations. 1989. Preprint.

[Esc91] J. Escher. *Über quasilineare parabolische Systeme.* PhD thesis, Univ. Zürich, Switzerland, 1991.

[Esc92] J. Escher. Nonlinear elliptic systems with dynamic boundary conditions. *Math. Z.*, **210** (1992), 413–439.

[Esc93a] J. Escher. On the qualitative behavior of some semilinear parabolic problems. *Diff. Int. Equ.*, (1993). To appear.

[Esc93b] J. Escher. Quasilinear parabolic systems with dynamical boundary conditions. *Comm. Partial Diff. Equ.*, (1993). To appear.

[Esc93c] J. Escher. Smooth solutions of nonlinear systems with dynamic boundary conditions. In *Proc. Semigroup Theory & Evol. Equ.*, Lecture Notes Pure Appl. Math., M. Dekker, New York, 1993. To appear.

[FasP77] A. Fasano, M. Primicerio. General free-boundary problems for the heat equation, III. *J. Math. Anal. Appl.*, **59** (1977), 1–14.

[FasP83] A. Fasano, M. Primicerio. Classical solutions of general two-phase parabolic free boundary problems in one dimension. In A. Fasano, M. Primicerio, editors, *Free Boundary Problems: Theory and Applications, vol. II*, pages 644–657, Pitman Research Notes in Math. #79, Pitman, London, 1983.

[FasPK77] A. Fasano, M. Primicerio, S. Kamin. Regularity of weak solutions to one-dimensional two-phase Stefan problems. *Ann. Mat. Pura Appl.* (4), **115** (1977), 341–348.

[Fat83] H.O. Fattorini. *The Cauchy Problem.* Addison-Wesley, Reading, Mass., 1983.

[Fra86] J. Franke. On the spaces $F^s_{p,q}$ of Triebel-Lizorkin type: Pointwise multipliers and spaces on domains. *Math. Nachr.*, **125** (1986), 29–68.

[Gey65] G. Geymonat. Su alcuni problemi ai limiti per i sistemi lineari elittici secondo Petrowsky. *Le Matematiche*, **20** (1965), 211–253.

[Gia83] M. Giaquinta. *Multiple Integrals in the Calculus of Variations and Nonlinear Elliptic Systems.* Princeton Univ. Press, Princeton, N.J., 1983.

[GiM87] M. Giaquinta, G. Modica. Local existence for quasilinear parabolic systems under nonlinear boundary conditions. *Ann. Mat. Pura Appl.*, **149** (1987), 41–59.

[GiS81] M. Giaquinta, M. Struwe. An optimal regularity result for a class of quasilinear parabolic systems. *Manuscripta Math.*, **36** (1981), 223–239.

[Goe92] M. Goebel. Continuity and Fréchet-differentiabilty of Nemytskii operators in Hölder spaces. *Mh. Math.*, **113** (1992), 107–119.

[GoKT92] H. Goldberg, W. Kampowsky, F. Tröltzsch. On Nemytskii operators in L_p-spaces of abstract functions. *Math. Nachr.*, **155** (1992), 127–140.

[Gol85] J.A. Goldstein. *Semigroups of Linear Operators and Applications.* Oxford Univ. Press, Oxford, 1985.

[Gri69] P. Grisvard. Équations différentielles abstraites. *Ann. Sc. Éc. Norm. Sup.*, **Sér. 4** (1969), 311–395.

[Gro87] M. Grobbelaar-Van Dalsen. Semilinear evolution equations and fractional powers of a closed pair of operators. *Proc. Royal Soc. Edinburgh*, **105A** (1987), 101–115.

[Grö89] K. Gröger. A $W^{1,p}$-estimate for solutions to mixed boundary value problems for second order elliptic differential equations. *Math. Ann.*, **283** (1989), 679–687.

[Grö92] K. Gröger. $W^{1,p}$-estimates of solutions to evolution equations corresponding to nonsmooth second order elliptic differential operators. *Nonlin. Anal., T.M.&A.*, **18** (1992), 569–577.

[GrR89] K. Gröger, J. Rehberg. Resolvent estimates in $W^{-1,p}$ for second order elliptic differential equations in case of mixed boundary conditions. *Math. Ann.*, **285** (1989), 105–113.

[Gui90] D. Guidetti. A maximal regularity result with applications to parabolic problems with nonhomogeneous boundary conditions. *Rend. Sem. Mat. Univ. Padova*, **84** (1990), 1–37.

[Gui91a] D. Guidetti. Abstract linear parabolic problems with nonhomogeneous boundary conditions. In Ph. Clément, E. Mitidieri, B. de Pagter, editors, *Semigroup Theory and Evolution Equations*, pages 227–242, M. Dekker, New York, 1991.

[Gui91b] D. Guidetti. On interpolation with boundary conditions. *Math. Z.*, **207** (1991), 439–460.

[Hen81] D. Henry. *Geometric Theory of Semilinear Parabolic Equations.* Lecture Notes in Math. #840, Springer Verlag, Berlin, 1981.

[Hin88] T. Hintermann. *Evolutionsgleichungen mit dynamischen Randbedingungen.* PhD thesis, Univ. Zürich, Switzerland, 1988.

[Hin89] T. Hintermann. Evolution equations with dynamic boundary conditions. *Proc. Royal Soc. Edinburgh*, **113A** (1989), 43–60.

[Jam70] G. Jameson. *Ordered Linear Spaces.* Lecture Notes in Math. #141, Springer Verlag, Berlin, 1970.

[JaM91] B. Jawerth, M. Milman. *Extrapolation Theory with Applications.* Memoirs AMS, Vol. 89, #440, 1991.

[Kač80] J. Kačur. Nonlinear parabolic equations with the mixed nonlinear and nonstationary boundary conditions. *Math. Slovaca*, **30** (1980), 213–237.

[Kim84] J.U. Kim. Smooth solutions to a quasilinear system of diffusion equations for a certain population model. *Nonlin. Anal., T.M.&A.*, **8** (1984), 1121–1144.

[Koc92] P. Koch Medina. *Feedback stabilization of time-periodic parabolic equations.* PhD thesis, Univ. Zürich, Switzerland, 1992.

[Kra64] M.A. Krasnosel'skii. *Topological Methods in the Theory of Nonlinear Integral Equations.* Pergamon Press, Oxford, 1964.

[KuJF77] A. Kufner, O. John, S. Fučik. *Function Spaces.* Academia, Prague, 1977.

[LadSU68] O.A. Ladyženskaja, V.A. Solonnikov, N.N. Ural'ceva. *Linear and Quasilinear Equations of Parabolic Type.* Amer. Math. Soc., Transl. Math. Monographs, Providence, R.I., 1968.

[LadU65] O.A. Ladyženskaja, N.N. Ural'ceva. A boundary value problem for linear and quasi-linear parabolic equations. *Amer. Math. Soc. Transl., Ser.* 2, **47** (1965), 217–299.

[LadU66] O.A. Ladyženskaja, N.N. Ural'ceva. A boundary value problem for linear and quasi-linear equations and systems of parabolic type. III. *Amer. Math. Soc. Transl., Ser.* 2, **56** (1966), 103–192.

[LadU67] O.A. Ladyženskaja, N.N. Ural'ceva. On the Hölder continuity of solutions and their derivatives for linear and quasi-linear equations of elliptic and parabolic types. *Amer. Math. Soc. Transl., Ser.* 2, **61** (1967), 207–269.

[Las80] I. Lasiecka. Unified theory for abstract parabolic boundary problems — a semigroup approach. *Appl. Math. Optim.*, **6** (1980), 287–333.

[Li92] S. Li. *Quasilinear evolution problems.* PhD thesis, TH Delft, Holland, 1992.

[Lio69] J.-L. Lions. *Quelques Méthodes de Résolution des Problèmes aux Limites Non Linéaires.* Dunod, 1969.

[LiM60] J.-L. Lions, E. Magenes. Problemi ai limiti non omogenei (I–VII). *Ann. Scuola Norm. Sup. Pisa*, (I) **14** (1960), 259–308; (III) **15** (1961), 39–101; (IV) **15** (1961), 311–326; (V) **16** (1962), 1–44. *Ann. Inst. Fourier*, (II) **11** (1961), 137–178. *J. d'Analyse Math.*, (VI) **11** (1963), 165–188. *Ann. Mat. Pura Appl.* 4, (VII) **63** (1963), 201–224.

[LiM72] J.-L. Lions, E. Magenes. *Non-Homogeneous Boundary Value Problems and Applications* I, II, III. Springer Verlag, Berlin, 1972, 1973.

[Liu89] Gui-Zhang Liu. *Evolution Equations and Scales of Banach Spaces.* PhD thesis, Univ. Eindhoven, Holland, 1989.

[Lun87] A. Lunardi. On the local dynamical system associated to a fully nonlinear abstract parabolic equation. In V. Lakshmikantam, editor, *Nonl. Analysis and Appl.*, pages 319–326, M. Dekker, New York, 1987.

[Lun89] A. Lunardi. Maximal space regularity in nonhomogeneous initial boundary value parabolic problems. *Num. Funct. Anal. and Optim.*, **10** (1989), 323–349.

[Lun91] A. Lunardi. An introduction to geometric theory of fully nonlinear parabolic equations. In T.T. Li, P. de Mottoni, editors, *Qualitative Aspects and Applications of Nonlinear Evolution Equations*, pages 107–131, World Scientific, Singapore, 1991.

[LuSW92] A. Lunardi, E. Sinestrari, W. von Wahl. A semigroup approach to time dependent parabolic initial boundary value problems. *Diff. Int. Equ.*, **5** (1992), 1275–1306.

[LuV91] A. Lunardi, V. Vespri. Hölder regularity in variational parabolic nonhomogeneous equations. *J. Diff. Equ.*, **94** (1991), 1–40.

[MaS85] V.G. Maz'ya, T.O. Shaposhnikova. *Theory of Multipliers in Spaces of Differentiable Functions.* Pitman, Boston, 1985.

[Nag83] R. Nagel. Sobolev spaces and semigroups. *Semesterbericht Funktionalanalysis Tübingen*, (1983), 1–20.

[Neč67] J. Nečas. *Les méthodes directes en théorie des équations elliptiques.* Academia, Prague, 1967.

[New68] J. Newman. Engineering design of electrochemical systems. *Indust. Engineering Chemistry*, **60** (1968), 12–27.

[Osw92] P. Oswald. On the boundedness of the mapping $f \mapsto |f|$ in Besov spaces. (1992). Preprint.

[Paz83] A. Pazy. *Semigroups of Linear Operators and Applications to Partial Differential Equations.* Springer Verlag, New York, 1983.

[PeF90] O. Penrose, P.C. Fife. Thermodynamically consistent models of phase-field type for kinetics of phase transitions. *Physica D*, **43** (1990), 44–62.

[PoT90] M.A. Pozio, A. Tesei. Global existence of solutions for a strongly coupled quasilinear parabolic system. *Nonlin. Anal., T.M.&A.*, **14** (1990), 657–689.

[PrS93] J. Prüss, H. Sohr. Imaginary powers of elliptic second order differential operators in L^p-spaces. *Hiroshima Math. J.*, (1993). To appear.

[Red87] R. Redlinger. Pointwise a-priori bounds for strongly coupled semilinear problems. *Indiana Univ. Math. J.*, **36** (1987), 441–454.

[Red89] R. Redlinger. Invariant sets for strongly coupled reaction-diffusion systems under general boundary conditions. *Arch. Rat. Mech. Anal.*, **108** (1989), 281–291.

[Red92] R. Redlinger. Existence of the global attractor for a strongly coupled parabolic system arising in population dynamics. 1992. Preprint.

[Rhe84] F. Rothe. *Global Solutions of Reaction-Diffusion Systems.* Lecture Notes in Math. #1072, Springer Verlag, Berlin, 1984.

[RMK86] I. Roušar, K. Micka, A. Kimla. *Electrochemical Engineering.* Elsevier, Amsterdam, 1986.

[Rub71] L.I. Rubinstein. *The Stefan Problem.* Amer. Math. Soc., Transl. Math. Monographs, Providence, R.I., 1971.

[Run86] Th. Runst. Mapping properties of non-linear operators in spaces of Triebel-Lizorkin and Besov type. *Anal. Math.*, **12** (1986), 313–346.

[Sad91] T. Sadamatsu. A property of an analytic semigroup. *J. Math. Kyoto Univ.*, **31** (1991), 931–936.

[Sch71] H.H. Schaefer. *Topological Vector Spaces.* Springer Verlag, New York, 1971.

[See71] R. Seeley. Norms and domains of the complex powers A_B^z. *Amer. J. Math.*, **93** (1971), 299–309.

[See72] R. Seeley. Interpolation in L^p with boundary conditions. *Stud. Math.*, **44** (1972), 47–60.

[ShKT79] N. Shigesada, K. Kawasaki, E. Teramoto. Spacial segregation of interacting species. *J. theor. Biology*, **79** (1979), 83–99.

[Sho74] R. Showalter. Degenerate evolution equations. *Indiana Univ. Math. J.*, **23** (1974), 655–677.

[Sic86] W. Sickel. On pointwise multipliers in Besov-Triebel-Lizorkin spaces. In B.W. Schulze, H. Triebel, editors, *Seminar Analysis*, pages 45–103, Teubner, Leipzig, 1986.

[Sic89] W. Sickel. On boundedness of superposition operators in spaces of Triebel-Lizorkin type. *Czechoslovak Math. J.*, **39** (1989), 323–347.

[Sic92] W. Sickel. Superposition of functions in Sobolev spaces of fractional order. A survey. *Banach Center Publ., Partial Diff. Equ.*, **27** (1992), 481–497.

[Sic93] W. Sickel. Pointwise multiplication in Triebel-Lizorkin spaces. *Forum Math.*, **5** (1993), 73–91.

[Sim72] C.G. Simader. *On Dirichlet's Boundary Value Problem*. Lecture Notes in Math. #268, Springer Verlag, 1972.

[Sit92] G. Simonett. *Zentrumsmannigfaltigkeiten für quasilineare parabolische Gleichungen*. PhD thesis, Univ. Zürich, Switzerland, 1992.

[Sit93a] G. Simonett. Center manifolds for quasilinear reaction-diffusion systems. 1993. Preprint.

[Sit93b] G. Simonett. Quasilinear parabolic equations and semiflows. In *Proc. Semigroup Theory & Evol. Equ.*, Lecture Notes Pure Appl. Math., M. Dekker, New York, 1993. To appear.

[SlH72] M.G. Slinko, K. Hartmann. *Methoden und Programme zur Berechnung chemischer Reaktoren*. Akademie Verlag, Berlin, 1972.

[Sob64] P.E. Sobolevskii. Coerciveness inequalities for abstract parabolic equations. *Soviet Math. (Doklady)*, **5** (1964), 894–897.

[Sob66] P.E. Sobolevskii. Equations of parabolic type in a Banach space. *Amer. Math. Soc. Transl., Ser.* 2, **49** (1966), 1–62.

[Sol65] V.A. Solonnikov. On boundary value problems for linear parabolic systems of differential equations of general form. *Proc. Steklov Inst. Math.*, **83** (1965), 1–184.

[SpZ92] J. Sprekels, S. Zheng. Global smooth solutions to a thermodynamically consistent model of phase-field type in higher space dimensions. *J. Math. Anal. Appl.*, (1992). To appear.

[StJM86] J. Stará, O. John, J. Malý. Counterexample to the regularity of weak solution of the quasilinear parabolic system. *Commentationes Math. Univ. Carolinae*, **27** (1986), 123–136.

[Str84] M. Struwe. A counterexample in regularity theory for parabolic systems. *Czechoslovak Math. J.*, **34** (1984), 183–188.

[Tai91] K. Taira. *Boundary Value Problems and Markov Processes*. Lecture Notes in Math. #1499, Springer Verlag, Berlin, 1991.

[Tan60] H. Tanabe. On the equation of evolution in a Banach space. *Osaka Math. J.*, **12** (1960), 363–376.

[Tan79] H. Tanabe. *Equations of Evolution*. Pitman, London, 1979.

[Tar88] D.A. Tarzia. *A bibliography on moving-free boundary problems for the heat-diffusion equation. The Stefan problem.* Progetto Nazionale M.P.I.: Equ. evol. e appl. fisico-mat., Firenze, 1988.

[Ter89] B. Terreni. Non-homogeneous initial-boundary value problems for linear parabolic systems. *Stud. Math.*, **92** (1989), 141–175.

[Tri78] H. Triebel. *Interpolation Theory, Function Spaces, Differential Operators.* North Holland, Amsterdam, 1978.

[Tri83] H. Triebel. *Theory of Function Spaces.* Birkhäuser, Basel, 1983.

[Tri92] H. Triebel. *Function Spaces* II. Birkhäuser, Basel, 1992.

[Trig80] R. Triggiani. Boundary feedback stabilizability of parabolic equations. *Appl. Math. Optim.*, **6** (1980), 201–220.

[Vai64] M.M. Vainberg. *Variational Methods for the Study of Nonlinear Operators.* Holden-Day, San Francisco, 1964.

[Ves88] V. Vespri. The functional space $C^{-1,\alpha}$ and analytic semigroups. *Diff. Int. Equ.*, **1** (1988), 473–493.

[Ves89] V. Vespri. Analytic semigroups generated in $H^{-m,p}$ by elliptic variational operators and applications to linear Cauchy problems. In Clément et al., editors, *Semigroup Theory and Applications*, pages 419–431, Lecture Notes Pure Appl. Math. 116, M. Dekker, New York, 1989.

[Ves91] V. Vespri. Abstract quasilinear parabolic equations with variable domains. *Diff. Int. Equ.*, **4** (1991), 1041–1072.

[vW85a] W. von Wahl. *The Equations of Navier-Stokes and Abstract Parabolic Equations.* Vieweg & Sohn, Braunschweig, 1985.

[vW85b] W. von Wahl. On the Cahn-Hilliard equation $u' + \Delta^2 u - \Delta f(u) = 0$. *Delft Progress Report*, **10** (1985), 291–310.

[vW90] W. von Wahl. Abschätzungen für das Neumann-Problem und die Helmholtz-Zerlegung von L^p. *Nachr. Akad. Wiss. Göttingen, II. Math.-Phys. Klasse*, (1990), 9–37.

[Wal86] Th. Walther. *Abstrakte Sobolev-Räume und ihre Anwendungen auf die Störungstheorie für Generatoren von C_0-Halbgruppen.* PhD thesis, Univ. Tübingen, Germany, 1986.

[Wei80] F.B. Weissler. Local existence and nonexistence for semilinear parabolic equations in L^p. *Indiana Univ. Math. J.*, **29** (1980), 79–102.

[Wie92] M. Wiegner. Global solutions to a class of strongly coupled parabolic systems. *Math. Ann.*, **292** (1992), 711–727.

[Yag88] A. Yagi. Fractional powers of operators and evolution equations of parabolic type. *Proc. Japan Acad., Ser. A*, **64** (1988), 227–230.

[Yag90] A. Yagi. Parabolic evolution equations in which the coefficients are the generators of infinitely differentiable semigroups. II. *Funkcial Ekvac.*, **33** (1990), 139–150.

[Yag91] A. Yagi. Abstract quasilinear evolution equations of parabolic type in Banach spaces. *Boll. UMI*, **5-B** (1991), 341–368.

[Yag92] A. Yagi. Global solution of some quasilinear parabolic systems in population dynamics. 1992. Preprint.

[Zhe92] S. Zheng. Global existence for a thermodynamically consistent model of phase field type. *Diff. Int. Equ.*, **5** (1992), 241–253.

Universität Zürich
Rämistr. 74
CH–8001 Zürich
Switzerland

The Functional Calculus in Sobolev Spaces

Gérard Bourdaud

The present exposition is a synthesis of various papers by Dalila Kateb, Djalil Kateb, Yves Meyer and the author, where the reader will find more detailed proofs.

I - Introduction

Let E be a real-valued function space. The *functional calculus problem* for E consists in finding necessary and sufficient conditions - as simple as possible - such that a real variable function G acts, via left composition, on E; in other words, that $G \circ f \in E$ for each $f \in E$. We say that the functional calculus is *non trivial* if for every "reasonable" function (say $G \in S(\mathbb{R}^n)$ with $G(0) = 0$) acts on E.
The functional calculus in the classical Sobolev spaces $W^{s,p}(\mathbb{R}^n)$, ($s \in N^*$, $1 \leq p \leq +\infty$) was entirely characterized by the author [1]; it is far from being known in case of *fractional* Sobolev spaces. Two families of fractional spaces are currently used: the Besov and the Lizorkin-Triebel spaces. Let us recall the definitions of these spaces in the wavelet framework.
A multiresolution analysis (V_j) on $\mathbb{R}^n$ (cf. Y.Meyer's book) leads to a Hilbert orthogonal decomposition:

$$L_2(\mathbb{R}^n) = V_0 \oplus W_0 \oplus W_1 \oplus W_2 \oplus \dots .$$

Then exists a function w^0 (scale function) and functions w^ε (wavelets) such that $\{w^0(x - k) \,/\, k \in \mathbb{Z}^n\}$ is an orthonormal basis of V_0 and $\{w^\varepsilon(x-k) \,/\, k \in \mathbb{Z}^n, \varepsilon \in \{0,1\}^n \setminus \{(0,\dots,0)\}\}$ an orthonormal basis of W_0. The regularity of w^0 and w^ε is supposed to be larger than $|s|$. Every reasonable distribution on $\mathbb{R}^n$ admits a wavelet expansion

$$f(x) = \sum_{k \in \mathbb{Z}^n} \beta_k \, w^0(x-k) + \sum_{\varepsilon} \sum_{j \geq 0} \sum_{k \in \mathbb{Z}^n} \alpha_{j,k,\varepsilon} \, 2^{jn/2} \, w^\varepsilon(2^j x - k).$$

The Besov space $B_p^{s,q}(\mathbb{R}^n)$ ($s \in \mathbb{R}$, $p, q \in [1, +\infty]$) is defined by the conditions

$$\left(\sum_{k \in \mathbb{Z}^n} |\beta_k|^p \right)^{1/p} < +\infty,$$

$$\sum_{\varepsilon} \left\{ \sum_{j \geq 0} \left(2^{nj(\frac{1}{2} - \frac{1}{p})} \, 2^{js} \Big(\sum_{k \in \mathbb{Z}^n} |\alpha_{j,k,\varepsilon}|^p \Big)^{1/p} \right)^q \right\}^{1/q} < +\infty.$$

The Besov scale *does not contain* the classical Sobolev spaces $W^{s,p}$ $(s \in \mathbf{N}, p \neq 2)$ for integers s, $B_p^{s,q}(\mathbb{R}^n)$ are somewhat strange spaces whose prototype is the Zygmund class $B_\infty^{1,\infty}(\mathbb{R}^n)$. The Triebel-Lizorkin space $F_p^{s,q}(\mathbb{R}^n)$ $(s \in \mathbb{R},\ p \in [1,+\infty[,\ q \in [1,+\infty])$ is defined by the condition

$$\left(\sum_{k\in\mathbb{Z}^n} |\beta_k|^p\right)^{1/p} < +\infty,$$

$$\sum_\varepsilon \| \left\{ \sum_{j\geq 0} \sum_{k\in\mathbb{Z}^n} (\,|\alpha_{j,k,\varepsilon}|\, 2^{nj/2}\, 2^{js} \chi_{j,k})^q \right\}^{1/q} \|_p < +\infty,$$

where $\chi_{j,k}$ is the characteristic function of the cube $2^j x - k \in [0,1[^n$.
For $p > 1$, it is known that $F_p^{s,2}(\mathbb{R}^n)$ is the (Bessel potential) space $(1-\Delta)^{-s/2}(L_p(\mathbb{R}^n))$; in particular, we have $F_p^{s,2}(\mathbb{R}^n) = W^{s,p}(\mathbb{R}^n)$ for each *integer* s. Similarly, for an integer s, $F_1^{s,2}(\mathbb{R}^n)$ is the Sobolev space "built" on the local Hardy space (Goldberg space) $F_1^{0,2}(\mathbb{R}^n)$. The Sobolev spaces $W^{s,1}$ and $W^{s,\infty}$ are not Besov nor Lizorkin-Triebel spaces; as a matter of fact, they do not admit inconditional basis; hence they cannot be described by conditions on the *modulus* of the wavelet coefficients β_k and $\alpha_{j,k,\varepsilon}$.
The differences between the various types of fractional spaces being not very significant - as far as the functional calculus is concerned - we shall adopt the *generic* notation $E_p^s(\mathbb{R}^n)$. $\|f\|$ is the norm of f in $E_p^s(\mathbb{R}^n)$, when there is no risk of mistake. Henceforth, it is assumed that $s > 0$.
Let us add - for the reader's stillness - that we seldom use the precise definitions of these spaces. Most of the time, it suffices to observe $E_p^s(\mathbb{R}^n)$ is embedded into $B_p^{s,\infty}(\mathbb{R}^n)$, for which a reasonable norm is known, namely $\|f\|_p + N_1(f)$, with

$$N_1(f) = \sup_{h\neq 0} |h|^{-s} \left(\int |(\Delta_h)^m f(x)|^p\, dx\right)^{1/p},$$

where Δ_h is the finite difference operator $\Delta_h f(x) = f(x+h) - f(x)$ and m is the first integer such that $m > s$.
We say that $E_p^s(\mathbb{R}^n)$ is *supercritical* if it is embedded into $L^\infty(\mathbb{R}^n)$, *subcritical* if not. It is well-known that $E_p^s(\mathbb{R}^n)$ is subcritical for $s < n/p$, supercritical for $s > n/p$. Here is the *exhaustive* list of supercritical $E_p^{n/p}(\mathbb{R}^n)$'s :

$$B_p^{\frac{n}{p},1}(\mathbb{R}^n), \quad F_1^{n,q}(\mathbb{R}^n) \quad \text{and} \quad W_1^n(\mathbb{R}^n).$$

Recall the fundamental estimate

$$\|f(\frac{\cdot}{\lambda})\| \leq \lambda^\varrho \|f\| \quad (\forall\lambda \in]0,1]), \tag{1}$$

where $\varrho = n/p - s$; the hypothesis $s > 0$ plays an essential role in (1); for instance, we have

$$\|f(\frac{\cdot}{\lambda})\|_{B_1^{0,1}(\mathbb{R}^n)} \leq c\,\lambda^n\, |\log(\lambda)|\, \|f\|_{B_1^{0,1}(\mathbb{R}^n)} \quad (\forall\lambda \in]0,\frac{1}{2}]), \tag{2}$$

and this estimate is optimal (see [3]).
Let us end our introduction with a remark : if we except the case $p = +\infty$, 0 is the only constant function which belongs to $E_p^s(\mathbb{R}^n)$; hence $G(0) = 0$ is a necessary condition for G to act on $E_p^s(\mathbb{R}^n)$; it will be implicit in the sequel.

II - Summary of the results

Theorem 1 *(Necessary and sufficient conditions). The following conditions are necessary and sufficient for the function G to act on the space $E = E_p^s$:*

(i) $G \in W^{s,p}(\mathbb{R})_{loc}$, *for s integer,* $s \geq 2$, *and* $E = W^{s,p}$, *if it is supercritical.*

(ii) $G' \in W^{s-1,p}(\mathbb{R})_{loc-unif}$, *for s integer,* $s = n/p \geq 2$ *and* $E = W^{s,p}(\mathbb{R}^n)$, *if it is subcritical.*

(iii) $G'' \in L^1(\mathbb{R})$, *if* $E = W^{2,1}(\mathbb{R}^n)$, *with* $n \geq 3$.

(iv) $G(t) = ct$, *if* $1+1/p \leq s < n/p$ *(except for* $E = W^{2,1}$ *and - perhaps - for* $E = B_p^{1+\frac{1}{p},1}$ *and* $F_1^{2,q}$*).*

(v) $G' \in L^\infty(\mathbb{R}^n)_{loc}$, *for* $0 < s < 1$, *if E is supercritical (as just as* $W^{1,p}$, *if it is supercritical).*

(vi) $G' \in L^\infty(\mathbb{R}^n)$, *for* $0 < s < 1$, *if E is subcritical (as just as* $W^{1,p}$, *if it is subcritical).*

Let us recall the difference between E_{loc} and $E_{loc-unif}$. If E is a function space in $\mathbb{R}^n$ and g a given non-zero function in $\mathcal{D}(\mathbb{R}^n)$, we say that f belongs to E_{loc} if all the functions

$$x \longrightarrow f(x)\, g(x-t),$$

for $t \in \mathbb{R}^n$, belong to E; and that f belongs to $E_{loc-unif}$ if, moreover,

$$\sup_{t \in \mathbb{R}^n} \|f(x)\, g(x-t)\|_E < +\infty.$$

The Theorem 1 solves the functional calculus problem in the three following cases:

- $0 < s < 1$,
- s integer (for the *classical* Sobolev spaces),
- $1 + 1/p \leq s \leq n/p$ (except for $B_p^{1+\frac{1}{p},1}$ and $F_1^{2,q}$).

In the other cases, we know necessary conditions which are not sufficient, and vice versa:

Theorem 2 *If a function G acts on* $E_p^s(\mathbb{R}^n)$ *then*

(i) *G is Lipschitz, if $E_p^s(\mathbb{R}^n)$ is subcritical,*

(ii) *G is locally Lipschitz, if $E_p^s(\mathbb{R}^n)$ is supercritical.*

These conditions are not sufficient, as soon as $s > 1$.

Theorem 3 *If $s \geq n/p$, every function G such that $G^{(j)} \in L^\infty$ $(j = 1, \ldots, [s]+1)$ acts on $E_p^s(\mathbb{R}^n)$.*

Theorem 4 *If $1 \leq s < 1 + 1/p$, every function G such that G'' is a bounded measure acts on $B_p^{s,q}(\mathbb{R}^n)$.*

Theorem 5 *If $1 \leq s < 1 + 1/p$, every function G such that $G' \in L^\infty(\mathbb{R}^n)$ and $\sup_{t>0} t^{p-1} \int_{|s|\geq t} |G''(s)|^p\, ds < +\infty$ acts on $B_p^{s,q}(\mathbb{R}^n)$.*

The Theorem 4 has been extended to Lizorkin-Triebel spaces under restrictive conditions $(1 < s < 1 + 1/p,\ p,\ q \in]1, +\infty[$ - see [5]).

V.Maz'ya has proved the non triviality of the functional calculus on the positive cone $(W^{2,p})^+$, whatever be the dimension n. This the functional calculus on $(E_p^s)^+$ may strongly differ from the one on E_p^s. Here is a first result of this type:

Theorem 6 *If $1 < p < n/2$ the following condition is necessary and sufficient for a function G, defined on $\mathbb{R}^+$, to act on $W^{2,p}(\mathbb{R}^n)^+$:*

$$G' \in L^\infty(\mathbb{R}^+) \quad \textit{and} \quad \sup_{t>0} t^{p-1} \int_t^{+\infty} |G''(s)|^p\, ds < +\infty.$$

Up to now, we have not been able to prove (or disprove) the triviality of the functional calculus on $B_p^{1+\frac{1}{p},1}(\mathbb{R}^n)$ $(1 \leq p < n-1)$ and $F_1^{2,q}(\mathbb{R}^n)$ $(n \geq 3)$.

III - Proofs: sufficient conditions for the classical spaces

The Lipschitz condition is classically known to be sufficient in the $W^{1,p}$ spaces. Let us study the case of $W^{m,p}$, for an integer $m \geq 2$.

We assume that $G^{(m)}$ belongs to L^p locally uniformly; this leads us to introduce a partition of unity

$$\sum_{k\in\mathbb{Z}} \psi^2(t-k) = 1 \quad (\forall t \in \mathbb{R}),$$

where ψ is a positive, compactly supported C^1 function;

the functions $H_k(t) = \int_t^\infty |G^{(m)}(s)|^p\, \psi(s-k)\, ds$ satisfy the condition

$$M = \sup_{k\in\mathbb{Z}} \|H_k\|_\infty < +\infty.$$

We have to prove that, for every $f \in W^{m,p}$, the integral

$$\mathcal{J} = \int |G^{(m)} \circ f|^p \, |D_j f|^{mp} \, dx \quad (j = 1, \ldots, n)$$

if finite. We use the partition of unity in order to obtain

$$\begin{aligned} \mathcal{J} &= \int |G^{(m)}(f(x))|^p \, \psi^2(f(x) - k) \, |D_j f|^{mp} \, dx \\ &= \int D_j(H_k \circ f) \, \psi(f - k) \, (D_j f) \, |D_j f|^{mp-2} \, dx; \end{aligned}$$

then an integration by parts implies $\mathcal{J} \leq M \, (\mathcal{J}_1 + \mathcal{J}_2)$, where

$$\begin{aligned} \mathcal{J}_1 &= (mp - 1) \sum_{k \in \mathbb{Z}} \int \psi(f(x) - k) \, |D_j^2 f(x)| \, |D_j f|^{mp-2} \, dx, \\ \mathcal{J}_2 &= \sum_{k \in \mathbb{Z}} \int |\psi'(f(x) - k)| \, |D_j f|^{mp} \, dx. \end{aligned}$$

The hypothesis $m \geq n/p$ and the Sobolev theorem - applied to $W^{m-1,p}$ and $W^{m-2,p}$ - give

$$\mathcal{J}_1 \leq C \, \|f\|^{mp-1}, \quad \mathcal{J}_2 \leq C \, \|f\|^{mp}.$$

Of course, the partition of unity is superfluous if $m > n/p$ (or $m = 2$ and $p = 1$).Let us point out that the sufficiency of the condition $G^{(p)} \in L^p$, when $n = 1$ was first proved by F.Szigeti. The functional calculus on the positive cone $(W^{2,p})^+$ relies on the following generalization of an inequality of Maz'ya:

Lemma 1 *Let $p \in]1, +\infty[$ and G a function, defined on $]0, +\infty$, such that*

$$A = \sup_{t>0} t^{p-1} \int_t^\infty |G''(s)|^p \, ds < +\infty.$$

Then for all $f \in (W^{2,p})^+$, it holds

$$\int_{\mathbb{R}^n} |G'' \circ f|^p \, |D_j f|^{2p} \, dx \leq (2p - 1)^p \, (p - 1)^{p-1} \, A \left(\int_{\mathbb{R}^n} |D_j^2 f|^p \, dx \right).$$

It suffices to prove the lemma for $n = 1$ (an integration with respect to the other variables will provide the general case). We set

$$H(t) = \int_t^\infty |G''(s)|^p \, ds;$$

that gives

$$\begin{aligned} &\int_{\mathbb{R}} |G'' \circ f|^p \, |f'|^{2p} \, dx \\ &= (2p - 1) \int_{\mathbb{R}} (H \circ f) \, |f'|^{2p-2} \, f'' \, dx \end{aligned}$$

$$\leq (2p-1)\,A \int_{\mathbb{R}} |f'|^{2p-2} f^{1-p}\,|f''|\,dx$$

$$\leq (2p-1)\,A \left(\int_{\mathbb{R}} |f'|^{2p} f^{-p}\,dx\right)^{1-1/p} \left(\int_{\mathbb{R}} |f''|^{p}\,dx\right)^{1/p}.$$

We conclude the proof with the aid of the following inequality

$$\int_{\mathbb{R}} |f'|^{2p} f^{-p}\,dx \leq (2p-1)^p\,(p-1)^{-p} \left(\int_{\mathbb{R}} |f''|^{p}\,dx\right)$$

(see Maz'ya, p. 363, or Adams).

IV - Proofs: sufficient conditions for the fractional spaces

The Theorem 3 is more or less classical: a first proof - based upon non-linear interpolation - appeared in an old paper of J.Peetre; the method of linearization of Bony and Meyer applies to Besov as well as to Lizorkin-Triebel spaces (see the paper of T.Runst).
Theorems 4 and 5 will follow from the

Proposition 1 *Let $1 < s < 1 + 1/p$. If E_p^s is the Besov space $B_p^{s,q}$ ($1 \leq p < +\infty$, $1 \leq q \leq +\infty$) or the Lizorkin-Triebel space $F_p^{s,q}$ ($1 < p < +\infty, 1 < q < +\infty$), there exists a constant $M = M(s,p,q,n) > 0$ such that*

$$\|\,|f-\tau| - |\tau|\,\| \leq M\,\|f\|, \tag{3}$$

for all real-valued $f \in E_p^s$ and all $\tau \in R$.

Let us outline the proof of the proposition for the Besov space $B_p^{s,q}(\mathbb{R})$. The extension to $\mathbb{R}^n$ will follow by the *Fubini property* enjoyed by $B_p^{s,p}(\mathbb{R}^n)$ (see the book of H.Triebel) and the case of $B_p^{s,q}(\mathbb{R}^n)$ by non-linear interpolation. The adaption to $F_p^{s,q}(\mathbb{R}^n)$ requires only minor modifications [5].
The inequality (3) reads as

$$\int\int_{\mathbb{R}\times\mathbb{R}} |\mathrm{sgn}\,(f(x)-\tau)\,f'(x) - \mathrm{sgn}\,(f(y)-\tau)\,f'(y)|^p\,|x-y|^{-1-p\sigma}\,dx\,dy \tag{4}$$

$$\leq M \int\int_{\mathbb{R}\times\mathbb{R}} |f'(x)-f'(y)|^p\,|x-y|^{-1-p\sigma}\,dx\,dy,$$

where $\sigma = s-1$. We assume that f is real-analytic, rapidly decreasing to infinity and introduce the open set Ω, where $f(x) > \tau$. The inequality (4) is then reduced to the majorization of

$$A_1 = \int\int_{\Omega\times\Omega^c} |f'(x)|^p\,|x-y|^{-1-p\sigma}\,dx\,dy$$

by the second member of (4). After an integration with respect to y, A_1 is estimated by

$$A_2 = \int_{\Omega} |f'(x)|^p\,\mathrm{dist}\,(x,\Omega^c)^{-p\sigma}\,dx.$$

Writing Ω as a disjoint union of open intervals J yields

$$A_2 \leq \sum_J \int_J |f'(x)|^p \,\mathrm{dist}\,(x, J^c)^{-p\sigma}\, dx.$$

We conclude the proof of the proposition with the aid of the following

Lemma 2 *If $p \in [1, +\infty[$ and $0 < \sigma < 1/p$, there exists $C = C(\sigma, p) > 0$ such that*

$$\int_J \int_J |f'(x)|^p \, dist(x, J^c)^{-p\sigma}\, dx \leq C \int\int_{J\times J} |f(x) - f(y)|^p \, |x-y|^{-1-p\sigma}\, dx\, dy \qquad (5)$$

for all open interval J and all C^1 function f on $\mathbb{R}$ satisfying :

(i) $f(x) = O(x^{-2})$ $(|x| \to +\infty)$, *if J is not bounded,*

(ii) $\int_J f(x)\, dx = 0$, *if J is bounded.*

Proof of Lemma 2. We observe that the inequality (5) is invariant by affine transforms, so that it suffices to prove

$$\int_0^\infty |f(x)|^p \, x^{-\sigma p}\, dx \leq C \int_0^\infty \int_0^\infty |f(x) - f(y)|^p \, |x-y|^{-\sigma p - 1}\, dx\, dy, \qquad (6)$$

$$\int_0^1 |f(x)|^p \, x^{-\sigma p}\, dx \leq C \int_0^1 \int_0^1 |f(x) - f(y)|^p \, |x-y|^{-\sigma p - 1}\, dx\, dy, \qquad (7)$$

under the respective hypothesis on f.

First we set

$$g(x) = f(x) - x^{-1} \int_0^x f(y)\, dy \quad (\forall x > 0),$$

so that

$$f(x) = g(x) - \int_x^\infty g(y)\, y^{-1}\, dy.$$

An easy calculation shows that $\int_0^\infty |g(x)|^p \, x^{-\sigma p}\, dx$ is estimated by the second member of (6); then we conclude the proof of (6) with the aid of the classical Hardy inequality

$$\int_0^\infty \left| \int_x^\infty g(y)\, y^{-1}\, dy \right|^p x^{-\sigma p}\, dx \leq C \int_0^\infty |g(x)|^p \, x^{-\sigma p}\, dx$$

(where the hypothesis $\sigma p < 1$ plays an essential role).

Let ζ be a C^1 function such that $0 \leq \zeta \leq 1$, $\zeta(x) = 1$ for $0 \leq x \leq 1/2$, $\zeta(x) = 0$ for $x \geq 3/4$. The application of (6) to the function ζf leads to

$$\int_0^1 |f(x)|^p \, x^{-\sigma p}\, dx \leq C \left\{ \int_0^1 |f(x)|^p \, dx + \int_0^1 \int_0^1 |f(x) - f(y)|^p \, |x-y|^{-\sigma p - 1}\, dx\, dy \right\};$$

the condition $\int_0^1 f(x)\, dx = 0$ allows us to write

$$f(x) = \int_0^1 (f(x) - f(y))\, dy,$$

which implies that $\int_0^1 |f(x)|^p\, dx$ is estimated by the second member of (7).

Let us point out that an independent proof of the proposition is due to P.Oswald.

Under the hypothesis of Theorem 4, the function G can be written

$$G(x) = \frac{1}{2}\int_{\mathbb{R}} (|x-\tau| - |\tau|)\, d\mu(\tau) + ax,$$

where $\mu = G''$; thus we obtain

$$\|G \circ f\| \leq \left\{ \frac{M}{2}\|\mu\| + |a| \right\} \|f\|.$$

Under the hypothesis of Theorem 5 the two functions

$$G_1(t) = G(t), \quad G_2(t) = G(-t) \quad (\forall t \in \mathbb{R}^+)$$

act on $(W^{2,p})^+$; by non-linear interpolation they also act on $(B_p^{s,q})^+$, for all $s \in]0,2[$. The identity

$$G \circ f = G_1 \circ f^+ + G_2 \circ f^-,$$

and the proposition provide us the wished conclusion.

V - Proofs: necessary conditions for the subcritical spaces

Most of the proofs of necessary conditions begin with the following lemma, which says that, in some sense, each function which acts on $E_p^s(\mathbb{R}^n)$, acts *boundedly* on $E_p^s(\mathbb{R}^n)$.

Lemma 3 *If a function G acts on $E_p^s(\mathbb{R}^n)$, there exist constant $C_j > 0$ such that, for all function f with support in the unit cube $Q = [-1/2, +1/2]^n$, $\|f\| \leq C_1$ implies $\|G \circ f\| \leq C_2$.*

This lemma is more or less classical in harmonic analysis. Suppose that, for every cube R and all numbers C_j, it is possible to find a function f, carried by R, so that $\|f\| \leq C_1$ and $\|G \circ f\| > C_2$.

We take a sequence R_j of disjoint cubes and functions $\varphi_j \in \mathcal{D}(\mathbb{R}^n)$ such that $\varphi_j(x) = 1$ on $R_j/2$ and $\varphi_j(x) = 0$ out of R_j. Let M_j be the norm of the operator $f \to f\varphi_j$, acting on $E_p^s(\mathbb{R}^n)$; we choose functions f_j such that

$$\mathrm{supp} f_j \subset R_j/2, \quad \|f_j\| \leq 2^{-j}, \quad \|G \circ f_j\| > j\, M_j.$$

Then the function $f = \sum_{j\geq 0} f_j$ belongs to $E_p^s(\mathbb{R}^n)$ and $(G \circ f_j)\,\varphi_j = G \circ f_j$; hence

$$j\, M_j < \|(G \circ f)\,\varphi_j\| \leq M_j\, \|G \circ f\|,$$

which is impossible.

Thus we have found *one* cube R and some $C_j > 0$ such that, for all f carried by R, $\|f\| \leq C_1$ implies $\|G \circ f\| \leq C_2$; by a suitable homothety, it is possible to replace R by *any* other cube.

A - Necessity of the Lipschitz condition

The first proofs of the necessity of the Lipschitz condition appeared in the papers of Igari and Marcus-Mizel.

Lemma 4 *In the subcritical case, there exists a sequence $(\Theta_\nu)_{\nu\geq 1}$ of C^∞ functions defined on $\mathbb{R}^n$, carried by Q, such that $\Theta_\nu(x) = 1$ on $2^{-\nu}Q$ and $\lim_{\nu\to+\infty} \|\Theta_\nu\| = 0$.*

In order to proof the lemma, let us give ourselves a function $\varphi \in \mathcal{D}(\mathbb{R}^n)$ such that $\varphi(x) = 1$ in Q and $\varphi(x) = 0$ out of $2Q$.

For $s < n/p$, it suffices to set $\Theta_\nu(x) = \varphi(2^\nu x)$; the inequality (1) gives the result. Assume now $s = n/p$ and set

$$\Theta_\nu(x) = \nu^{-1} \sum_{1\leq j\leq \nu} \varphi(2^j x);$$

the above sum is - roughly speaking - a wavelet expansion (in fact an atomic decomposition, in the sense of Frazier and Jawerth); this leads to the estimates

$$\|\Theta_\nu\|_{B_p^{\frac{n}{p},q}} \leq C\,\nu^{\frac{1}{q}-1},$$

$$\|\Theta_\nu\|_{F_p^{\frac{n}{p},q}} \leq C\,\nu^{\frac{1}{p}-1}.$$

Let us give now the proof of Theorem 2(i). We shall exhibit two constants $\sigma > 0$ and $K > 0$ such that $|a-b| \leq \sigma$ implies $|G(a) - G(b)| \leq K\,|a-b|$, whatever be the numbers a and b. Let $g(x) = \varphi(\tau^{-1}x)$, where φ is the smooth function of Lemma 4 and $\tau = \tau(s)$ a small positive number. Then we set

$$f(x) = (b-a) \sum_{|k_j|\leq\mu} g(r^{-1}x - k) + a\Theta_\nu(x)$$

(we sum over all $k \in \mathbb{Z}^n$ such that $|k_j| \leq \mu$ for $j = 1,\ldots,n$; the positive integers μ, ν and the number $r \in [0,1]$ will be precised in a while). We propose to call the function f an "Igari's comb". Indeed, S.Igari used a rudimentary version of this function in order to describe the functional calculus on $H^s(\mathbb{R})$, for $0 < s < 1/2$; then S.Janson adapted this method to the Hardy-Sobolev space $F_1^{1,2}(\mathbb{R}^n)$.

Let C_1 and C_2 be the constants associated to G according to Lemma 3; then the Lemma 4 allows us to choose an integer ν such that $2\,|a|\,\|\Theta_\nu\| \leq C_1$; the theorems of Frazier and Jawerth and the inequality (1) lead to

$$\|\sum_{|k_j|\leq\mu} g(r^{-1}(\cdot) - k)\| \leq C_3\, r^{\varrho}\, \mu^{n/p}.$$

I claim that a suitable choice of r and μ provides the inequalities

$$C_1(3|a-b|)^{-1} \leq C_3 r^{\varrho}\mu^{n/p} \leq C_1(2|a-b|)^{-1}, \tag{8}$$

$$r\mu \leq 2^{-\nu-2}. \tag{9}$$

As a matter of fact, for $\varrho > 0$, it suffices to define r by the *equality* $C_3 r^\varrho \mu^{n/p} = C_1(2|a-b|)^{-1}$; $r\mu$ is then an $O(\mu^{1-\frac{n\varrho}{p}})$; as $s > 0$ implies $1 - \frac{n\varrho}{p} < 0$, it is possible to realize (9) for a large integer μ.

If $\varrho = 0$, we assume that $C_1(3|a-b|)^{-1}$ is large enough; then one can find an integer $\mu \geq 1$ which satisfies (8); r is determined by (9).

The relations (8), (9) and a suitable choice of τ imply $\|f\| \leq C_1$ together with

$$f(x) = b \quad \text{on} \quad r(\tau Q + k), \quad f(x) = a \quad \text{on} \quad r(Q+k) \setminus r(2\tau Q + k),$$

for all k such that $|k_j| \leq \mu$, $(j = 1, \ldots, n)$. Thus the norm of $G \circ f$ in $E^s_p(\mathbb{R}^n)$ is majorized by C_2; this leads to

$$\begin{aligned} &|G(a) - G(b)|^p \, (2\mu + 1)^n \operatorname{vol}(r\tau Q^+) \\ &\leq \sum_{|k_j| \leq \mu} \int_{r(\tau Q^+ + k)} |(\Delta^m_{r\tau e_1}(G \circ f)(x)|^p \, dx \\ &\leq r^{sp} \, \tau^{sp} C_2^p, \end{aligned}$$

where $Q^+ = [0, 1/2]^n$; from the relation (8), it follows that

$$|G(a) - G(b)| \leq C_3 \, r^{-\varrho} \, \mu^{-n/p} \leq C_4 |a - b|.$$

B - The triviality case

We are going to prove part (iv) of Theorem 1. First we shall treat the case $s > 1 + 1/p$ (due to Dahlberg, Runst and the author). According to an idea of T.Runst, we embed $E^s_p(\mathbb{R}^n)$ in a Besov space $B^{t,\infty}_k$, with $1 < t < 2$.

Of course, if s itself is less than 2, it suffices to set $t = s$ and $k = p$. For $s \geq 2$, it is possible to find t and k such that

(i) $\varrho = \frac{n}{p} - s = \frac{n}{k} - t$,

(ii) $k \geq p$,

(iii) $1 + \frac{1}{k} < t < 2$;

this leads to the wished embedding.

Let u be a C^∞ function, carried by Q, such that $u(x) = x_1$ on $Q/2$. Set

$$f(x) = a \, u(r^{-1}x), \tag{10}$$

where $r \in]0, 1]$ and $a \geq 1$ satisfy

$$a \, \|u\| \, r^\varrho = C_1. \tag{11}$$

(The hypothesis $\varrho > 0$ makes this possible, as soon as a is large enough). Let $h = (\tau a^{-1}r, 0, \ldots, 0)$, where $0 < \tau \leq 1/8$; for any $x \in \frac{r}{4}Q$, we have

$$G(f(x+h)) - G(f(x)) = G(ar^{-1}x_1 + \tau) - G(ar^{-1}x_1).$$

The inequality $\|G \circ f\|_{B_k^{t,\infty}} \leq C_2$ leads to

$$\int_{\frac{r}{4}Q} |D_1(G \circ f)(x+h) - D_1(G \circ f)(x)|^k\, dx \leq C_2^k\, |h|^{k(t-1)}$$

and, after a change of variable,

$$\int_{|v| \leq \frac{a}{8}} |G'(v+\tau) - G'(v)|^k\, dv \leq C_3 r^{-k\varrho}\, a^{1-tk}. \tag{12}$$

According to (11), the second member of (12) is an $O(a^{1-tk+k})$; letting $a \to +\infty$ gives $G'(v+\tau) - G'(v) = 0$ for all v and τ; this means that G' is a constant.

Quite recently [4], we succeeded in adapt this proof to the limit case $s = 1 + 1/p$ (except for $q = 1$ in the Besov case, and $p = 1$ in the Lizorkin-Triebel case). We need the following

Lemma 5 *There exists a sequence $(u_\nu)_{\nu \geq 1}$ of C^∞ functions, carried by Q, such that*

(i) $u_\nu(x) = x_1$, *for* $|x_1| \leq 2^{-\nu-1}$ *and* $|x_j| \leq 1/4$, $(j = 2, \ldots, n)$;

(ii) $\lim_{\nu \to +\infty} u_\nu = 0$ *in* $E_p^{1+\frac{1}{p}}(\mathbb{R}^n)$ *(except in* $B_p^{1+\frac{1}{p},1}(\mathbb{R}^n)$ *and* $F_1^{2,q}(\mathbb{R}^n)$*).*

Proof of Lemma 5. First we consider the sequence (Θ_ν) constructed in the Lemma 4, such that $\lim_{\nu \to +\infty} \Theta_\nu = 0$ in $E_p^{1/p}(\mathbb{R})$, with the additional conditions that φ is even and $\int \varphi(x)\, dx = 0$; then it suffices to set $u_\nu(x) = \int_{-\infty}^{x} \Theta_\nu(t)\, dt$. For $n \geq 2$, the functions $u_\nu^{(n)}$ will be defined as

$$u_\nu^{(n)} = u_\nu(x_1)\, \varphi(x_2) \ldots \varphi(x_n).$$

Let us return to the proof of the triviality of the functional calculus on $E_p^{1+1/p}(\mathbb{R}^n)$. f is chosen again according to (10) and (11), but u is replaced by u_ν, for a large integer ν, and we assume $a = 2^\nu \|u_\nu\|^{-1}$.

If $p > 1$, it holds

$$\int_{|v| \leq 4^{-1}\|u_\nu\|^{-1}} |G'(v+\tau) - G'(v)|^p\, dv \leq C_3\, \|u_\nu\|^p; \tag{13}$$

letting $\nu \to +\infty$ leads to $G' = Ct$.

If $p = 1$ (i.e. if $E_p^s = B_1^{2,q}$, with $q > 1$), the inequality (13) is replaced by

$$\int_{|v| \leq 4^{-1}\|u_\nu\|^{-1}} |G'(v+\tau) + G'(v-\tau) - 2G'(v)|\, dv \leq C_3\, \|u_\nu\|;$$

then G is a polynomial of degree less than 3 and a classical reasoning allows to conclude again G(t) = ct.

C - Other necessary conditions

First we assume that G acts on $W^{2,1}(\mathbb{R}^n)$, with $n \geq 3$. Then we test G on the function f given by (10) and (11), where $\varrho = n-2$. The inequality $\|D_1^2(G \circ f)\|_1 \leq C_2$ yields

$$\int_{|v|\leq a/4} |G''(v)|\, dv \leq C_3$$

for any large $a > 0$; this means $G'' \in L^1$.

Next we suppose that G acts on $W^{2,p}(\mathbb{R}^n)^+$, with $1 < p < n/2$. The function f is again given by (10) and (11), but u is now replaced by a positive C^∞ function $\tilde{u}$, carried by Q, with $\tilde{u}(x) = x_1$, for $1/8 \leq x_1 \leq 1/4$ and $|x_j| \leq 1/4\,(j = 2, \ldots, n)$. The inequality $\|D_1^2(G \circ f)\|_p \leq C_2$ leads to

$$\int_{a/8}^{a/4} |G''(v)|^p\, dv \leq C_3\, a^{1-p},$$

which easily implies

$$\int_a^{+\infty} |G''(v)|^p\, dv \leq C_4\, a^{1-p}.$$

In order to prove the necessity of $G' \in L^\infty$, we observe that the Igari's comb can be assumed to be a positive function.

Let us finally study the case where G acts on $W^{m,p}(\mathbb{R}^n)$, for m integer, $m = n/p \geq 2$ and $p > 1$. We set

$$f(x) = \lambda\, u(2^{\nu-1}x) + a\Theta_\nu(x).$$

λ is fixed in such a way that

$$2\,\lambda\,\|u\| = C_1$$

and the integer $\nu = \nu(a) \geq 1$ in order to satisfy

$$2\,|a|\,\|\Theta_\nu\| \leq C_1.$$

The integration of $|D_1^m(G \circ f)|^p$ on the cube $2^{-\nu}Q$ leads to

$$\int_{|v-a|\leq\lambda/4} |G^{(m)}(v)|^p\, dv \leq C_3,$$

which exactly means that $G^{(m)}$ belongs to L^p locally uniformly ; together with $G' \in L^\infty$, this implies $G' \in W^{m-1,p}(\mathbb{R})_{loc-unif}$.

VI - Proofs : necessary conditions for the supercritical spaces

In this case, $G' \in L^\infty(\mathbb{R})_{loc}$ is the only non obvious necessary condition. We fix a number a and - according to Igari's method - set

$$S\,f(x) = \varphi(x)\,(G(a + f(x)) - G(a))\,.$$

If G acts on E_p^s, the same is true for the non-linear operator S ; moreover S acts boundedly, in the following sense : there exists constants $C_j > 0$ and a cube $Q' \subset Q$ such that, for all f carried by Q', $||f|| \leq C_1$ implies $||Sf|| \leq C_2$ (The proof of this property, similar to that of Lemma 3, is left to the reader).

Lemma 6 *There exists a family of functions $\psi_\varepsilon \in \mathcal{D}(\mathbb{R}^n)$ $(0 < \varepsilon \leq 1/2)$ such that $\psi_\varepsilon(x) = 0$ out of $(1+\varepsilon)Q$ and*

(i) for $s > 1/p$, the norm of ψ_ε in E_p^s is an $O(\varepsilon^{\frac{1}{p}-s})$,

(ii) the norms of ψ_ε in $B_p^{1/p,1}(\mathbb{R})$ and $F_1^{1,q}(\mathbb{R})$ are $O(|\log \varepsilon|)$.

Proof of Lemma 6. We construct ψ_ε for $n = 1$ (in higher dimension, it suffices to set $\psi_\varepsilon^{(n)}(x) = \psi_\varepsilon(x_1) \cdot \ldots \cdot \psi_\varepsilon(x_n)$). If ω is a positive C^∞ function, with support in the segment $[0,1]$, such that $\int \omega(x)\, dx = 1$, the function

$$\psi_\varepsilon(x) = \int_{-\infty}^{x} 2\,\varepsilon^{-1} \left\{ \omega(-\varepsilon^{-1}(2t+1)) - \omega(\varepsilon^{-1}(2t-1)) \right\} dt$$

has the required properties (the estimate of $||\psi_\varepsilon||$ follows from (1) and (2)) (Q.E.D.).

Let ξ and ℓ be the center and the half-side of Q'; we set

$$f(x) = (b' - b)\,\psi_\varepsilon(r^{-1}(x - \xi)) + b\,\varphi(\ell^{-1}(x - \xi)).$$

The Lemma 6 and the inequality (1) lead to

$$||\psi_\varepsilon(r^{-1}((\cdot) - \xi))|| \leq C_3\, \chi(\varepsilon)\, r^\varrho \quad (0 < \varepsilon \leq 1/2),$$

where $\chi(\varepsilon) = \varepsilon^{\frac{1}{p}-s}$ is $s > 1/p$ and $\chi(\varepsilon) = |\log \varepsilon|$ if $s = 1/p$ and $n = 1$. The parameters ε and r are required to satisfy $2r \leq \ell$ and

$$2\,C_3\,|b' - b|\, r^\varrho\, \chi(\varepsilon) = C_1. \tag{14}$$

For $\varrho < 0$, $\varepsilon = \varepsilon(s)$ is a fixed small number and r is determined by the equality (14) ; for $\varrho = 0$, we set $r = \ell/2$ and define ε by the equality (14). If

$$2\,|b|\,\ell^\varrho\, ||\varphi|| \leq C_1,$$

we get $||f|| \leq C_1$; if ε is small enough, one also has

$$S\,f(x) = G(a + b') \quad \text{on} \quad \xi + r\,Q,$$

$$S\,f(x) = G(a + b) \quad \text{on} \quad (\xi + 2r\,Q) \setminus (\xi + r\,(1+\varepsilon)\,Q).$$

The inequality $||Sf|| \leq C_2$, the identity (14) and the following lemmata lead to

$$|G(a+b) - G(a+b')| \leq C_4\,|b - b'|;$$

in other words, G is Lipschitz in a neighborhood of a.

Lemma 7 *Let $\varepsilon > 0$, a sufficiently small number, $r > 0$ and $\xi \in \mathbb{R}^n$. If a function f belongs to $B_p^{s,\infty}(\mathbb{R}^n)$ and satisfies*

$$f(x) = \beta \quad on \quad \xi + rQ, \quad f(x) = \beta' \quad on \quad (\xi + 2rQ) \setminus (\xi + r(1+\varepsilon)Q), \tag{15}$$

the norm of f in $B_p^{s,\infty}(\mathbb{R}^n)$ is greater than

$$C\,|\beta - \beta'|\, r^{\varrho}\, \varepsilon^{\frac{1}{p}-s}.$$

Proof. We assume $\varepsilon \leq \frac{1}{2m}$. In the case $\xi = 0$ and $r = 1$, one integrates $|(\Delta^m_{\varepsilon e_1} f(x)|^p$ on the box P_ε defined by

$$\frac{1-\varepsilon}{2} \leq x_1 \leq \frac{1}{2}, \quad |x_j| \leq \frac{1}{2} \quad (j = 1, 2, \ldots, n);$$

hence

$$N_1(f) \geq \varepsilon^{-s}\, |\beta - \beta'|\, (\text{vol}\, P_\varepsilon)^{1/p} = 2^{-1/p}\, |\beta - \beta'|\, \varepsilon^{\frac{1}{p}-s}.$$

To obtain the general case, we set $f(x) = f_0(r^{-1}(x - \xi))$, which yields

$$(\Delta_h)^m f(x) = (\Delta_{h/r})^m f_0(r^{-1}(x - \xi)),$$

so that $N_1(f) = r^{\varrho}\, N_1(f_0)$.

Lemma 8 *Let $\varepsilon > 0$, a sufficiently small number, $r > 0$ and $\xi \in \mathbb{R}$. If a function f belongs to $B_p^{1/p,1}(\mathbb{R})$ or $F_1^{1,\infty}(\mathbb{R})$ and satisfies (15), the norm of f is greater than $C\,|\beta - \beta'|\,|\log \varepsilon|$.*

Proof. As in the preceding lemma, we reduce ourselves to the case $\xi = 0$ and $r = 1$. The $B_p^{1/p,1}(\mathbb{R})$-norm is equivalent to $\|f\|_p + N_2(f)$, where

$$N_2(f) = \int_{\mathbb{R}} \left(\int_{\mathbb{R}} |(\Delta_h)^2 f(x)|^p \, dx \right)^{1/p} |h|^{-1-1/p}\, dh.$$

Suppose $\varepsilon \leq h \leq 1/4$ and integrate $|(\Delta_h)^2 f(x)|^p$ on the interval $[((1+\varepsilon)/2) - h, 1/2]$; it holds

$$\begin{aligned} N_2(f) &\geq |\beta - \beta'| \int_\varepsilon^{1/4} (h - \frac{\varepsilon}{2})^{1/p}\, h^{-1-1/p}\, dh \\ &\geq |\beta - \beta'|\, 2^{-1/p} \int_\varepsilon^{1/4} h^{-1}\, dh \\ &= |\beta - \beta'|\, 2^{-1/p} \log \frac{1}{4\varepsilon}. \end{aligned}$$

The $F_1^{1,\infty}(\mathbb{R})$-norm is equivalent to $\|f\|_p + N_3(f)$, where

$$N_3(f) = \|\sup_{t>0}\, t^{-1} \int_{[-1,+1]} |(\Delta_{th})^2 f(\cdot)|\, dh\|_1$$

(see the paper H.Triebel).

One assumes $\varepsilon \leq 1/16$ and $3/8 \leq x \leq 1/2$; for all $h \in [1/2, 1]$ and $t \in [1+\varepsilon-2x, (1-x)/2]$, one has $(\Delta_{th})^2 f(x) = |\beta - \beta'|$; hence

$$\sup_{t>0} t^{-1} \int_{[-1,+1]} |(\Delta_{th})^2 f(x)|\, dh \geq \frac{1}{2} |\beta - \beta'| (1+\varepsilon-2x)^{-1};$$

integrating on the interval $[3/8, 1/2]$ yields

$$N_3(f) \geq \frac{1}{2} |\beta - \beta'| \log(1 + \frac{1}{4\varepsilon}).$$

References

1 - Papers summarized in the exposition

[1] *Bourdaud, G.* . Le calcul fonctionnel dans les espaces de Sobolev. Invent. Math. 104 (1991) , 435-446.

[2] - . Le calcul fonctionnel dans lspace de Besov critique. Proc. Amer. Math. Soc. (to appear).

[3] - . Fonctions qui opèrent sur les espaces de Besov et de Triebel. Annales I.H.P. - Analyse non linéaire - 10 (1993) (to appear).

[4] - . La trivialité du calcul fonctionell dans $H^{3/2}(\mathbb{R}^4)$. C.R. Acad. Sci. Paris, t.314, Série 1, 187-190.

[5] - . Calcul fonctionnel et propriété de Fubini dans les espaces de Triebel -Lizorkin. Preprint, Paris VII (1992).

[6] *Bourdaud, G., Kateb, Dalila* . Fonctions qui opérent sur certains espaces de Besov. Ann. Inst. Fourier, Grenoble, 40 (1990), 153-162.

[7] - . Fonctions qui opérent sur les espaces de Besov. Proc. Amer. Math. Soc. 112 (1991), 1067-1076.

[8] *Bourdaud, G., Kateb, Djalil* . Calcul fonctionnel dans l'espace de Sobolev fractionnaire. Math. Z. 210 (1992), 607-613.

[9] *Bourdaud, G., Meyer, Y.* . Fonctions qui opérent sur les espaces de Sobolev. J. Funct. Anal. 97 (1991), 351-360.

2 - Previous works on the functional calculus

[10] *Adams, D.R.* . On the existence of capacitary strong type estimates in $\mathbb{R}^n$. Ark. Mat. 14 (1976), 125-140.

[11] *Bourdaud, G.* . Fonctions qui opèrent sur les espaces de Sobolev. Sém. Anal. Harm., Orsay (1980-1981) , 6-17.

[12] *Dahlberg, B.E.J.* . A note on Sobolev spaces. Proc. Symp. Pure Math. 35, 1 (1979),

183-185.

[13] *Igari, S.* . Sur les fonctions qui opèrent sur l'espace $\hat{A}^2$. Ann. Inst. Fourier, Grenoble, 15 (1965), 525-536.

[14] *Janson, S* . Harmonic analysis and partial differential equations (El Escorial, 1987), 193-201, Lect. Notes in Math. 1384, Springer 1989.

[15] *Marcus, M., Mizel, J.* . Complete characterization of functions which act via superposition on Sobolev spaces. Trans. Amer. Math. Soc. 251 (1979), 187-218.

[16] *Maz'ya, V.* . Sobolev spaces. Springer 1985.

[17] *Oswald, P.* . On the boundedness of the mapping $f \longrightarrow |f|$ in Besov spaces. Comment. Univ. Carolinae 33 (1992), 57-66.

[18] *Peetre, J.* . Interpolation of Lipschitz operators and metric spaces. Mathematica (Cluj) 12 (35) (1970), 325-334.

[19] *Runst, T.* . Mapping properties of non-linear operators in spaces of Triebel-Lizorkin and Besov type. Anal. Math. 12 (1986), 313-346.

[20] *Szigeti, F.* . On the Nemitzki operators in Sobolev spaces. Z. Angew. Math. Mech. 63 (1983), T 332.

3 - Works on wavelets and function spaces

[21] *Bourdaud, G.* . Analyse fonctionnelle dans l'espace Euclidien. Publ. Math. Paris VII, 23 (1987).

[22] *Frazier, M., Jawerth, B.* . Decomposition of Besov spaces. Indiana Univ. Math. J. 84 (1985), 777-799.

[23] . A discrete transform and applications to distribution spaces. J. Funct. Anal. 93 (1990), 34-170.

[24] *Goldberg, D.* . A local version of real Hardy spaces. Duke Math. J. 46 (1979), 27-42.

[25] *Meyer, Y.* . Ondelettes et Opèrateurs. Hermann, Paris 1990/91.

[26] *Triebel, H.* . Theory of function spaces. Birkhäuser, Basel, Boston, Stuttgart 1983.

[27] - . Local approximation spaces. Z. Anal. Anwendungen 8 (1989), 261-288.

Gérard Bourdaud
Univ. Paris VII - Denis Diderot
C.N.R.S.U.A. 212
Tour 45 – 55 – 5° étage
2, place Jussieu
75251 Paris Cedex 05
France

Solvability of Strongly Nonlinear Degenerated Elliptic Problems

Pavel Drábek

1 Introduction.

Let us consider *quasilinear elliptic boundary value problem* (BVP)

$$\begin{aligned} -\operatorname{div}(a(x,u)|\nabla u|^{p-2}\nabla u) &= \lambda b(x,u)|u|^{p-2}u + f(x,u) \quad \text{in } \Omega, && (1.1a)\\ u &= 0 \quad \text{on } \partial\Omega, && (1.1b) \end{aligned}$$

where Ω is a bounded domain, $p > 1$ is a real number λ is a real parameter and $a(x,s), b(x,s)$, $f(x,s)$ satisfy appropriate growth conditions. We will assume that the coefficient $a(x,s)$ contains a *degeneration* or a *singularity* to a certain extent. The purpose of this note is to present results concerning the existence of the weak solution of the BVP (1.1). Under some additional assumptions on $f(x,s)$ we prove the existence of nonnegative weak solution of the BVP (1.1). In order to deal with the degenerate (or singular) coefficient $a(x,s)$ in the equation (1.1a) we work in a suitable weighted Sobolev space with the weight which controls the degeneration (or singularity) of $a(x,s)$ in a certain sense.

We use a functional–theoretical approach based on the degree theory for a generalized monotone mappings (see BROWDER, PETRYSHIN [2] and SKRYPNIK [9]), apriori estimates and some recent results of the author concerning the existence of the least eigenvalue and a corresponding nonnegative eigenfunction of the quasilinear eigenvalue problem

$$\begin{aligned} -\operatorname{div}(a(x,u)|\nabla u|^{p-2}\nabla u) &= \lambda b(x,u)|u|^{p-2}u \quad \text{in } \Omega, && (1.2a)\\ u &= 0 \quad \text{on } \partial\Omega, && (1.2b) \end{aligned}$$

(see DRÁBEK [4]).

The results of this paper generalize some existence theorems for p-Laplacian (see DRÁBEK [3], FLECKINGER, HERNÁNDEZ, de THÉLIN [5] and the references therein).

2 Weak formulation, maximum principle.

Let us suppose that Ω is an open bounded subset of n-dimensional Euclidean space $\mathbb{R}^n, p > 1$, is an arbitrary real number and w is a *weight function* (i.e. positive and measurable) in Ω. Assume that

$$w \in L^1_{loc}(\Omega) \text{ and } \frac{1}{w} \in L^{1/(p-1)}_{loc}(\Omega). \tag{2.1}$$

Let us define the *weighted Sobolev space* $W^{1,p}(w,\Omega)$ as the set of all real valued functions u defined in Ω for which

$$\|u\|_{1,p,w} = (\int_\Omega |u|^p dx + \int_\Omega w|\nabla u|^p dx)^{1/p} < \infty. \tag{2.2}$$

It follows from (2) that $W^{1,p}(w,\Omega)$ is a *reflexive Banach space* and that $W^{1,p}_0(w,\Omega)$ is well defined as a closure of $C^\infty_0(\Omega)$ in $W^{1,p}(w,\Omega)$ with respect to the norm $\|\cdot\|_{1,p,w}$ (see e.g. KUFNER, SÄNDIG [7]).

Let $s \geq \frac{1}{p-1}$ be a real number. A simple application of the Hölder inequality yields that the *continuous imbedding*

$$W^{1,p}(w,\Omega) \hookrightarrow W^{1,p_1}(\Omega) \tag{2.3}$$

holds provided

$$\frac{1}{w} \in L^s(\Omega) \text{ and } p_1 = \frac{ps}{s+1}.$$

It follows from (2) and from the Sobolev imbedding theorem (see e.g. ADAMS [1], KUFNER, JOHN, FUČÍK [6]) that for $s+1 \leq ps < n(s+1)$ we have

$$W^{1,p}_0(w,\Omega) \hookrightarrow W^{1,p_1}_0(\Omega) \hookrightarrow L^q(\Omega), \tag{2.4}$$

where $1 \leq q = \dfrac{np_1}{n-p_1} = \dfrac{nps}{n(s+1)-ps}$, and for $ps \geq n(s+1)$ the imbedding (2) holds with arbitrary $1 \leq q < \infty$.

Mereover, the *compact imbedding*

$$W^{1,p}_0(w,\Omega) \hookrightarrow\hookrightarrow L^r(\Omega)$$

holds provided $1 \leq r < q$.

An easy calculation yields that $s > \frac{n}{p}$ implies $q > p$. In particular, we have

$$W^{1,p}_0(w,\Omega) \hookrightarrow\hookrightarrow L^{p+\eta}(\Omega) \tag{2.5}$$

for $0 \leq \eta < q - p$, provided

$$\frac{1}{w} \in L^s(\Omega) \text{ and } s \in (\frac{n}{p}, +\infty) \cap [\frac{1}{p-1}, +\infty). \tag{2.6}$$

In what follows we will always assume that (2) is fulfilled.

Simple application of the Friedrichs inequality in $W_0^{1,p_1}(\Omega)$, the Hölder inequality and the assumption (2) yield that the norm

$$\|u\|_w = (\int_\Omega w|\nabla u|^p dx)^{1/p}.$$

on the space $W_0^{1,p}(w,\Omega)$ is equivalent to the norm $\|\cdot\|_{1,p,w}$ defined by (2). Moreover, $W_0^{1,p}(w,\Omega)$ equipped with the norm $\|\cdot\|_w$ is *uniformly convex Banach space.*

Let $g : [0,\infty) \to [1,\infty)$ be a nondecreasing function, $\alpha(x) \in L^\infty(\Omega)$. We assume that $a(x,s), b(x,s)$ are the Carathéodory functions satisfying

$$\frac{w(x)}{c_1} \leq a(x,s) \leq c_1 g(|s|)w(x), \tag{2.7}$$

$$0 \leq b(x,s) \leq \alpha(x), \tag{2.8}$$

for a.e. $x \in \Omega$ and for all $s \in \mathbb{R}$ with some $c_1 > 1$. Moreover, assume that

$$\operatorname{meas}\{x \in \Omega; b(x,v(x)) > 0\} > 0$$

for any $v \in L^p(\Omega), v \not\equiv 0$. We will also assume that $f(x,s)$ is the Carathéodory function satisfying the assumption

$$|f(x,s)| \leq \beta(x)|s|^{p-1} + \gamma(x), \tag{2.9}$$

for a.e. $x \in \Omega$ and for all $s \in \mathbb{R}$, where $\beta(x), \gamma(x) \in L^\infty(\Omega)$.

Definition 2.1 We will say that $u \in W_0^{1,p}(w,\Omega)$ is *the weak solution* of the BVP (1.1) if

$$\begin{aligned} &\int_\Omega a(x,u(x))|\nabla u|^{p-2}\nabla u \nabla\varphi dx = \\ &= \lambda \int_\Omega b(x,u(x))|u|^{p-2}u\varphi dx + \int_\Omega f(x,u(x))\varphi dx \end{aligned} \tag{2.10}$$

holds for any $\varphi \in W_0^{1,p}(w,\Omega)$.

Definition 2.2 We will say that $\lambda \in \mathbb{R}$ is the *eigenvalue* and $u \in W_0^{1,p}(w,\Omega)$, $u \neq 0$, is the corresponding *eigenfunction* of the eigenvalue problem (1.2) if

$$\int_\Omega a(x,u(x))|\nabla u|^{p-2}\nabla u \nabla\varphi dx = \lambda \int_\Omega b(x,u(x))|u|^{p-2}u\varphi dx \tag{2.11}$$

holds for any $\varphi \in W_0^{1,p}(w,\Omega)$.

The following result concerning the existence of the first (the least) eigenvalue and the corresponding nonnegative eigenfunction of (1.2) is proved in DRÁBEK [4] (see Theorem 4.10 in [4]).

Theorem 2.1 *Let us assume* (2), (2). *Then for a given real number* $R > 0$ *there exists the least eigenvalue* $\lambda_R > 0$ *and a corresponding eigenfunction* $u_R \geq 0$ *a.e. in* Ω *of the eigenvalue problem* (1.2) *such that* $\|u_R\|_p = R$.

Due to Theorem 2.1 we have a function

$$R \mapsto \lambda_R$$

mapping $(0,\infty)$ into $(0,\infty)$. Using the assumptions (2) and (2) we prove the *estimate of* λ_R *from below* uniformly with respect to $R>0$. Set

$$\lambda^{\#} = \frac{1}{c_1 c_2 \|\alpha(x)\|_\infty},$$

where $c_2 > 0$ is the constant of the imbedding $W_0^{1,p}(w,\Omega) \hookrightarrow L^p(\Omega)$ (cf. (2)), i.e.

$$\|u\|_p \le c_2 \|u\|_w \tag{2.12}$$

for any $u \in W_0^{1,p}(w,\Omega)$. Let us assume that $w \in W_0^{1,p}(w,\Omega)$ is such that both integrals

$$\int_\Omega a(x,v(x))|\nabla v|^p dx \text{ and } \int_\Omega b(x,v(x))|v|^p dx$$

are finite.

Lemma 2.1 *Let $\lambda \le \lambda^{\#}$. Then*

$$\int_\Omega a(x,v(x))|\nabla v|^p dx - \lambda \int_\Omega b(x,v(x))|v|^p dx \ge 0. \tag{2.13}$$

Proof The assertion is clear if $\lambda \le 0$. Let $\lambda > 0$. It follows from (2) that

$$\int_\Omega a(x,v(x))|\nabla v|^p dx \ge \frac{1}{c_1}\int_\Omega w(x)|\nabla v|^p dx = \frac{1}{c_1}\|v\|_w^p. \tag{2.14}$$

On the other hand applying (2) and (2) we obtain

$$\int_\Omega b(x,v(x))|v|^p dx \le \|\alpha(x)\|_\infty \int_\Omega |v|^p dx \le c_2\|\alpha(x)\|_\infty \|v\|_w^p. \tag{2.15}$$

Combining () and () we get

$$\int_\Omega a(x,v(x))|\nabla v|^p dx - \lambda \int_\Omega b(x,v(x))|v|^p dx \ge \left[\frac{1}{c_1} - \lambda c_2 \|\alpha(x)\|_\infty\right] \|v\|_w^p. \tag{2.16}$$

Since $\lambda \le \lambda^{\#}$ implies $\frac{1}{c_1} - \lambda c_2 \|\lambda(x)\|_\infty \ge 0$ the assertion follows from ().

■

Lemma 2.2 *Let $\lambda^* = \inf_{R>0} \lambda_R$. Then $\lambda^* \ge \lambda^{\#}$.*

Proof Let $R>0$ be arbitrary. Let λ_R be the eigenvalue and $u_R \in W_0^{1,p}(w,\Omega) \cap L^\infty(\Omega)$, $\|u_R\|_p = R$ be the corresponding eigenfunction of (1.2). Choosing $\varphi = u_R$ as a test function in (2.2) we obtain

$$\int_\Omega a(x,u_R(x))|\nabla u_R|^p dx = \lambda_R \int_\Omega b(x,u_R(x))|u_R|^p dx. \tag{2.17}$$

The estimates () and () with v replaced by u_R yield

$$\frac{1}{c_1} \le \lambda_R c_2 \|\alpha(x)\|_\infty, \text{ i.e. } \lambda_R \ge \frac{1}{c_1 c_2 \|\alpha(x)\|_\infty} = \lambda^\#.$$

Since $R > 0$ is arbitrary we have $\lambda^* \ge \lambda$.

■

Let us consider BVP

$$\begin{aligned} -\operatorname{div}(a(x,u)|\nabla u|^{p-2}\nabla u) &= \vartheta b(x,u)|u|^{p-2}u + h \quad \text{in } \Omega, && (2.1a)\\ u &= 0 \quad \text{on } \partial\Omega, && (2.1b) \end{aligned}$$

with a real parameter ϑ and with $h(x) \in L^{p'}(\Omega), \frac{1}{p} + \frac{1}{p'} = 1$.

Definition 2.3 We will say that the BVP (2.1) satisfies *the maximum principle* if $h \ge 0$ in Ω implies $u \ge 0$ a.e. in Ω for all possible weak solutions $u \in W_0^{1,p}(w,\Omega)$ of (2.1).

Lemma 2.3 (Sufficient condition.) *Let $b(x,s) > 0$ for a.e. $x \in \Omega$ and for all $s \in \mathbf{R}$. If*

$$\vartheta < \lambda^\# \qquad (2.19)$$

then the BVP (2.1) *satisfies the maximum principle.*

Proof Let $u \in W_0^{1,p}(w,\Omega)$ be the weak solution of (2.1) corresponding to $h \in L^{p'}(\Omega), h \ge 0$ a.e. in Ω. Then

$$\int_\Omega a(x,u)|\nabla u|^{p-2}\nabla u \nabla\varphi dx = \vartheta \int_\Omega b(x,u)|u|^{p-2}u\varphi dx + \int_\Omega h\varphi dx \qquad (2.20)$$

holds for any $\varphi \in W_0^{1,p}(w,\Omega)$. Choose $\varphi = u^- = \max\{-u(x), 0\}$ as a test function in (). We obtain

$$-\int_\Omega a(x,u)|\nabla u^-|^p dx = -\vartheta \int_\Omega b(x,u)|u^-|^p dx + \int_\Omega hu^- dx. \qquad (2.21)$$

It follows from () and from the fact $h \ge 0$ a.e. in Ω that

$$\int_\Omega a(x,u^-)|\nabla u^-|^p dx - \vartheta \int_\Omega b(x,u^-)|u^-|^p dx \le 0. \qquad (2.22)$$

On the other hand it follows from Lemma 2.1 (see (2.1)) that

$$\int_\Omega a(x,u^-)|\nabla u^-|^p dx - \lambda^\# \int_\Omega b(x,u^-)|u^-|^p dx \ge 0. \qquad (2.23)$$

The inequalities () and () imply

$$(\lambda^\# - \vartheta) \int_\Omega b(x,u^-)|u^-|^p dx \le 0.$$

Due to (2.3) and the assumption on $b(x,s)$ we get $u^- = 0$ a.e. in Ω, i.e. $u \ge 0$ a.e. in Ω. Hence the BVP (2.1) satisfies the maximum principle.

■

Lemma 2.4 (Necessary condition.) *Let both $a(x,s)$ and $b(x,s)$ be even in s, i.e.*

$$a(x,s) = a(x,-s) \quad \textit{and} \quad b(x,s) = b(x,-s)$$

for all $s \in \mathbf{R}$ and for a.e. $x \in \Omega$. Let the BVP (2.1) *satisfy the maximum principle. Then*

$$\vartheta < \lambda_R$$

for any $R > 0$.

Proof We will proceed via contradiction. Let us assume that there exists $R > 0$ such that $\vartheta \geq \lambda_R$. Let u_R be the corresponding nonnegative eigenfunction of (1.2) satisfying $||u_R||_p = R$ (see Theorem 2.1). Set

$$h(x) = (\vartheta - \lambda_R) b(x, u_R(x)) |u_R(x)|^{p-2} u_R(x). \tag{2.24}$$

Then $h(x) \geq 0$ a.e. in Ω. Since both $a(x,s)$ and $b(x,s)$ are even in s, the function $-u_R$ is the solution of the BVP (2.1) with h given by (). But $u_R \not\equiv 0$ and $-u_R \leq 0$ a.e. in Ω. This is a contradiction with the fact that the BVP (2.1) satisfies the maximum principle.

■

Remark 2.1 Let us consider the case of p-Laplacian:

$$a(x,s) \equiv 1 \quad \text{and} \quad b(x,s) \equiv 1. \tag{2.25}$$

Then

$$\lambda_R \equiv \lambda^* = \lambda^\# = \lambda_1 \quad \text{for} \quad R > 0,$$

where $\lambda_1 > 0$ is the first eigenvalue of p-Laplacian (see DRÁBEK [3] and the references therein). The functions $a(x,s), b(x,s)$ given by (2.1) satisfy the assumptions of Lemmas 2.3 and 2.4. It follows from here that the BVP

$$\begin{aligned} -\mathrm{div}(|\nabla u|^{p-2}\nabla u) &= \vartheta |u|^{p-2}u + h \quad \text{in } \Omega, \\ u &= 0 \quad \text{on } \partial\Omega, \end{aligned}$$

satisfies the maximum principle *if and only if*

$$\vartheta < \lambda_1$$

(cf. FLECKINGER, HERNÁNDEZ, de THÉLIN [5]).

3 Main results

In this section we will formulate and prove the existence results for the BVP (1.1).

Theorem 3.1 *Let us assume* (2)–(2) *and*

$$\lambda < \frac{1 - c_1 c_2 ||\beta(x)||_\infty}{c_1 c_2 ||\alpha(x)||_\infty} \; (= \lambda^\# (1 - c_1 c_2 ||\beta||_\infty)). \tag{3.1}$$

Then the BVP (1.1) *has at least one weak solution $u \in W_0^{1,p} \cap L^\infty(\Omega)$.*

Proof We will prove the assertion in four steps. In the first step we prove an apriori estimate in $L^r(\Omega)$ for any $1 \le r < \infty$, and in the second step an apriori estimate in $L^\infty(\Omega)$ for any possible solution of the BVP (1.1). In the third step we find a suitable operator representation of the BVP (1.1). The fourth step is devoted to the application of the degree theory in order to prove the existence of the weak solution.

■

Step 1 (apriori estimate in $\mathbf{L^r(\Omega), 1 \le r < \infty}$). Let us suppose that $u \in W_0^{1,p}(w,\Omega)$ is the weak solution of the BVP (1.1). For $\varphi \in W_0^{1,p}(w,\Omega)$ set formally

$$\langle T(u),\varphi\rangle = \int_\Omega a(x,u)|\nabla u|^{p-2}\nabla u\nabla\varphi dx - \lambda\int_\Omega b(x,u)|u|^{p-2}u\varphi dx - $$

$$-\int_\Omega f(x,u)\varphi dx.$$

Choosing $\varphi = u$ as a test function in (2.1) we obtain due to (2)-(2),(2),(2.1) that

$$\begin{aligned}
0 = \langle T(u),u\rangle &= \frac{\lambda}{\lambda\#}\Big[\int_\Omega a(x,u)|\nabla u|^p dx - \lambda\#\int_\Omega b(x,u)|u|^p dx\Big] + \\
&+\frac{\lambda\# - \lambda}{\lambda\#}\int_\Omega a(x,u)|\nabla u|^p dx - \int_\Omega f(x,u)u dx \ge \\
&\ge \frac{\lambda\# - \lambda}{c_1\lambda\#}\|u\|_w^p - c_2\|\beta(x)\|_\infty\|u\|_w^p - \|\gamma(x)\|_\infty\|u\|_1 \ge \\
&\ge \frac{\lambda\#(1 - c_1c_2\|\beta(x)\|_\infty) - \lambda}{c_1\lambda\#}\|u\|_w^p - c_2(\text{meas}\,\Omega)^{(p-1)/p}\|\gamma(x)\|_\infty\|u\|_w.
\end{aligned} \tag{3.2}$$

It follows from (3.1) and (1) that $\|u\|_w \le c_3$ for any possible weak solution u of the BVP (1.1). By the imbedding (2) we have $\|u\|_q \le c_4$. In particular, we also have

$$\|u^+\|_q \le c_4 \tag{3.3}$$

(here $u^+ = \max\{u(x),0\}$). For a given real $M > 0$ set

$$\begin{aligned}
v_M(x) &= \inf\{u(x),M\} \quad &&\text{on} \quad \{x\in\Omega; u(x)\ge 0\},\\
v_M(x) &= 0 &&\text{on} \quad \{x\in\Omega, u(x) < 0\}.
\end{aligned}$$

For a real $\kappa \ge 0$ set $\varphi = v_M^{\kappa p+1} \in W_0^{1,p}(w,\Omega)\cap L^\infty(\Omega)$ and choose this φ as a test function in (2.1). We obtain

$$\begin{aligned}
(\kappa p+1)\int_{\Omega(u>0)} a(x,u)v_M^{\kappa p}|\nabla v_M|^p dx = \\
= \lambda\int_{\Omega(u>0)} b(x,u)u^{p-1}v^{\kappa p+1}dx + \int_{\Omega(u>0)} f(x,u)v_M^{\kappa p+1}dx)
\end{aligned} \tag{3.4}$$

The left hand side of (1) is estimated by using (2) and (2):

$$
\begin{aligned}
(\kappa p+1)\int_{\Omega(u>0)} a(x,u)v_M^{\kappa p}|\nabla v_M|^p dx \geq \\
\geq \frac{\kappa p+1}{c_1}\int_{\Omega(u>0)} w(x)v_M^{\kappa p}|\nabla v_M|^p dx = \\
= \frac{\kappa p+1}{c_1(\kappa+1)^p}\int_{\Omega(u>0)} w(x)|\nabla(v_M^{\kappa+1})|^p dx \geq \\
\geq \frac{\kappa p+1}{c_1 c_2^p(\kappa+1)^p}(\int_{\Omega(u>0)} (v_M^{\kappa+1})^q dx)^{p/q}.
\end{aligned} \tag{3.5}
$$

The right hand side of (1) is estimated by using (2), (2) and the Hölder inequality:

$$
\begin{aligned}
\lambda\int_{\Omega(u>0)} b(x,u)u^{p-1}v_M^{\kappa p+1}dx + \int_{\Omega(u>0)} f(x,u)v_M^{(\kappa p+1)}dx \leq \\
\leq \lambda\|\alpha(x)\|_\infty \int_{\Omega(u>0)} u^{(\kappa+1)p}dx + \|\beta(x)\|_\infty \int_{\Omega(u>0)} u^{(\kappa+1)p}dx + \\
+\|\gamma(x)\|_\infty (\operatorname{meas}\Omega)^{(\kappa p+1)/((\kappa+1)p)} \int_{\Omega(u>0)} u^{(\kappa+1)p}dx.
\end{aligned} \tag{3.6}
$$

It follows from (1) - (1) that

$$
(\int_{\Omega(u>0)} v_M^{(\kappa+1)q}dx)^{p/q} \leq c_5 \int_{\Omega(u>0)} u^{(\kappa+1)p}dx \tag{3.7}
$$

with some $c_5 > 0$. Let κ be such that $(\kappa+1)p = q$. Due to (1) and (1) we get

$$
\int_{\Omega(u>0)} v_M^{(\kappa+1)q}dx \leq c_6. \tag{3.1}
$$

Hence $v_M \in L^{q_1}(\Omega)$ for any $M > 0$, where $q_1 = (\kappa+1)q = \frac{q}{p}q$. Moreover, the Fatou lemma together with (3.1) yield

$$
\int_\Omega (u^+)^{q_1}dx \;\leq\; \liminf_{M\to\infty}\int_{\Omega(u>0)} v_M^{q_1}dx \;\leq\; c_6,
$$

i.e.

$$
\|u^+\|_{q_1} \leq c_7. \tag{3.9}
$$

Thus we can choose the number κ in (1) such that $(\kappa+1)p = q_1$. Then it follows from (1) and (1) that

$$
\int_{\Omega(u>0)} v_M^{(\kappa+1)q}dx \leq c_8.
$$

Applying the Fatou lemma as above we get

$$
\|u^+\|_{q_2} \leq c_9,
$$

where $q_2 = (\kappa+1)q = (\frac{q}{p})^2 q$. Proceeding further we obtain $\|u^+\|_{q_k} \le c(q_k)$ with $q_k = (\frac{q}{q})^k q, k = 1, 2, \cdots$. Similarly we can handle with u^- and hence for any $r \in [0,\infty)$ there exists a constant $c = c(r) > 0$ such that

$$\|u\|_r \le c(r) \tag{3.10}$$

for any possible weak solution $u \in W_0^{1,p}(w,\Omega)$ of the BVP (1.1).

Step 2 (apriori estimate in $\mathbf{L}^\infty(\Omega)$). Let $k \ge 0$ be a real number. Define

$$\begin{aligned} \varphi(x) &= \sup\{u(x), k\} - k \quad && \text{on } \{x \in \Omega; u(x) \ge 0\} \\ \varphi(x) &= 0 && \text{on } \{x \in \Omega, u(x) < 0\}. \end{aligned}$$

Substitute this function φ into (2.1). We obtain

$$\begin{aligned} &\int_{\Omega(u>k)} a(x,u)|\nabla\varphi|^p dx = \\ &= \lambda \int_{\Omega(u>k)} b(x,u)(\varphi+k)^{p-1}\varphi dx + \int_{\Omega(u>k)} f(x,u)\varphi dx. \end{aligned} \tag{3.11}$$

Let us fix a real number $r > 0$ such that

$$r > \max\{\frac{(p-1)q}{q-p}, q\}.$$

Now, we will estimate the left hand side and the right hand side of (2) from below and from above, respectively. We have

$$\int_{\Omega(u>k)} a(x,u)|\nabla\varphi|^p dx \ge \frac{1}{c_1}\int_{\Omega(u>k)} w(x)|\nabla\varphi|^p dx \ge c_{10}\|\varphi\|_q^p \tag{3.12}$$

with some c_{10}. We apply the Hölder inequality (two times), (2) and (1) in order to get

$$\begin{aligned} &\int_{\Omega(u>k)} b(x,u)(\varphi+k)^{p-1}\varphi dx \le \|\alpha(x)\|_\infty \int_{\Omega(u>k)} (\varphi+k)^{p-1}\varphi dx \le \\ &\le \|\alpha(x)\|_\infty \Big(\int_{\Omega(u>k)} (\varphi+k)^p dx\Big)^{(p-1)/p} \Big(\int_{\Omega(u>k)} \varphi^p dx\Big)^{1/p} \le \\ &\le \|\alpha(x)\|_\infty \Big(\int_\Omega |u|^r dx\Big)^{(p-1)/r} [\text{meas}\,\Omega(u>k)]^{((p-1)/p)(1-p/r)} \cdot \\ &\quad \cdot \Big(\int_\Omega \varphi^q dx\Big)^{1/q} [\text{meas}\,\Omega(u>k)]^{(1/p)(1-p/q)} \le \\ &\le \|\alpha(x)\|_\infty c(r)^{p-1}\|\varphi\|_q [\text{meas}\,\Omega(u>k)]^{((p-1)/p)(1-p/r)+(1/p)(1-p/q)}. \end{aligned} \tag{3.13}$$

Similarly using (2) and (1) we obtain

$$
\begin{aligned}
\int_{\Omega(u>k)} f(x,u)\varphi dx &\le \|\beta(x)\|_\infty \int_{\Omega(u>k)} |u|^{p-1}\varphi dx + \|\gamma(x)\|_\infty \int_{\Omega(u>k)} \varphi dx \le \\
&\le \|\beta(x)\|_\infty \Big(\int_{\Omega(u>k)} |u|^p dx\Big)^{(p-1)/p} \Big(\int_{\Omega(u>k)} \varphi^p dx\Big)^{1/p} + \\
&+ \|\gamma(x)\|_\infty \Big(\int_{\Omega(u>k)} dx\Big)^{(p-1)/p} \Big(\int_{\Omega(u>k)} \varphi^p dx\Big)^{1/p} \le \\
&\le \|\beta(x)\|_\infty c(r)^{p-1} \|\varphi\|_q [\operatorname{meas}\Omega(u>k)]^{((p-1)/p)(1-p/r)+(1/p)(1-p/q)} + \\
&+ \|\gamma(x)\|_\infty [\operatorname{meas}\Omega(u>k)]^{(p-1)/r} \|\varphi\|_q [\operatorname{meas}\Omega(u>k)]^{((p-1)/p)(1-p/r)+(1/p)(1-p/q)}.
\end{aligned}
\tag{3.14}
$$

It follows from (2) - (2) that

$$
\|\varphi\|_q^{p-1} \le c_{11} [\operatorname{meas}\Omega(u>k)]^{((p-1)/p)(1-p/r)+(1/p)(1-p/q)} \tag{3.15}
$$

with some $c_{11} = c_{11}(r) > 0$. On the other hand, for $h > h$ we obtain

$$
\begin{aligned}
\|\varphi\|_q^{p-1} = \Big(\int_{\Omega(u>k)} |u-k|^q dx\Big)^{(p-1)/q} &\ge \Big(\int_{\Omega(u>h)} |u-k|^q dx\Big)^{(p-1)/q} \ge \\
&\ge (h-k)^{p-1} [\operatorname{meas}\Omega(u>h)]^{(p-1)/q}.
\end{aligned}
\tag{3.16}
$$

Set $\xi(t) = \operatorname{meas}\Omega(u>t)$. Then it follows from (2) and (2) that nonnegative and nonincreasing function $\xi(t)$ satisfies

$$
\xi(h) \le c_{12}(h-k)^{-q}(\xi(k))^\delta
$$

for any $h > k \ge 0$, where

$$
\delta = \frac{q}{p-1}\Big[\frac{p-1}{p}(1-\frac{p}{r}) + \frac{1}{p}(1-\frac{p}{q})\Big].
$$

An easy calculation yields that $r > \frac{(p-1)q}{q-p}$ implies $\delta > 1$. Due to the result of MURTHY, STAMPACCHIA [8] (cf Lemma 2.5 in DRÁBEK [4]) there is $d^+ = d^+(r) > 0$ such that $\xi(d^+) = 0$, i.e. $u^+(x) \le d^+$ a.e. in Ω. Similarly (working with u^-) we get $u^-(x) \le d^-$ a.e. in Ω for some $d^- = d^-(r) > 0$. Since r was fixed, we have

$$
\|u\|_\infty \le d, \; d = \max\{d^+, d^-\}. \tag{3.17}
$$

for all possible weak solutions $u \in W_0^{1,p}(w,\Omega)$ of the BVP (1.1).

Step 3 (operator representation of the BVP (1.1)). Let us define the function

$$
\tilde{a}(x,s) = \begin{cases} a(x,s) & \text{for } x \in \Omega, |s| \le d, \\ a(x,d) & \text{for } x \in \Omega, s > d, \\ a(x,-d) & \text{for } x \in \Omega, s < -d. \end{cases} \tag{3.18}
$$

where $d > 0$ is the real number from (2). Then due to Step 2 the function $u \in W_0^{1,p}(w,\Omega)$ is the weak solution of the BVP (1.1) if and only if u satisfies (2) and it is the weak solution of the BVP

$$\begin{aligned} -\operatorname{div}(\tilde{a}(x,u)|\nabla u|^{p-2}\nabla u) &= \lambda b(x,u)|u|^{p-2}u + f(x,u) \quad \text{in } \Omega, & (3.2a)\\ u &= 0 \quad \text{on } \partial\Omega. & (3.2b)\end{aligned}$$

Define the operators

$$A, B, F : W_0^{1,p}(w,\Omega) \to [W_0^{1,p}(w,\Omega)]^*$$

by the following way

$$\begin{aligned} \langle A(u), \varphi\rangle &= \int_\Omega \tilde{a}(x,u)|\nabla u|^{p-2}\nabla u \nabla\varphi dx,\\ \langle B(u), \varphi\rangle &= \int_\Omega b(x,u)|u|^{p-2}u\varphi dx,\\ \langle F(u), \varphi\rangle &= \int_\Omega f(x,u)\varphi dx\end{aligned}$$

for any $u, \varphi \in W_0^{1,p}(w,\Omega)$ (where $\langle\cdot,\cdot\rangle$ denotes the pairing between the dual space $[W_0^{1,p}(w,\Omega)]^*$ and $W_0^{1,p}(w,\Omega)$). Then (2) - (2) and (3) guarantee that A, B and F are well defined operators. Set

$$\tilde{T}(u) = A(u) - \lambda B(u) - F(u)$$

for any $u \in W_0^{1,p}(w,\Omega)$. It is easy to see that $u \in W_0^{1,p}(w,\Omega)$ is the weak solution of the BVP (3.2) if and only if u is the solution of the *operator equation*

$$\tilde{T}(u) = 0. \tag{3.20}$$

Step 4 (application of the degree theory). The operator

$$\tilde{T} : W_0^{1,p}(w,\Omega) \to [W_0^{1,p}(w,\Omega)]^*$$

defined in Step 3 satisfies the condition $\alpha(W_0^{1,p}(w,\Omega))$ (see SKRYPNIK [9], cf. the condition (S_+) in BROWDER, PETRYSHIN [2]): *Let u_n converge weakly to u_0 in $W_0^{1,p}(w,\Omega)$ and*

$$\limsup_{n\to\infty}\langle\tilde{T}(u_n), u_n - u_0\rangle \le 0 \tag{3.21}$$

hold. Then u_n converge strongly to u_0 in $W_0^{1,p}(w,\Omega)$.

Let us verify this condition. The weak convergence of u_n to u_0 in $W_0^{1,p}(w,\Omega)$ together with (2) imply the strong convergence of u_n to u_0 in $L^p(\Omega)$. Hence applying the Hölder inequality we obtain

$$\begin{aligned} |\langle B(u_n), u_n - u_0\rangle| &= |\int_\Omega b(x,u_n)|u_n|^{p-2}u_n(u_n-u_0)dx| \le\\ &\le \|\alpha(x)\|_\infty\Big(\int_\Omega |u_n|^p dx\Big)^{(p-1)/p}\Big(\int_\Omega |u_n-u_0|^p dx\Big)^{1/p} \le\\ &\le c_{13}\|\alpha(x)\|_\infty\|u_n-u_0\|_p \to 0,\end{aligned} \tag{3.22}$$

and

$$
\begin{aligned}
|\langle F(u_n), u_n - u_0\rangle| = |\int_\Omega f(x,u_n)(u_n - u_0)dx| \le \\
\le \|\beta(x)\|_\infty \Big(\int_\Omega |u_n|^p dx\Big)^{(p-1)/p} \Big(\int_\Omega |u_n - u_0|^p dx\Big)^{1/p} + \\
+ \|\gamma(x)\|_\infty \int_\Omega |u_n - u_0| dx \le \\
\le c_{13}\|\beta(x)\|_\infty \|u_n - u_0\|_p + c_{14}\|\gamma(x)\|_\infty \|u_n - u_0\|_p \to 0
\end{aligned}
\tag{3.23}
$$

for $n \to \infty$. It follows from (4) - (4) that

$$
\begin{aligned}
0 \ge \limsup_{n\to\infty}\langle \tilde{T}(u_n), u_n - u_0\rangle = \\
= \limsup_{n\to\infty}\langle A(u_n) - \lambda B(u_n) - F(u_n), u_n - u_0\rangle = \\
= \limsup_{n\to\infty}\langle A(u_n), u_n - u_0\rangle = \limsup_{n\to\infty}\langle A(un) - A(u_0), u_n - u_0\rangle.
\end{aligned}
\tag{3.24}
$$

The continuity of the Nemytskii operator

$$G : (u, z_1, \cdots, z_1) \mapsto \tilde{a}(x,u)(z_1^2 + \cdots + z_n^2)^{(p-1)/2}$$

from $L^p(\Omega) \times L^p(w,\Omega) \times \cdots \times L^p(w,\Omega)$ into $L^{p/(p-1)}(w^{-1/(p-1)}, \Omega)$, the boundedness of u_n in $W_0^{1,p}(w,\Omega)$ and the Hölder inequality imply

$$
\begin{aligned}
\int_\Omega [\tilde{a}(x,u_n) - \tilde{a}(x,u_0)]|\nabla u_0|^{p-2}\nabla u_0(\nabla u_n - \nabla u_0)dx \le \\
\le \Big(\int_\Omega w^{-1/(p-1)}|\tilde{a}(x,u_n)|\nabla u_0|^{p-2}\nabla u_0 - \tilde{a}(x,u_0)|\nabla u_0|^{p-2}\nabla u_0|^{p/(p-1)}dx\Big)^{(p-1)/p} \cdot \\
\cdot \Big(\int_\Omega w|\nabla u_n - \nabla u_0|^p dx\Big)^{1/p} \le \\
\le c_{15}\|G(u_n, \nabla u_0) - G(u_0, \nabla u_0)\|_{L^{p/(p-1)}(w^{-1/(p-1)},\Omega)} \to 0
\end{aligned}
\tag{3.25}
$$

for $n \to \infty$. It follows from the Hölder inequality, (2), (4) and (4) that

$$
\begin{aligned}
0 \ge \limsup_{h\to\infty}\langle A(u_n) - A(u_0), u_n - u_0\rangle = \\
= \limsup_{n\to\infty}\Big\{\int_\Omega [\tilde{a}(x,u_n) - \tilde{a}(x,u_0)]|\nabla u_0|^{p-2}\nabla u_0(\nabla u_n - \nabla u_0)dx + \\
+ \int_\Omega \tilde{a}(x,u_n)[|\nabla u_n|^{p-2}\nabla u_n - |\nabla u_0|^{p-2}\nabla u_0](\nabla u_n - \nabla u_0)dx\Big\} \ge \\
\ge \frac{1}{c_1}\limsup_{n\to\infty}\int_\Omega w[|\nabla u_n|^{p-2}\nabla u_n - |\nabla u_0|^{p-2}\nabla u_0](\nabla u_n - \nabla u_0)dx \ge \\
\ge \frac{1}{c_1}\limsup_{n\to\infty}(\|u_n\|_w^{p-1} - \|u_0\|_w^{p-1})(\|u_n\|_w - \|u_0\|_w).
\end{aligned}
\tag{3.26}
$$

We obtain from (4) that

$$\| u_n\|_w \to \|u_0\|_w$$

for $n \to \infty$. The uniform convexity of $W_0^{1,p}(w,\Omega)$ then implies that u_n converge strongly to u_0 in $W_0^{1,p}(w,\Omega)$. Thus the condition $\alpha(W_0^{1,p}(w,\Omega))$ is satisfied.

Similarly as in (1) we prove that

$$\langle \tilde{T}(u), u\rangle > 0 \tag{3.27}$$

for any $u \in \partial B_\rho(0) = \{u \in W_0^{1,p}(w,\Omega); \|u\|_w = \rho\}$ with $\rho > 0$ sufficiently large (it is sufficient to realize that $\tilde{a}(x,s)$ satisfies the same bound from below as $a(x,s)$). Hence the degreedeg $[\tilde{T}; B_\rho(0), 0]$ is well defined (see SKRYPNIK [9]) and due to (3.27) we have

$$\deg\,[\tilde{T}; B_\rho(0), 0] = 1. \tag{3.28}$$

The basic property of the degree and (4) yield that there is $u \in W_0^{1,p}(w,\Omega)$, $\|u\|_w < \rho$, satisfying (3). Due to the discussions in Steps 2 and 3 we have $u \in L^\infty(\Omega)$ and it is the weak solution of the BVP (1.1).

Theorem 3.2 *Let us assume* (2) - (2), (3.1), *and, moreover, suppose that* $b(x,s) > 0$, $f(x,s) \geq 0$ *for a.e.* $x \in \Omega$ *and for all* $s \in \mathbb{R}$. *Then the* BVP (1.1) *has at least one weak solution* $u \in W_0^{1,p}(w,\Omega) \cap L^\infty(\Omega)$ *satisfying* $u \geq 0$ *a.e. in* Ω.

Proof It follows from Theorem 3.1 that the BVP (1.1) has at least one weak solution $u \in W_0^{1,p}(w,\Omega) \cap L^\infty(\Omega)$. Let us apply Lemma 2.3 where we put $\vartheta = \lambda$ and $h(x) = f(x, u(x))$ in (2.1). We obtain immediately $u(x) \geq 0$ for a.e. $x \in \Omega$.

■

4 Applications

Example 4.1 Let us consider the BVP

$$\begin{aligned} -\operatorname{div}(w(x)e^{u^2}|\nabla u|^2\nabla u) &= \frac{1}{c\pi^2}u^3\arctan^2(u) + \frac{1}{3c}u^3 \quad \text{in } \Omega, && (4.1a) \\ u &= 0 \quad \text{on } \partial\Omega. && (4.1b) \end{aligned}$$

where $w(x)$ is the weight function satisfying (2), c is the constant of the imbedding $W_0^{1,4}(w,\Omega) \hookrightarrow L^4(\Omega)$. In this case we have $p = 4$, $\lambda = 1/c\pi^{-2}$, $na(x,s) = w(x)e^{s^2}$, $b(x,s) = s^3\arctan^2(s)$, $f(x,s) = \frac{1}{3c}s^3$, $\alpha(x) = \frac{\pi^2}{4}$, $\beta(x) \equiv \frac{1}{3c}$, $\gamma(x) \equiv 0$. It is possible to show that the hypotheses of Theorem 3.1 are satisfied. Hence the BVP (4.1) has at least one weak solution $u \in W_0^{1,4}(w,\Omega) \cap L^\infty(\Omega)$.

□

Example 4.2 Let $\Omega = (0,\pi) \times (0,\pi)$ be the plane domain, $x = (x_1, x_2) \in \mathbb{R}^2$. Define

$$w(x) = \left|\frac{x_1}{\pi}\right|^\mu \left|\frac{x_2}{\pi}\right|^\nu, \quad \mu,\nu \in \mathbb{R},\ x \in \mathbb{R}^2,$$

and consider the BVP

$$-\operatorname{div}(w(x)(1+u^4)\nabla u) = (\frac{u}{2\pi}+1) \text{ arccotan } (u) \quad \text{in } \Omega, \tag{4.2a}$$

$$u = 0 \quad \text{on } \partial\Omega. \tag{4.2b}$$

In this case we have $n = p = 2$, $\lambda = 1$, $a(x,s) = w(x)(1+s^4)$, $b(x,s) = \frac{s}{2\pi}\text{arccotan}(s)$, $f(x,s) = \text{arccotan}(s)$, $\alpha(x) \equiv \frac{1}{2}$, $\beta(x) \equiv 0$, $\gamma(x) \equiv \pi$. We will assume

$$\mu < 0, \nu < 0. \tag{4.3}$$

Then (4.2) guarantees the validity of (2). Since $\lambda_1 = 1$ is the first eigenvalue of the homogeneous Dirichlet problem for the Laplace operator on $\Omega = (0,\pi) \times (0,\pi)$, the condition (4.2) implies

$$\int_0^\pi \int_0^\pi u^2 dx_1 dx_2 \leq \int_0^\pi \int_0^\pi |\nabla u|^2 dx_1 dx_2 \leq \int_0^\pi \int_0^\pi w(x_1,x_2)|\nabla u|^2 dx_1 dx_2$$

for any $u \in W_0^{1,2}(w,\Omega)$. Hence we can put $c = 1$ for the imbedding constant $W_0^{1,5}(w,\Omega) \hookrightarrow L^2(\Omega)$. Thus $\lambda^\# \geq 2$ and the assumptions of Theorem 3.2 are fulfilled. Hence the BVP (4.2) has at least one weak solution $u \in W_0^{1,2}(w,\Omega) \cap L^\infty(\Omega)$ such that $u \geq 0$ a.e. in Ω.

□

References

[1] R. A. Adams: *Sobolev Spaces*, Academic Press, Inc., New York 1975.

[2] F. E. Browder, W. V. Petryshin: *Approximation methods and the generalized topological degree for nonlinear mappings in Banach spaces*, J. Func. Analysis 3, (1969), 217–245.

[3] P. Drábek: *Solvability and Bifurcations of Nonlinear Equations*, Pitman Research Notes 232, Longman, Essex 1992.

[4] P. Drábek: *The least eigenvalues of nonhomogeneous degenerated quasilinear eigenvalue problems*, Preprint no. 34, University of West Bohemia, Pilsen 1992.

[5] J. Fleckinger, J. Hernández, F. de Thélin: *Principe du maximum pour un systéme elliptique non linéaire*, C. R. Acad. Sci. Paris, t. 314, Sér. I (1992), 665–668.

[6] A. Kufner, O. John, S. Fučík: *Function Spaces*, Academia, Prague 1977.

[7] A. Kufner, A. M. Sändig: *Some Applications of Weighted Sobolev Spaces*, Teubner, Band 100, Leipzig 1987.

[8] M. K. V. Murthy, G. Stampacchia: *Boundary value problems for some degenerate elliptic operators,* Annali di Matematica (4), 80 (1968), 1 – 122.

[9] I. V. Skrypnik: *Nonlinear Elliptic Boundary Value Problems* (Russian), Naukovaja Dumka, Kyjev 1973 (English translation: Teubner, Leipzig 1986).

Pavel Drábek
Department of Mathematics
University of West Bohemia
Americká 42
30614 Plzeň, Czech Republic
Fax: 0042-19-220019
e-mail:
pdrabek@scylla.zcu.cs

Some Recent Results on Hardy Type Operators in Weighted Function Spaces and Related Topics

M. Krbec, B. Opic, L. Pick, J. Rákosník

1 PRELIMINARIES

1.1 INTRODUCTION

What follows is a survey of recent results in several areas of the theory of function spaces which have been intensively studied in last years. In particular, we concentrate on the boundedness and compactness of the Hardy and imbedding operators and on the Poincaré and Friedrichs inequalities in weighted spaces, taking the general setting of Banach function spaces as a common framework. We do not attempt to present an exhaustive survey of the topics. Our aim is to show the current state of the subjects in question with an emphasis on our contribution to their solution.

The Hardy operator T,

$$Tf(x) = \int_0^x f(t)\,\mathrm{d}t, \quad x > 0,$$

treated in Section 2 and 3, has been a subject of study by generations of mathematicians since first fundamental results proved in the twenties and thirties. It is a powerful tool with many applications and importance in various branches of analysis. Its behaviour in Lebesgue spaces with power weights had been known for nearly half a century before a general theorem was proved by G. Tomaselli [T], B. Muckenhoupt [MU], J.S. Bradley [BR], V.G. Maz'ya [M] and V. Kokilashvili [KO]. This together with results for the Lebesgue spaces L^p with $p < 1$ has practically completed the theory for the scale of Lebesgue spaces. However, the theory in more general spaces, such as Lorentz and Orlicz spaces, remained mostly open until recently. We present some results, but observe that challenges still do exist in this area. This concerns also the spaces L^∞ and BMO. A unified approach is offered by the notion of Banach function spaces, introduced by W.A.J. Luxemburg [L] for Banach spaces of locally integrable functions with norms possessing, among others, certain monotonicity and Fatou properties. Banach function spaces cover most of classical function spaces except BMO. Some results for operators related to the Hardy operator like the integral average or the geometric mean are added.

Another basic and important operator affording a characterization of function spaces is the imbedding operator. In Chapter 4 imbeddings between couples of Banach function spaces are established. Then we modify the definition of Banach function spaces so that their elements possess generalized derivatives of all orders, and it is natural to define abstract Sobolev spaces as classes of functions whose generalized derivatives of corresponding orders belong to certain Banach function spaces. We establish bounded and compact imbeddings of abstract Sobolev spaces into Banach function spaces.

The last chapter is devoted to a rather general approach to inequalities of Poincaré and Friedrichs type which are in close relation to the imbedding operators and have many important applications. We present a series of conditions equivalent to the validity of these inequalities in abstract Sobolev spaces.

The paper consists only of a survey of results, brief comments and a few examples. In order to keep a reasonable length we omit all proofs and the reader is referred to corresponding original papers. The references are by no means intended as comprehensive as such a list would be longer than the survey itself. It should serve as a key to further literature cited there.

We would like to thank the colleagues from FSU Jena for organizing a very enjoyable and stimulating conference and for their warm hospitality.

1.2 BASIC NOTATION

Let Ω be a domain in $\mathbb{R}^N$ and let μ be a complete, totally σ-finite non-negative measure on Ω (cf. [HA]). If Q is a μ-measurable subset of Ω, χ_Q represents the characteristic function of Q. By $M(Q,\mu)$ we denote the family of all complex-valued, μ-measurable functions on Q. The symbol μ is omitted provided that μ is the Lebesgue N-measure.

If Q is a measurable subset of Ω and $f \in M(Q)$ is integrable, we write $f(Q)$ for $\int_Q f(x)\,\mathrm{d}x$, and f_Q for $f(Q)/|Q|$, where $|Q|$ is the Lebesgue N-measure of Q. As usual, ∂Q stands for the boundary of Q.

Throughout, v, w, ρ, σ are used for weights on Ω, that is, for measurable and a.e. on Ω positive and finite functions.

The letters C, K, etc., will represent various constants allowed to change their values from one occurrence to another, but independent of the appropriate quantities.

We adopt the following convention: $\infty^0 = 0^0 = 1$, $1/\infty = 0$, $1/0 = \infty$. For $0 < p \le \infty$ we define p' by $1/p + 1/p' = 1$.

Given two complex Banach spaces X and Y, we write $X \hookrightarrow Y$ or $X \hookrightarrow\hookrightarrow Y$ if $X \subset Y$ and the natural imbedding of X in Y is continuous or compact, respectively. Denote by $\mathcal{B}(X,Y)$ the space of all bounded linear maps from X to Y; the subspace of $\mathcal{B}(X,Y)$ consisting of all compact maps is denoted by $\mathcal{K}(X,Y)$. Further, let X^* be the dual space of X : $X^* = \mathcal{B}(X,\mathbb{C})$.

By $C^{0,1}$ we shall understand the collection of all bounded domains in $\mathbb{R}^N$ whose boundaries can be locally described by Lipschitz functions; the precise definition is given, for example, in [KJF, Section 5.5.6].

1.3 BANACH FUNCTION SPACES

In this section we shall recall the general notion of a Banach function space. We follow the exposition of W. A. J. Luxemburg [L].

Definition 1 *A normed linear space*

$$X = X(\Omega,\mu) = \big(X(\Omega,\mu),\ \|\cdot\|_{X(\Omega,\mu)}\big)$$

is called a Banach function space *if the following conditions are satisfied:*

(i) *The norm* $\|f\|_{X(\Omega,\mu)}$ *is defined for all* $f \in M(\Omega,\mu)$ *and* $f \in X(\Omega,\mu)$ *if, and only*

(ii) $\|f\|_{X(\Omega,\mu)} = 0$ *if, and only if,* $f(x) = 0$ μ*-a.e. on* Ω.

(iii) $\|f\|_{X(\Omega,\mu)} = \|\,|f|\,\|_{X(\Omega,\mu)}$ *for all* $f \in X(\Omega,\mu)$.

(iv) *Given* $Q \subset \Omega$ *with* $\mu(Q) < \infty$, *there exists a constant* C_Q *such that for all* $f \in X(\Omega,\mu)$,

$$\int_Q |f(x)|\,\mathrm{d}\mu \le C_Q \|f\|_{X(\Omega,\mu)}.$$

(v) *For every* $Q \subset \Omega$ *with* $\mu(Q) < \infty$, *we have* $\chi_Q \in X(\Omega,\mu)$.

(vi) *If* $f, g \in M(\Omega,\mu)$ *and* $0 \le f(x) \le g(x)$ μ*-a.e. on* Ω, *then*

$$\|f\|_{X(\Omega,\mu)} \le \|g\|_{X(\Omega,\mu)}.$$

(vii) *If* $\{f_n\} \subset M(\Omega,\mu)$ *and* $0 \le f_n(x) \nearrow f(x)$ μ*-a.e. on* Ω, *then*

$$\|f_n\|_{X(\Omega,\mu)} \nearrow \|f\|_{X(\Omega,\mu)}.$$

Example 1 *Recall that every Banach function space is a Banach space.*

Definition 2 *Let* $X(\Omega,\mu)$ *be a Banach function space. The space* $X'(\Omega,\mu)$ *given by*

$$X'(\Omega,\mu) = \Big\{ f \in M(\Omega,\mu) : \int_\Omega f(x) g(x)\,\mathrm{d}\mu < \infty \quad \text{for every } g \in X(\Omega,\mu) \Big\}$$

and endowed with the norm

$$\|f\|_{X'(\Omega,\mu)} = \sup \Big\{ \int_\Omega fg\,\mathrm{d}\mu : \|g\|_{X(\Omega,\mu)} \le 1 \Big\}$$

is called the associate space to $X(\Omega,\mu)$.

Let us note that $X'(\Omega,\mu)$ is again a Banach function space and that the *Hölder inequality*

$$\int_\Omega fg\,\mathrm{d}\mu \le \|f\|_{X(\Omega,\mu)} \|g\|_{X'(\Omega,\mu)}$$

holds. Moreover,

$$\|f\|_{X(\Omega,\mu)} = \sup \Big\{ \int_\Omega fg\,\mathrm{d}\mu : \|g\|_{X'(\Omega,\mu)} \le 1 \Big\}.$$

Definition 3 *An element* f *in a Banach function space* $X(\Omega,\mu)$ *is said to have* absolutely continuous norm *(we will write* $f \in ACN(X)$*) if* $\lim_{n\to\infty} \|f\chi_{Q_n}\|_{X(\Omega,\mu)} = 0$ *for every sequence* $\{Q_n\}_{n\in\mathbb{N}}$ *satisfying* $Q_n \searrow \emptyset$ μ*-a.e. (that is,* $\chi_{Q_n} \searrow 0$ μ*-a.e.; cf. [BS]).*

We now give some examples of Banach function spaces.

Example 2 (i) Weighted Lebesgue space $L^p(\Omega, w)$, *where* $0 < p \le \infty$ *and* w *is a weight on* Ω, *is defined by*

$$L^p(\Omega,w) = \Big\{ f \in M(\Omega) : \|f\|_{p,w} = \Big(\int_\Omega |f|^p w\,\mathrm{d}x \Big)^{1/p} < \infty \Big\}, \quad 0 < p < \infty,$$

$$L^\infty(\Omega,w) \equiv L^\infty(\Omega) = \Big\{ f \in M(\Omega) : \|f\|_\infty = \operatorname*{ess\,sup}_{x\in\Omega} |f(x)| < \infty \Big\}.$$

For all $1 \le p \le \infty$, $L^p(\Omega,w)$ *is a Banach function space.*

(ii) Weighted Orlicz space $L_\Phi(\Omega,w)$, *where* Φ *is a* Young function *(that is,* $\Phi\mathbb{R} \to [0,\infty)$ *is even and continuous, increasing and convex function on* $[0,\infty)$ *with* $\Phi(0)=0$, $\lim_{t\to 0}\Phi(t)/t = \lim_{t\to\infty} t/\Phi(t) = 0$*), is defined as*

$$L_\Phi(\Omega,w) = \Big\{ f \in M(\Omega) : \int_\Omega \Phi\big(\lambda^{-1} f(x)\big) w(x)\,\mathrm{d}x < \infty \quad \text{for some } \lambda > 0 \Big\}.$$

The Orlicz space is equipped with the Luxemburg norm

$$\|f\|_{\Phi,w} = \inf\Big\{ \lambda > 0 : \int_\Omega \Phi\big(\lambda^{-1} f(x)\big) w(x)\,\mathrm{d}x \le 1 \Big\},$$

and also with the (equivalent) Orlicz norm

$$|||f|||_{\Phi,w} = \sup \int_\Omega fgw\,\mathrm{d}x,$$

where the supremum is taken over all functions g *with property* $\int_\Omega \widetilde{\Phi}(g) w\,\mathrm{d}x \le 1$*; here* $\widetilde{\Phi}$ *is the so called* complementary Young function to Φ *and is given by*

$$\widetilde{\Phi}(t) = \sup_{s>0}\big(s|t| - \Phi(s) \big), \quad t \in \mathbb{R}.$$

Every weighted Orlicz space is a Banach function space for any choice of Φ.

The assumption $\lim_{t\to 0}\Phi(t)/t = \lim_{t\to\infty} t/\Phi(t) = 0$ *for the Young function is not always substantial. We refer to the cited papers for details.*

(iii) Weighted Lorentz space $L^{p,q}(\Omega,w)$, *where* $0 < p < \infty$, $0 < q \le \infty$, *is defined by*

$$L^{p,q}(\Omega,w) = \Big\{ f \in M(\Omega) : \|f\|_{p,q,w} = \Big(\int_0^\infty q\lambda^{q-1} w\big(\{x : |f(x)| > \lambda\}\big)^{q/p}\,\mathrm{d}\lambda \Big)^{1/q} < \infty \Big\}$$

for $0 < q < \infty$, *and*

$$L^{p,\infty}(\Omega,w) = \Big\{ f \in M(\Omega) : \|f\|_{p,\infty,w} = \sup_{\lambda>0} \lambda\, w\big(\{x : |f(x)| > \lambda\}\big)^{1/p} < \infty \Big\}.$$

Every Lorentz space is a Banach function space provided that $1 \le q \le p$. *In the case* $p > 1$, $1 \le q \le \infty$ *it becomes a Banach function space by replacing the quantity* $\|\cdot\|_{p,q,w}$ *with an equivalent functional which is a norm for all* $q \ge 1$ *(cf. [BS]).*

(iv) *An example of a space playing an important role in analysis which is not a Banach function space is the weighted* BMO *denoted by* $\mathrm{BMO}(\Omega,w)$ *and defined as the set of all* $f \in M(\Omega)$ *such that*

$$\|f\|_{\mathrm{BMO},w} = \sup_{Q\subset\Omega} \frac{1}{w(Q)} \int_Q |f(x) - f_Q|\, w(x)\,\mathrm{d}x < \infty;$$

here Q *stands for a cube in* $\mathbb{R}^N$ *with sides parallel to coordinate axes. Note that the quantity* $\|\cdot\|_{\mathrm{BMO},w}$ *is only a seminorm on* $\mathrm{BMO}(\Omega,w)$ *as* $\|f\|_{\mathrm{BMO},w} = 0$ *for every constant function on* Ω. *Moreover, it is not difficult to show that, for example, this seminorm violates the condition (iii) of Definition 1.*

Convention 1 *If* $X(\Omega, \mu)$ *is a Banach function space with a measure* μ *given by* $\mathrm{d}\mu = w(x)\,\mathrm{d}x$, *where* w *is a weight on* Ω, *we write* $X(\Omega, w)$ *instead of* $X(\Omega, \mu)$. *In Sections 2.1–3.4 we deal with spaces of functions defined on* $\Omega = (0, \infty)$ *and thus we omit the domain* Ω *in the notation of spaces, and for* $Q = (a, b) \subset \Omega = (0, \infty)$ *and* $f \in M(\Omega)$, *we write* $f(a, b)$ *instead of* $f(Q)$.

2 BOUNDEDNESS OF THE HARDY OPERATOR

Six sections of this chapter are devoted to the problem of boundedness of the Hardy operator T,

$$Tf(x) = \int_0^x f(t)\,\mathrm{d}t, \quad x > 0,$$

and to related questions. Thus, throughout the domain Ω is the interval $(0, \infty)$.

Theorem 1 in Section 2.1 presents the Bradley and Maz'ya type conditions in terms of parameters $q, r \in (0, \infty)$ and weights v, w, which are equivalent to the boundedness of the operator $T\, L^r(v) \to L^q(w)$. In other words, Theorem 1 describes weight pairs v, w, for which the operator T is of strong type (r, q). The characterization of weights v, w for which T is of weak type (r, q) follows from Theorem 4.

The situation is much more complicated for Orlicz spaces (Section 2.2). There are at least five types of inequalities (5)–(9). The norm inequality of the type (5) with generally different Young functions on both sides is still an open problem.

The case of Lorentz spaces was almost completely solved by E. Sawyer in Theorems 4 and 5. His condition (see Theorem 5), restricted to Lebesgue spaces, is not entirely analogous to that of Maz'ya. We point out in Theorem 6 however that they are equivalent and that the Maz'ya type condition can be extended to the case $T\, L^r(v) \to L^{p,q}(w)$ as well.

The results known so far suggest the natural question whether there is a unifying condition equivalent to the boundedness of T as a mapping between two Banach function spaces. A partial answer concerning the weak-type inequality is given in Section 2.4. The case of strong-type inequality remains open. Theorem 8 deals with the boundedness of the operator $T_{\rho,\sigma}\, L^{r,s}(v) \to L^{p,q}(w)$,

$$T_{\rho,\sigma} f(x) = \rho(x) \int_0^x f(x)\sigma(y)\,\mathrm{d}y,$$

where σ, ρ are weights. The operator $T_{\rho,\sigma}$ is a generalization of the Hardy operator T, which appears naturally by investigating T in weighted spaces.

Section 2.5 deals with two particular target spaces: L^∞ whose importance stems from the fact that it is a cornerstone of the L^p-scale, and weighted BMO which sometimes replaces L^∞ in the interpolation theory.

In Section 2.6 the operators of integral average and geometric means are treated.

2.1 LEBESGUE SPACES

The problem of boundedness of the Hardy operator T,

$$Tf(x) = \int_0^x f(t)\,\mathrm{d}t, \quad x > 0,$$

on a weighted Lebesgue space has been completely solved.

Theorem 1 *Let $0 < q, r < \infty$ and let v, w be weights. Then*

$$\|Tf\|_{q,w} \le C\|f\|_{r,v} \quad \text{for all } f \in L^r(v) \tag{1}$$

if, and only if,

$$B = \sup_{x>0} \Big(\int_x^\infty w(t)\,\mathrm{d}t\Big)^{1/q} \Big(\int_0^x v(t)^{1-r'}\,\mathrm{d}t\Big)^{1/r'} < \infty \tag{2}$$

(with usual interpretation for $r = 1$) provided that $1 \le r \le q$, and if, and only if,

$$A = \Big[\int_0^\infty \Big(\int_x^\infty w(t)\,\mathrm{d}t\Big)^{\tau/q} \Big(\int_0^x v(t)^{1-r'}\,\mathrm{d}t\Big)^{\tau/q'} v(x)^{1-r'}\,\mathrm{d}x\Big]^{1/\tau} < \infty \tag{3}$$

(here $1/\tau = 1/q - 1/r$) provided that $q < r$ and $r > 1$.

This theorem was proved by G. Tomaselli [T], B. Muckenhoupt [MU] for $r = q$, by J.S. Bradley [BR], V.G. Maz'ya [M], V.M. Kokilashvili [KO] for $r \le q$, by V.G. Maz'ya [M] for $1 \le q < r$, and by G. Sinnamon [SI] for $0 < q < 1 < r$.

Several ways of generalization of Theorem 1 are of interest. We shall mention at least two of them:

(a) to extend the result to more general function spaces;

(b) to consider more general Hardy type operators.

2.2 ORLICZ SPACES

When dealing with weighted inequalities for integral operators in Orlicz spaces, the most difficulties we meet come from the fact that Young functions unlike powers are not homogeneous. This leads, for example, to non-equivalence of modular problems of type

$$\int_0^\infty \Phi\big(Tf(x)\big)w(x)\,\mathrm{d}x \le \int_0^\infty \Phi\big(Cf(x)\big)v(x)\,\mathrm{d}x \tag{4}$$

and norm problems

$$\|Tf\|_{\Phi,w} \le C\|f\|_{\Phi,v}. \tag{5}$$

It is easy to show that (4) always implies (5). Also it turns out that modular problems are much more easier to handle than their norm counterparts. For example, we can prove the following

Theorem 2 ([P1]) *Let Φ_1, Φ_2 be Young functions and let v, w be weights. Then the following statements are equivalent:*

(i) *There exists a constant C such that for all $f \in L_{\Phi_1}(v)$,*

$$\int_0^\infty \Phi_2\big(Tf(x)\big)w(x)\,\mathrm{d}x \le \int_0^\infty \Phi_1\big(Cf(x)\big)v(x)\,\mathrm{d}x. \tag{6}$$

(ii) *There exists a constant K such that for any $\lambda > 0$ and all $f \in L_{\Phi_1}(v)$,*

$$\Phi_2(\lambda)\,w\big(\{x : |Tf(x)| > \lambda\}\big) \le \int_0^\infty \Phi_1\big(Kf(x)\big)v(x)\,\mathrm{d}x. \tag{7}$$

(iii) *There exists a constant D such that*

$$\sup_{\alpha,x>0} \alpha\, w(x,\infty)\, R_{\Phi_2}\Big(D^{-1}\int_0^x S_{\Phi_1}\Big(\frac{1}{\alpha v(t)}\Big)\,\mathrm{d}t\Big) \le D,$$

where $R_{\Phi_2}(t)=\Phi_2(t)/t$, $S_{\Phi_1}(t)=\widetilde{\Phi}_1(t)/t$.

Recall that (7) is called weak-type inequality. It is a surprising property of the Hardy operator that (6), an inequality of strong type, is equivalent to the weak inequality (7).

In some sense more general inequalities,

$$\Phi_2^{-1}\Big(\int_0^\infty \Phi_2\big(\rho(x)Tf(x)\big)w(x)\,\mathrm{d}x\Big) \le \Phi_1^{-1}\Big(\int_0^\infty \Phi_1\big(C\sigma(x)f(x)\big)v(x)\,\mathrm{d}x\Big), \tag{8}$$

where ρ, w, σ, v are four possibly different weights, are subject of the recent preprint of S. Bloom and R. Kerman [BK2].

On the other hand, we have a characterization for the modified inequality of the type (7):

Theorem 3 ([P1]) *Let Φ be Young function and let v, w be weights. Then the inequality*

$$w\big(\{x: |Tf(x)|>\lambda\}\big) \le \int_0^\infty \Phi\big(C\lambda^{-1}|f(x)|\big)v(x)\,\mathrm{d}x \tag{9}$$

holds with a constant C independent of $f\in L_\Phi(v)$ and $\lambda>0$ if, and only if,

$$\sup_{x>0}\int_0^x S_\Phi\Big(\frac{w(x,\infty)}{Dv(t)}\Big)\,\mathrm{d}t \le D. \tag{10}$$

2.3 LORENTZ SPACES

A crucial step towards boundedness results for the Hardy operator in Lorentz spaces was done by E. Sawyer [S]. We shall recall some of his results.

Theorem 4 *Let $0<p<\infty$, $0<q\le\infty$, $1<r<\infty$, $1\le s\le\infty$, and let v, w be weights. Assume that*

$$\|Tf\|_{p,q,w} \le C\|f\|_{r,s,v} \quad \text{for all } f\in L^{r,s}(v). \tag{11}$$

Then

$$\sup_{x>0}\|\chi_{(x,\infty)}\|_{p,q,w}\,\Big\|\frac{\chi_{(0,x)}}{v}\Big\|_{r',s',v} < \infty. \tag{12}$$

Conversely, (12) *implies* (11) *if, and only if,* $q\ge\max\{r,s\}$.

Theorem 5 *Let $0<p<\infty$, $1\le r<\infty$, $0<q<r$, $1/\tau=1/q-1/r$ and let v, w be weights. Then*

$$\|Tf\|_{p,q,w} \le C\|f\|_{r,v} \quad \text{for all } f\in L^r(v) \tag{13}$$

if, and only if,

$$\sup_{\{x_k\}}\Big\{\sum_k\Big(\int_{x_k}^{x_{k+1}} w(x,\infty)^{(q/p)-1}w(x)\,\mathrm{d}x\Big)^{\tau/q}\Big(\int_{x_{k-1}}^{x_k} v(x)^{1-r'}\,\mathrm{d}x\Big)^{\tau/r'}\Big\}^{1/\tau} < \infty, \tag{14}$$

where the supremum is taken over all positive increasing sequences $\{x_k\}$.

Theorem th62 characterizes when the Hardy operator T is bounded from a Lebesgue space into a Lorentz space. This Sawyer's result leaves open a question, whether there is a characterizing condition for (13) similar to that of (3). This question seems to be quite natural and the following recent observation of B. Opic and L. Pick (not yet published) shows that the answer is affirmative.

Theorem 6 *Let* $0 < p < \infty$, $1 < r < \infty$, $0 < q < r$, $1/\tau = 1/q - 1/r$ *and let* v, w *be weights. Then* (13) *holds if, and only if,*

$$\Big\{\int_0^\infty \Big(\int_x^\infty w(t)\,\mathrm{d}t\Big)^{\tau/p}\Big(\int_0^x v(t)^{1-r'}\,\mathrm{d}t\Big)^{\tau/q'} v(x)^{1-r'}\,\mathrm{d}x\Big\}^{1/\tau} < \infty. \tag{15}$$

2.4 GENERAL REMARKS TO BOUNDEDNESS

Theorems 1, 2 and 4 suggest that there might be a general condition equivalent to the boundedness of the operator T between two weighted Banach function spaces $X(v)$ and $Y(w)$. However, there is no such known criterion in such a general context. On the other hand, there are some partial results of independent interest.

First, it turns out that the Bradley type condition (2) is always equivalent to the weak-type boundedness.

Theorem 7 ([B]) *Suppose that* v, w *are weights and* $X(v)$, $Y(w)$ *two Banach function spaces. Then the Hardy operator* T *satisfies the weak-type inequality*

$$\sup_{\lambda>0} \lambda\,\|\chi_{\{x:|Tf(x)|>\lambda\}}\|_{Y(w)} \le C\|f\|_{X(v)} \tag{16}$$

for all $f \in X(v)$ *if, and only if,*

$$\sup_{x>0} \|\chi_{(x,\infty)}\|_{Y(w)} \left\|\frac{\chi_{(0,x)}}{v}\right\|_{X'(v)} < \infty. \tag{17}$$

Since (16) is obviously weaker than the strong boundedness

$$\|Tf\|_{Y(w)} \le C\|f\|_{X(v)} \quad \text{for all } f \in X(v), \tag{18}$$

we obtain

Corollary 1 *For* (18) *it is always necessary that* (17) *holds.*

However, we know that in certain cases (cf. Theorems 1 and 4) the condition (17) is also sufficient. Let us recall that:

(i) If $X(v) = L^r(v)$ and $Y(w) = L^q(w)$ with $1 \le r, q < \infty$, then (17) is sufficient for (18) if, and only if, $r \le q$.

(ii) If $X(v) = L^{r,s}(v)$ and $Y(w) = L^{p,q}(w)$ with $0 < p < \infty$, $0 < q \le \infty$, $1 < r < \infty$, $1 \le s \le \infty$, then (17) is sufficient for (18) if, and only if, $q \ge \max\{r, s\}$.

The last mentioned result will be of particular interest when we switch to more general integral operators

$$T_{\rho,\sigma}f(x) = \rho(x)\int_0^x f(y)\sigma(y)\,\mathrm{d}y, \tag{19}$$

where ρ, σ are other weights. For such operators, an analogue of Theorem 7 need not be true in the sense that the analogue of the condition (17), that is,

$$\sup_{x>0} \|\rho\chi_{(x,\infty)}\|_{Y(w)} \left\| \frac{\sigma\chi_{(0,x)}}{v} \right\|_{X'(v)} \leq C, \tag{20}$$

need not be necessary even for weak-type boundedness analogous to (16). To see this, take $X(v) = Y(w) = L^1$, $v = w = \sigma = 1$, and $\rho(x) = 1/x$. Obviously, within frames of Lebesgue spaces, replacing of T by $T_{\rho,\sigma}$ does not play any role since we can just combine the weights. However, this is not true, for example, for Lorentz spaces, and it is interesting that the result mentioned above as (ii) is no longer valid. Details can be found in a forthcoming paper by D. E. Edmunds, P. Gurka and L. Pick, from which the following theorem is taken.

Theorem 8 *Let $1 < p, r < \infty$, $1 \leq q, s \leq \infty$, $X(v) = L^{r,s}(v)$, $Y(w) = L^{p,q}(w)$. Then* (20) *is sufficient for the inequality*

$$\|T_{\rho,\sigma} f\|_{p,q,w} \leq C\|f\|_{r,s,v} \tag{21}$$

to hold on $L^{r,s}(v)$ if

$$\max\{r, s\} \leq \min\{p, q\}. \tag{22}$$

Conversely, if (22) *is not valid, then there exist such ρ and σ that* (20) *is true and* (21) *is not.*

Theorem 8 shows that not all operators $T_{\rho,\sigma}$ preserve boundedness properties of T.

2.5 SPACES L^∞ AND BMO(w)

Spaces BMO(w) appear as a natural replacement of L^∞ in the interpolation theory. Obviously, $L^\infty \subset$ BMO where BMO := BMO(w) with $w \equiv 1$. We shall mention some results of [LP] where the Hardy operator is investigated as an operator with target in L^∞ or BMO(w).

Theorem 9 *Let $X(v)$ be a weighted Banach function space. Then*

$$\|Tf\|_\infty \leq C\|f\|_{X(v)} \quad \text{for all } f \in X(v) \tag{23}$$

if, and only if,

$$\frac{1}{v} \in X'(v). \tag{24}$$

A characterization of the analogue of (23) with L^∞ replaced by BMO(w) is much more difficult. The key is in finding a suitable replacement for the function $1/v$.

Define for $[a, b] \subset (0, \infty)$

$$G_{ab}(x) = \frac{w(a,x)\, w(x,b)}{(b-a)\, w(a,b)\, v(x)} \chi_{(a,b)}(x). \tag{25}$$

Using monotonicity of Tf for $f \geq 0$, we can prove a nice representation result.

Lemma 1 *Let $f \geq 0$. Then $\|Tf\|_{\text{BMO},w}$ is equivalent to*

$$\sup_{[a,b]\subset(0,\infty)} \int_a^b G_{ab}(x) f(x) w(x)\, dx. \tag{26}$$

The representation formula (26) yields

Theorem 10 *The following statements are equivalent:*

(i) *There exists $K > 0$ such that for all $f \in X(v)$,*

$$\|Tf\|_{\mathrm{BMO},w} \leq K\|f\|_{X(v)}. \tag{27}$$

(ii) *There exists $C > 0$ such that*

$$\sup_{[a,b]\subset(0,\infty)} \|G_{ab}\|_{X'(v)} \leq C. \tag{28}$$

Moreover, if K and C are the best constants in (27) and (28), respectively, then $C \leq K \leq 4C$.

It is of interest to discuss the case when $w \equiv 1$. Then G_{ab} turns to

$$H_{ab}(x) = \frac{(b-x)(x-a)}{(b-a)^2}\frac{\chi_{(a,b)}(x)}{v(x)}, \tag{29}$$

and it is quite easy to prove that

$$H_{ab}(x) \leq \frac{\chi_{(a,b)}(x)}{4v(x)}. \tag{30}$$

Hence, (24) implies (28) with $w \equiv 1$, and in turn (23) implies (27) with $w \equiv 1$, which just reflects the fact that $L^\infty \subset \mathrm{BMO}$.

We shall demonstrate with two examples that in certain cases of $X(v)$ it can happen that (23) is equivalent to (27) with $w \equiv 1$, but that this equivalence is false in general. Our first example works with $X(v) = L^1(v)$.

Example 3 *The following statements are equivalent:*

(i) *There exists $K > 0$ such that*

$$\|Tf\|_{\mathrm{BMO},w} \leq K\|f\|_{1,v} \quad \textit{for all } f \in L^1(v).$$

(ii) *There exists $C > 0$ such that*

$$\|Tf\|_\infty \leq C\|f\|_{1,v} \quad \textit{for all } f \in L^1(v).$$

(iii) *The function $1/v$ belongs to L^∞.*

The second example uses $X(v) = L^p(v)$ with $p > 1$.

Example 4 *Let $w \equiv 1$, $v(x) = x^{p-1}$ for $x > 0$ and let $X(v) = L^p(v)$, $p > 1$. Then, $1/v \notin X'(v) = L^{p'}(v)$. On the other hand,*

$$\|H_{ab}\|_{p',v} \leq \left[B(p', p'+1)\right]^{1/p'} < \infty,$$

where B is the Beta function. Therefore, (27) with $w = 1$ holds, but (23) is not true.

2.6 BOUNDEDNESS OF RELATED OPERATORS

One of the most important and also most interesting operators is the averaging operator

$$Af(x) = \frac{1}{x}\int_0^x f(t)\,\mathrm{d}t, \quad x > 0, \tag{31}$$

and the geometric mean operator

$$Gf(x) = \exp\Big(\frac{1}{x}\int_0^x \log f(t)\,\mathrm{d}t\Big), \quad f > 0 \text{ a.e. on } \mathbb{R}_+, \tag{32}$$

which can be considered as a limit of

$$A_p f(x) = \Big(\frac{1}{x}\int_0^x f(t)^{1/p}\,\mathrm{d}t\Big)^p, \quad x > 0,$$

(see [HLP]).

As already mentioned in Section 2.4, it is easy to get $L^p(v) - L^q(w)$ boundedness results for the operator A by means of combining weights and using boundedness results for the Hardy operator T. However, there are unpleasant obstacles when switching from Lebesgue to Orlicz or Lorentz spaces.

As to the geometric mean operator, Knopp's result [KN] states that $G\,L^1 \to L^1$ is bounded. (This can be also obtained by taking limits of both sides of the Hardy inequality. The behaviour of constants on the right hand side, however, is not known in the general (weighted) case so that another approach has to be used (cf. [H])). Full characterization of weights v, w for which $G\,L^1(v) \to L^1(w)$ is bounded is due to H. P. Heinig, R. Kerman and M. Krbec (cf. [H]). Finally, $L^p(v) - L^q(w)$ boundedness of G was characterized in [OG] (for $0 < p \le q < \infty$) and then in [PO] (for $0 < q < p \le \infty$) and independently in [HKK] (for $0 < p \le q < \infty$ and for $0 < q < p \le \infty$, $p > 1$). We mention here some results from [OG], [PO] and [HKK].

Theorem 11 *Assume that $0 < p, q \le \infty$ and that v, w are weights. Let*

$$\|Gf\|_{q,w} \le C\|f\|_{p,v} \quad \textit{for all } u \in L^p(v). \tag{33}$$

Then for every $\alpha > 1$

$$B(\alpha) = \sup_{x>0} x^{(\alpha-1)/p}\Big\|\chi_{(x,\infty)}(s)\,u(s)^{1/q}\Big(\frac{G(1/v)(s)}{s^\alpha}\Big)^{1/p}\Big\|_q < \infty. \tag{34}$$

If $0 < p \le q \le \infty$, and (34) *is true for some $\alpha > 1$, then* (33) *holds, and*

$$\sup_{\alpha>1}\Big(\frac{\alpha-1}{\alpha}\Big)^{-1/p} B(\alpha) \le \|G\| \le \inf_{\alpha>1}\exp\Big(\frac{\alpha-1}{q}\Big)B(\alpha).$$

Remark 1 *In [OG] also some other characterization of* (33) *can be found. More general operators that* (31) *and* (32) *have been considered in [HKK].*

Theorem 11 left open the question of characterizing (33) for $q < p$. Here the crucial role is played by a surprising observation that the boundedness of A is equivalent to the boundedness of G provided that $p > 1$ and the source space is non-weighted. This is an unexpected fact since G is bounded on L^1 (see [KN]) while A is not.

More precisely, we have

Lemma 2 *Let $1 < p < \infty$, $0 < q < \infty$, and let w be a weight. Then $A\ L^p \to L^q(w)$ is bounded if, and only if, $G\ L^p \to L^q(w)$ is bounded.*

The result for $q < p$ reads as follows:

Theorem 12 *Let $0 < q < p \le \infty$, let v, w be weights and define the weight ρ by $\rho(x) = w(x)\big[G(1/v)(x)\big]^{q/p}$, $x > 0$. Then the following statements are equivalent:*
(i) *$G\ L^p(v) \to L^q(w)$ is bounded.*
(ii) *There exists $s > 1/p$ such that $G\ L^{p,s} \to L^{q,s}(\rho)$ is bounded.*
(iii) *$A \in \mathcal{B}\big(L^{p,s}, L^{q,s}(\rho)\big)$ for some $s > 1/p$.*
(iv) *There exists $s > 1/p$ such that*

$$\int_0^\infty x^{rs/(sq)'}\Big(\int_x^\infty \frac{\rho(t)}{t^{sq}}\,\mathrm{d}t\Big)^{r/q}\,\mathrm{d}x < \infty,$$

where $1/r = 1/q - 1/p$.

This theorem has the following corollary of independent interest.

Theorem 13 *Let $0 < p, q < \infty$ and let w be a weight. Then*

$$A \in \mathcal{B}\big(L^\alpha, L^{\alpha q/p}(w)\big)$$

either for every $\alpha > 1$, or for no $\alpha > 1$.

Remark 2 *In other words, given $0 < q < \infty$, $1 < p < \infty$, and a weight w, then the membership $A \in \mathcal{B}\big(L^p, L^q(w)\big)$ depends only on q/p, and not on p, q separately. This is not obvious and perhaps not known.*

Example 5 *If $0 < p \le q < \infty$ and $w(x) = x^\delta$, $v(x) = x^\gamma$ for $x > 0$ where $\delta, \gamma \in \mathbb{R}$ are such that $q(1+\gamma) = p(1+\delta)$, then (9.3) holds. Observe that here the result again depends only on q/p.*

3 COMPACTNESS OF THE HARDY OPERATOR

In this chapter we present some results on compactness of Hardy type operators in weighted spaces. Again the most simple and thoroughly solved is the case of Lebesgue spaces — see Section 3.1. Section 3.2 contains the recent and not yet published result by D. E. Edmunds, P. Gurka and L. Pick characterizing the compactness of the operator $T_{\rho,\sigma}\ L^{r,s}(v) \to L^{p,q}(w)$ for $1 < p, q, r, s < \infty$ with $\max\{r, s\} \le \min\{p, q\}$. Theorems 17 and 18 describe Banach function spaces $X(v)$ such that $T \in \mathcal{K}(X(v), L^\infty)$ and $T \in \mathcal{K}(X(v), \mathrm{BMO})$, respectively. Contrary to Section 2.5, the space BMO is unweighted here.

When dealing with a non-compact operator it is often useful to know "how much" non-compact the operator is. This can be expressed by means of the distance of the operator from the class of compact operators. In Section 3.4 the problem is discussed for a generalization of T which covers also the Riemann-Louville fractional operators of orders equal or greater then 1.

As in Chapter 2, the domain Ω is always the interval $(0, \infty)$.

3.1 LEBESGUE SPACES

Let us recall the "classical results" on compactness of the Hardy operator T (see [OK]).

Theorem 14 *Let $1 < r \leq q < \infty$ and let v, w be weights. Then $T \in \mathcal{K}(L^r(v), L^q(w))$ if, and only if, the number B from (2) is finite, and*

$$\lim_{x \to 0_+} F(x) = \lim_{x \to \infty} F(x) = 0, \tag{35}$$

where

$$F(x) = \Big(\int_x^\infty w(t)\,\mathrm{d}t \Big)^{1/q} \Big(\int_0^x v(t)^{1-r'}\,\mathrm{d}t \Big)^{1/r'} = \|\chi_{(x,\infty)}\|_{q,w} \left\| \frac{\chi_{(0,x)}}{v} \right\|_{r',v}. \tag{36}$$

Comparing Theorems 1 and 14, we see that in the case $1 < r \leq q < \infty$ there is an essential difference between conditions characterizing boundedness and compactness of T. This difference is expressed in demanding additionally the condition (35) in the case of compactness. The following theorem shows that there is no difference between boundedness and compactness of T in the case $1 < q < r < \infty$.

Theorem 15 *Assume that $1 < q < r < \infty$ and let v, w be weights. Then $T \in \mathcal{K}(L^r(v), L^q(w))$ if, and only if, the condition (3) holds.*

3.2 LORENTZ SPACES

Let us consider the operator $T_{\rho,\sigma}$ given by (19). The following theorem, recently proved by D. E. Edmunds, P. Gurka and L. Pick, has not yet been published.

Theorem 16 *Let $p, q, r, s \in (1, \infty)$ and let v, w be weights. Suppose that*

$$\max\{r, s\} \leq \min\{p, q\}.$$

Then $T_{\rho,\sigma} \in \mathcal{K}\big(L^{r,s}(v), L^{p,q}(w)\big)$ if, and only if,

$$\sup_{R>0} A(R) = \sup_{R>0} \|\rho\chi_{(R,\infty)}\|_{p,q,w} \left\| \frac{\sigma}{v}\chi_{(0,R)} \right\|_{r',s',v} < \infty$$

and

$$\lim_{R \to 0_+} A(R) = \lim_{R \to \infty} A(R) = 0.$$

3.3 COMPACTNESS RESULTS IN L^∞ AND BMO

The results of Section 3.1 (and 3.2), in particular the comparison of Theorem 14 to Theorem 1 suggests that the difference between boundedness conditions and those for compactness should be expressed in a certain sense by means of continuity of norm of proper weights in proper spaces (cf. (35) and (36)).

For this reason we introduce the definition of continuous norm of an element f from a Banach function space $X(v)$.

Definition 4 *Let v be a weight and let $X(v)$ be the corresponding Banach function space. We will say that $f \in X(v)$ has a* continuous norm *(and we will write $f \in \mathrm{CN}(X(v))$) if for any $a \in [0, \infty)$, $b \in (0, \infty]$ and for any sequences $\{x_n\}$, $\{y_n\} \subset \mathbb{R}$ such that either $x_n \searrow a$ or $y_n \nearrow b$ we have*

$$\lim_{n\to\infty} \|f\chi_{(a,x_n)}\|_{X(v)} = \lim_{n\to\infty} \|f\chi_{(y_n,b)}\|_{X(v)} = 0.$$

Remark 3 *Definition 4 is taken from [LP]. Obviously,* $\mathrm{ACN}(X(v))$ *(see Definition 3) is a subset of* $\mathrm{CN}(X(v))$.

Theorem 17 ([LP]) *The following statements are equivalent:*

(i) $T \in \mathcal{K}(X(v), L^\infty)$;

(ii) $1/v \in \mathrm{CN}(X'(v))$.

As usual (see Section 2.6), turning to BMO we find much more difficult situation.

Theorem 18 ([LP]) *The operator T is compact from $X(v)$ into* BMO *if, and only if,*

(i) $\sup_{[a,b]\subset(0,\infty)} \|H_{ab}\|_{X'(v)} < \infty$;

(ii) $\chi_{(r,R)}/v \in \mathrm{CN}(X'(v))$ *for every* $0 < r < R < \infty$;

(iii) $\lim_{n\to\infty} \sup_{[a,b]\subset(0,\infty)} \|H_{ab}\,\chi_{I_n}\|_{X'(v)} = 0$ *whenever* $I_n = (\alpha_n, \beta_n)$ *and either* $\alpha_n \nearrow \infty$ *or* $\beta_n \searrow 0$.

3.4 DISTANCE FROM COMPACT OPERATORS

Let $1 < p \le q < \infty$ and let v, w be weights. In this section we consider a *generalized Hardy operator*

$$T_\varphi f(x) = \int_0^x \varphi(x,y) f(y)\,\mathrm{d}y, \qquad x > 0, \tag{37}$$

acting between two weighted spaces $L^p(v)$, $L^q(w)$. Observe that $T_{\rho,\sigma}$ in (19) is a special case of T_φ. Our aim is to present lower and upper bounds on the distance of T_φ from the space of all compact linear operators $P\ L^p(v) \to L^q(w)$. The results are taken from [O].

Several authors (cf. [MS], [ST1], [BK1]) have established necessary and sufficient conditions for the inequality

$$\|T_\varphi f\|_{q,w} \le C\|f\|_{p,v} \tag{38}$$

to hold on $L^p(v)$, where the kernel of the operator T_φ satisfies various assumptions. Here we consider the kernel φ with the following properties (cf. [BK1]):

$$\begin{array}{ll} \text{(i)} & \varphi\ \{(x,y)\in\mathbb{R}^2 : 0<y<x<\infty\} \to (0,\infty). \\ \text{(ii)} & \varphi(x,y) \text{ is non-decreasing in } x \text{ and non-increasing in } y. \\ \text{(iii)} & \text{There exists } D>0 \text{ such that } \varphi(x,y) \le D[\varphi(x,z)+\varphi(z,y)] \text{ if } y<z<x. \end{array} \tag{39}$$

The Riemann–Liouville operator T_s, given by

$$T_s f(x) = \int_0^x (x-y)^{s-1} f(y)\,\mathrm{d}y, \qquad x>0,\ s\geq 1, \tag{40}$$

is an example of an operator T_φ whose kernel satisfies (39–i)–(39–iii). In particular, taking $s=1$ in (40), we obtain the Hardy operator from preceding sections. Note that in all papers mentioned above the kernel φ is such that conditions (39–i)–(39–iii) are fulfilled.

Let $0\leq\alpha<\beta\leq\infty$, $t\in(\alpha,\beta)$, $i\in\{0,1\}$. Set

$$B^{(i)}_{(\alpha,\beta)}(t) = \Big(\int_t^\beta \varphi(x,t)^{qi} w(x)\,\mathrm{d}x\Big)^{1/q}\Big(\int_\alpha^t \varphi(t,x)^{p'(1-i)} v(x)^{1-p'}\,\mathrm{d}x\Big)^{1/p'},$$

$$B^{(i)}_{(\alpha,\beta)} = \sup_{\alpha<t<\beta} B^{(i)}_{(\alpha,\beta)}(t), \quad B_{(\alpha,\beta)} = \max_{i=0,1} B^{(i)}_{(\alpha,\beta)}.$$

Applying the method of [MS] one can prove

Theorem 19 *Let $1<p\leq q<\infty$ and let the function φ satisfy the conditions (39–i)–(39–iii). Then there exists a positive constant C such that the inequality (38) holds for all $f\in L^p(v)$ if, and only if, $B_{(0,\infty)}<\infty$.*

Moreover, if C is the least constant for which (38) holds on $L^p(v)$, then

$$B_{(0,\infty)} \leq C \leq k(q,p,D)\, B_{(0,\infty)},$$

where

$$k(q,p,D) = 2^{1/q'}(D+1)^2\Big[D^q\Big(1+\frac{q}{p'}\Big)\Big(1+\frac{p'}{q}\Big)^{q/p'}+1\Big]^{1/q} \tag{41}$$

with D from (39–iii).

In other words, Theorem 19 provides a necessary and sufficient condition for the relation $T_\varphi\in\mathcal{B}(L^p(v),L^q(w))$. In the following theorem we establish lower and upper bounds on the distance of T_φ from $\mathcal{K}(L^p(v),L^q(w))$ provided $T_\varphi\in\mathcal{B}(L^p(v),L^q(w))$ and the function φ, in addition to conditions (39–i)–(39–iii), satisfies

(39–iv) For any $\gamma\in(0,\infty)$ the function φ is uniformly continuous on the set
$$\{(x,y)\in\mathbb{R}^2 : 0<y<x<\gamma\}$$

(Note that for the Riemann–Liouville operator T_s from (40) the condition (39–iv) is again fulfilled.) In particular, from such bounds one easily obtains a characterization of those operators T_φ that are compact.

Theorem 20 *Suppose that $1<p\leq q<\infty$, $X=L^p(v)$, $Y=L^q(w)$. Let T_φ be the operator (37) with the kernel φ satisfying (39–i)–(39–iv). Moreover, let $B_{(0,\infty)}<\infty$. Then*

$$\frac{1}{2}J \leq \operatorname{dist}(T_\varphi,\mathcal{K}(X,Y)) \leq k(q,p,D)J,$$

where

$$J = \lim_{\alpha\to 0_+} B_{(0,\alpha)} + \lim_{\beta\to\infty} B_{(\beta,\infty)},$$

and $k(q,p,D)$ is given by (41).

Corollary 2 *Let all assumptions of Theorem 20 be fulfilled. Then*

$$T_\varphi \in \mathcal{K}(X,Y) \Leftrightarrow J = 0.$$

Remark 4 *If we suppose that*

$$\lim_{r\to 0_+} B^{(i)}_{(0,\infty)}(r) = 0 = \lim_{r\to\infty} B^{(i)}_{(0,\infty)}(r), \qquad i = 0,1, \tag{42}$$

then it is easy to show that $J = 0$. Thus, the assumptions of Theorem 20 and (42) *imply that $T_\varphi \in \mathcal{K}(X,Y)$.*

On the other hand, if, for example, the function φ satisfying (39–i)–(39–iii) *is bounded, then one can prove that $J = 0$ implies* (42) *(cf. the proof of Corollary (ii) from [J]).*

Moreover, let us note that, using the arguments from [ST2], it is possible to prove the implication $T_\varphi \in \mathcal{K}(X,Y) \Rightarrow$ (42) *provided that the function φ has the properties* (39–i)–(39–iii).

Remark 4 and Corollary 2 imply

Corollary 3 *Suppose that all the assumptions of Theorem 20 are fulfilled. Then $T_\varphi \in \mathcal{K}(X,Y)$ if, and only if,* (42) *is true.*

V. D. Stepanov proved the statement of Corollary 3 under assumption $\varphi(x,y) = k(x-y)$ with a suitable function k (cf. [ST1]).

4 IMBEDDINGS

Our aim in this chapter is to present conditions which guarantee continuous or compact imbeddings of given function spaces. We start with imbeddings of various Banach function spaces in Section 4.1. The most general situation, the imbedding $X(\Omega,v) \hookrightarrow Y(\Omega,w)$, is the subject of Theorem 24.

In Section 4.2 we turn to the imbeddings of abstract Sobolev spaces defined on the basis of Banach function spaces. For this purpose we have to modify slightly the notion of Banach function spaces (see Definition 5). The idea of the imbedding theorem is based on the assumption of an existence of local imbeddings and on the control of the norms in narrow strips along the boundary.

4.1 IMBEDDINGS BETWEEN BANACH FUNCTION SPACES

In this section we assume that all functions are defined on a domain Ω in $\mathbb{R}^N$. We suppose that v, w are weights on Ω, $X(\Omega,v)$, $Y(\Omega,w)$ are Banach function spaces and we study the continuous imbedding

$$X(\Omega,v) \hookrightarrow Y(\Omega,w), \tag{43}$$

that is, the weighted inequality of the form

$$\|f\|_{Y(\Omega,w)} \le C\|f\|_{X(\Omega,v)} \tag{44}$$

with C independent of $f \in X(\Omega,v)$.

In the case $X(\Omega,v) = L^p(\Omega,v)$, $Y(\Omega,w) = L^q(\Omega,w)$, $0 < p,q < \infty$, the imbedding (43) was characterized by Kabaila [K] (for a partial result see [AV]).

Theorem 21 *Let $0 < q < p < \infty$. Then*

$$L^p(\Omega, v) \hookrightarrow L^q(\Omega, w)$$

if, and only if,

$$w^{\frac{p}{p-q}} v^{-\frac{q}{p-q}} \in L^1(\Omega).$$

In [KP], the "limiting approach" was used to prove a similar theorem in a scale of special Orlicz spaces.

Theorem 22 *For every $t \geq 0$ and $K > 1$ set $\Phi_K(t) = t(1 + \log_+ t)^K$ and $F_K(t) = \exp t^{1/K}$. Let C_p be the norm of the imbedding $L^p(\Omega, v) \hookrightarrow L^1(\Omega, w)$, that is,*

$$C_p = \sup_{0 \neq f \in L^p(\Omega, v)} \|f\|_{1,w} / \|f\|_{p,v}.$$

Then the following statements are equivalent.

(i) $L_{\Phi_K}(\Omega, v) \hookrightarrow L^1(\Omega, w)$.

(ii) *There exists $C > 0$ such that*

$$C_p \leq C(p-1)^{-K} \qquad \text{for all } p \in (1, 2).$$

(iii) $w/v \in L_{F_K}(\Omega, v)$.

This theorem leads to an imbedding theorem for quite large set of weighted Orlicz spaces:

Theorem 23 ([KP]) *Let Φ_1, Φ_2 be two Young functions such that $\Phi_2 \circ \Phi_1^{-1}$ is a Young function. Let N be the complementary Young function to $\Phi_2 \circ \Phi_1^{-1}$. Assume that at least one of Φ_1, Φ_2 satisfies the Δ_2 condition. Then the following statements are equivalent:*

(i) $L_{\Phi_2}(\Omega, v) \hookrightarrow L_{\Phi_1}(\Omega, w)$,

(ii) $w/v \in L_N(\Omega, v)$.

Using ideas from [K] and [KP], the following imbeddings of general Banach function spaces can be proved:

Theorem 24 ([P2]) *Let v, w be weights on Ω and let $X(\Omega, v)$, $Y(\Omega, w)$ be Banach function spaces. Then $X(\Omega, v) \hookrightarrow Y(\Omega, w)$ if, and only if,*

$$\sup_{\|g\|_{Y'(\Omega, w)} \leq 1} \left\| g \frac{w}{v} \right\|_{X'(\Omega, v)} < \infty. \tag{45}$$

This theorem yields results, for example, for imbeddings between weighted Orlicz spaces and for certain imbeddings between weighted Lorentz and Lebesgue spaces. We shall sketch the argument.

First, setting $Y(\Omega, w) = L^1(\Omega)$, then (45) easily turns to $w/v \in X'(\Omega, v)$. By a suitable homogeneity argument, this gives another proof of slightly more general version of Theorem 23 Also, it yields

Corollary 4 *Let $1 \leq r \leq \min\{p, q\}$. Then the imbedding*

$$L^{p,q}(\Omega, v) \hookrightarrow L^r(\Omega, w)$$

holds if, and only if, $w/v \in L^{(p/r)', (q/r)'}(\Omega, v)$.

4.2 CONTINUOUS AND COMPACT IMBEDDINGS OF SOBOLEV TYPE SPACES

The aim of this section is to present conditions which guarantee that the identical operator from an abstract Sobolev space into a Banach function space is continuous or compact. As examples some results for weighted Sobolev spaces will be included. The notion of an abstract Sobolev space will be based on the following modified definition of a Banach function space (cf. [EOP]).

Throughout this section Ω will be a domain in $\mathbb{R}^N$ and $\{\Omega_n\}$ a fixed sequence of bounded domains such that

$$\Omega_n \subset \overline{\Omega}_n \subset \Omega_{n+1} \subset \Omega \quad \text{for each } n \in \mathbb{N}, \tag{46}$$

and

$$\Omega = \bigcup_{n=1}^{\infty} \Omega_n. \tag{47}$$

For each $n \in \mathbb{N}$ we put

$$\Omega^n = \Omega \setminus \Omega_n. \tag{48}$$

Definition 5 *(A modified definition of Banach function spaces.) A normed linear space $X = X(\Omega) = (X(\Omega), \|\cdot\|_{X(\Omega)})$, where $X(\Omega) \subset M(\Omega)$, is called a* Banach function space *if the following conditions are satisfied:*

(i) *The norm $\|f\|_{X(\Omega)}$ is defined for all $f \in M(\Omega)$ and $f \in X(\Omega)$ if, and only if,*

$$\|f\|_{X(\Omega)} < \infty.$$

(ii) *$\|f\|_{X(\Omega)} = 0$ if, and only if, $f(x) = 0$, a.e. on Ω.*

(iii) *$\|f\|_{X(\Omega)} = \|\,|f|\,\|_{X(\Omega)}$ for all $f \in X(\Omega)$.*

(iv) *Given any $n \in \mathbb{N}$, there is a constant $C_n = C_n(X)$ such that for all $f \in X(\Omega)$*

$$\int_{\Omega_n} |f(x)|\,\mathrm{d}x \le C_n \|f\|_{X(\Omega)}.$$

(v) *For all $n \in \mathbb{N}$, $\chi_{\Omega_n} \in X(\Omega)$.*

(vi) *If $f, g \in M(\Omega)$ and $0 \le f(x) \le g(x)$ a.e. on Ω, then*

$$\|f\|_{X(\Omega)} \le \|g\|_{X(\Omega)}.$$

(vii) *The space $X(\Omega) = (X(\Omega), \|\cdot\|_{X(\Omega)})$ is complete.*

Remark 5 *(i) Condition (iv) implies that $X(\Omega) \subset L^1_{\mathrm{loc}}(\Omega)$. Thus any function $f \in X(\Omega)$ has distributional derivatives on Ω of all orders.*

(ii) Conditions (v) and (vi) imply that $C_0^\infty(\Omega) \subset X(\Omega)$.

Example 6 *Note that a weighted Lebesgue space $L^p(\Omega, w)$, $1 \le p \le \infty$, is a Banach function space (in the sense of Definition 5) provided that w, $w^{1-p'} \in L^1_{\mathrm{loc}}(\Omega)$.*

Similarly, a weighted Orlicz space $L_\Phi(\Omega, w)$ is a Banach function space for any Young function Φ if $w(\Omega_n) < \infty$ and $\|\frac{1}{w}\chi_{\Omega_n}\|_{\widetilde{\Phi},w} < \infty$ for every $n \in \mathbb{N}$.

Similar assertion holds for a weighted Lorentz space $L^{p,q}(\Omega, w)$ *(after replacing the quantity* $\|\cdot\|_{p,q,w}$ *with an equivalent norm if necessary) provided that* $1 < p < \infty$, $1 \le q \le \infty$ *and* $w(\Omega_n) < \infty$, $\|\frac{1}{w}\chi_{\Omega_n}\|_{p',q',w} < \infty$ *for all* $n \in \mathbb{N}$.

Definition 6 *Suppose that* $X(\Omega)$ *is a Banach function space and let* $n \in \mathbb{N}$. *For any* $f \in M(\Omega_n)$, *we define a function* $E_n f \in M(\Omega)$ *(*the extension of f to Ω*) by*

$$(E_n f)(x) = \begin{cases} f(x), & x \in \Omega_n, \\ 0, & x \in \Omega^n. \end{cases}$$

The space $X(\Omega_n)$ *given by*

$$X(\Omega_n) = \{f \in M(\Omega_n) : \|E_n f\|_{X(\Omega)} < \infty\}$$

is endowed with the norm

$$\|f\|_{X(\Omega_n)} := \|E_n f\|_{X(\Omega)}.$$

The space $X(\Omega_n) = (X(\Omega_n), \|\cdot\|_{X(\Omega_n)})$ *is a Banach function space.*

Definition 7 *Let* $X = X(\Omega)$ *and* $Y = Y(\Omega)$ *be Banach function spaces with norms* $\|\cdot\|_{X(\Omega)}$ *and* $\|\cdot\|_{Y(\Omega)}$. *By Remark 5(i), any* $f \in X(\Omega)$ *has distributional derivatives* $\partial f/\partial x_i$ *(*$i = 1, \dots, N$*). Accordingly, we define the* abstract Sobolev space $W(X, Y) = W(X(\Omega), Y(\Omega))$ *by*

$$W(X,Y) = \{f \in X(\Omega) : \frac{\partial f}{\partial x_i} \in Y(\Omega) \ \ \textit{for}\ i = 1, \dots, N\}$$

and equip this space with the norm

$$\|f\|_W = \|f\|_{W(X,Y)} := \|f\|_X + \|\nabla f\|_Y,$$

where by $\|\nabla f\|_Y$ *we mean* $\sum_{i=1}^{N} \|\frac{\partial f}{\partial x_i}\|_Y$.

Remark 6 *rem156 (i)* $W(X,Y)$ *is not a Banach function space, nevertheless* $W(X,Y)$ *is a Banach space.*

(ii) In addition to the abstract Sobolev space $W(X(\Omega), Y(\Omega))$ *we have at our disposal abstract Sobolev spaces* $W(X(\Omega_n), Y(\Omega_n))$, $n \in \mathbb{N}$, *(cf. Definition 6) which are Banach spaces.*

(iii) If $f \in C_0^\infty(\Omega)$ *then, of course,* $\partial f/\partial x_i \in C_0^\infty(\Omega)$ *for* $i = 1, \dots, N$, *and hence (using Remark 5(ii)),* $C_0^\infty(\Omega) \subset W(X(\Omega), Y(\Omega))$. *This enables to define the subspace*

$$W_0(X(\Omega), Y(\Omega))$$

of $W(X(\Omega), Y(\Omega))$ *as the closure of* $C_0^\infty(\Omega)$ *with respect to* $\|\cdot\|_{W(X,Y)}$.

Let $X(\Omega)$, $Y(\Omega)$ and $Z(\Omega)$ be Banach function spaces. Our goal is to establish some relations between the "local embeddings"

$$W(X(\Omega_n), Y(\Omega_n)) \hookrightarrow Z(\Omega_n), \qquad n \in \mathbb{N}, \tag{49}$$

and the "global imbeddings"

$$W(X(\Omega), Y(\Omega)) \hookrightarrow Z(\Omega). \tag{50}$$

We have

Theorem 25 *Suppose that* (49) *is fulfilled. Then* (50) *holds if, and only if,*

$$A := \lim_{n\to\infty} \sup_{\|f\|_{W(X,Y)}\le 1} \|f\chi_{\Omega^n}\|_{Z(\Omega)} < \infty. \tag{51}$$

Note that the limit in (51) always exists.

Assuming that the "local imbeddings" (49) are compact, we look for some additional conditions that would guarantee the compactness of the "global imbedding" (50). Our result reads as follows:

Theorem 26 *Suppose that*

$$\|f\chi_{\Omega^n}\|_{Z(\Omega)} \to 0 \quad as \quad n\to\infty \qquad for\ each\ f\in Z(\Omega) \tag{52}$$

(cf. Definition 3). Let the imbeddings (49) *be compact. Then*

$$W(X(\Omega), Y(\Omega)) \hookrightarrow\hookrightarrow Z(\Omega)$$

if, and only if, $A = 0$, *where* A *is given by* (51).

Remark 7 *Suppose that* (50) *holds. Let* I *be the identical operator from* $W(X(\Omega), Y(\Omega))$ *into* $Z(\Omega)$. *For each* $n \in \mathbb{N}$ *define operators* I_n $W(X(\Omega), Y(\Omega)) \to Z(\Omega)$ *by* $I_n u = u\chi_{\Omega_n}$. *Denote by* $\beta(I)$ *the ball measure of non-compactness of* I *(see [EE2]). Using some ideas from [CG–F] one can prove that*

$$\beta(I) = A = \lim_{n\to\infty} \|I - I_n\|$$

provided (52) *is fulfilled.*

Example 7 *Taking* $X(\Omega) = Y(\Omega) = L^p(\Omega; d^\beta)$, *where* $\beta \in \mathbb{R}$ *and* $d(x) = \operatorname{dist}(x, \partial\Omega)$, $x \in \Omega$, *the abstract Sobolev space* $W(X(\Omega), Y(\Omega))$ *is just the* weighted Sobolev space *usually denoted by* $W^{1,p}(\Omega; d^\beta, d^\beta)$. *The space* $W_0(X(\Omega), Y(\Omega))$ *is then written as* $W_0^{1,p}(\Omega; d^\beta, d^\beta)$. *Using Theorems 25 and 26 (together with their variants for the space* $W_0(X(\Omega), Y(\Omega)))$ *one can prove (cf. [GO II] or [OK], Theorems 19.16 and 19.17)):*

Let $1 \le p \le q \le \infty$, $\Omega \in C^{0,1}$, $\alpha, \beta \in \mathbb{R}$ *with* $\beta > p-1$ *or* $\beta \ne p-1$. *Then*

$$W^{1,p}(\Omega; d^\beta, d^\beta) \hookrightarrow L^q(\Omega; d^\alpha)$$

or

$$W_0^{1,p}(\Omega; d^\beta, d^\beta) \hookrightarrow L^q(\Omega; d^\alpha)$$

if, and only if

$$\frac{N}{q} - \frac{N}{p} + 1 \ge 0 \qquad and \qquad \frac{\alpha}{q} - \frac{\beta}{p} + \frac{N}{q} - \frac{N}{p} + 1 \ge 0. \tag{53}$$

These imbeddings are compact if, and only if, the inequalities (53) *are strict.*

For other examples we refer to [GO], [OK], [OR], [KOP].

5 POINCARÉ AND FRIEDRICHS INEQUALITIES

Our aim in this chapter is to present results recently obtained by the authors in collaboration with D. E. Edmunds on inequalities of Poincaré and Friedrichs type. The results can be found in [EO], [EOP], [EOR].

5.1 POINCARÉ INEQUALITY

We start with recalling the important result of C. J. Amick [AM].

Theorem 27 *Let Ω be a bounded domain in $\mathbb{R}^N$. Then there exists a constant C such that the inequality*

$$\int_\Omega |f|^2 \,\mathrm{d}x \le C\left(\left|\int_\Omega f\,\mathrm{d}x\right|^2 + \sum_{i=1}^n \int_\Omega \left|\frac{\partial f}{\partial x_i}\right|^2 \mathrm{d}x\right)$$

holds for all $f \in W^{1,2}(\Omega)$ if, and only if, $\beta(I)$, the ball measure of non-compactness of the imbedding $I\ W^{1,2}(\Omega) \to L^2(\Omega)$, satisfies $\beta(I) < 1$.

The Amick's result was generalized and further investigated by many authors, we refer to [EE1], [EH] or [CG–F]. The above mentioned collaboration of the authors with D. E. Edmunds lead finally to quite a complete theory some results of which we are going to present.

Convention 2 *From this point on the symbols $\Omega, \Omega_n, \Omega^n$ have the same meaning as in Section 4.2 (cf.* (46)–(48)*). Also the notion of a Banach function space is understood as in Section 4.2.*

Definition 8 *Suppose that $X = X(\Omega)$, $Y = Y(\Omega)$ are Banach function spaces, $W(X,Y)$ is the corresponding abstract Sobolev space. Let $F \in W(X,Y)^*$. We say that [F,X,Y]* supports the Poincaré inequality *if there is a positive constant K such that for all $u \in W(X,Y)$,*

$$\|u\|_{X(\Omega)} \le K\big(|F(u)| + \|\nabla u\|_{Y(\Omega)}\big). \tag{54}$$

Define

$$A = \lim_{n\to\infty} \sup_{\|u\|_{W(X,Y)}\le 1} \|u\chi_{\Omega^n}\|_{X(\Omega)} \tag{55}$$

and observe that this limit always exists and that $A \in [0,1]$.

We have the following results:

Theorem 28 *Let $\chi_\Omega \in W(X,Y)$, $F \in W(X,Y)^*$, and $F(\chi_\Omega) \ne 0$. Suppose that the following conditions are satisfied:*

$$W(X(\Omega_n), Y(\Omega_n)) \hookrightarrow\hookrightarrow X(\Omega_n) \qquad \textit{for all} \quad n \in \mathbb{N}. \tag{56}$$

$$\textit{There exists } n \in \mathbb{N} \textit{ such that } \quad \|\chi_{\Omega^n}\|_{X(\Omega)} < \|\chi_\Omega\|_{X(\Omega)}. \tag{57}$$

Then [F,X,Y] supports the Poincaré inequality if, and only if, $A < 1$.

Remark 8 (i) *The characterizing condition $A < 1$ is independent of F. Thus, if the assumptions of Theorem 28 are fulfilled with $F_1, F_2 \in W(X,Y)^*$, then either both $[F_1, X, Y]$ and $[F_2, X, Y]$ support the Poincaré inequality or neither of them.*

(ii) *The condition* (56) *is quite natural. For example, it is fulfilled for $X(\Omega) = L^p(\Omega; v_0)$, $Y(\Omega) = L^p(\Omega; v_1)$ provided that $\Omega_n \in C^{0,1}$ for all $n \in N$ and weights v_0, v_1 are locally bounded from below and from above on Ω by positive constants.*

(iii) *An interesting peculiarity occurs when $X(\Omega)$ is an Orlicz space $L_\Phi(\Omega)$. Then* (57) *is satisfied provided that $L_\Phi(\Omega)$ is equipped with the Orlicz norm, however it need not be satisfied when the Orlicz norm is replaced by the Luxemburg one (cf. [EOP, Lemma 7.6 and Example 7.7]).*

(iv) *Assume that*

$$\|f\chi_{\Omega^n}\|_{X(\Omega)} \to 0 \quad as \quad n \to \infty \qquad for\ each\ f \in X(\Omega). \tag{58}$$

Then one can prove (cf. Remark 7) that $A = \beta(I)$, where $\beta(I)$ stands for the ball measure of non-compactness of the identical operator I from $W(X(\Omega), Y(\Omega))$ into $X(\Omega)$.

Another characterization of the Poincaré inequality is given by

Theorem 29 *Let $\chi_\Omega \in W(X,Y)$, $F \in W(X,Y)^*$, $F(\chi_\Omega) \neq 0$, and let L be a projection from $W(X,Y)$ onto $\mathcal{P}_0 = \{c\chi_\Omega : c \in \mathbb{C}\}$. Suppose that $Z = Z(\Omega)$ is another Banach function space such that*

$$W(X,Y) \hookrightarrow Z \hookrightarrow X. \tag{59}$$

Then the following statements are equivalent:

(i) *$[F, X, Y]$ supports the Poincaré inequality.*

(ii) *There exists a constant C_1 such that for all $u \in W(X,Y)$,*

$$\|u - Lu\|_{Z(\Omega)} \le C_1 \|\nabla u\|_{Y(\Omega)}.$$

(iii) *There exists a constant C_2 such that for all $u \in W(X,Y)$,*

$$\inf_{p \in \mathcal{P}_0} \|u - p\|_{Z(\Omega)} \le C_2 \|\nabla u\|_{Y(\Omega)}.$$

(iv) *The operator $T\, W(X,Y) \to [Y(\Omega)]^N$ defined by $Tf = \nabla f$ has closed range.*

The following theorem deals with the case when $\chi_\Omega \notin W(X,Y)$.

Theorem 30 *Suppose that $\chi_\Omega \notin W(X,Y)$. Denote by Θ the zero functional on $W(X,Y)$. Let symbols $Z(\Omega)$ and T have the same meaning as in Theorem 29. Then the following statements are equivalent:*

(i) *$[\Theta, X, Y]$ supports the Poincaré inequality.*

(ii) *There exists a constant C such that for all $u \in W(X,Y)$,*

$$\|u\|_{Z(\Omega)} \le C \|\nabla u\|_{Y(\Omega)}.$$

(iii) *The operator T has closed range. Moreover, let condition* (56) *hold. Then each of statements* (i)–(iii) *is equivalent to*

(iv) *$A < 1$.*

Remark 9 *For the Poincaré inequality in an abstract Sobolev space of higher order (and even in an anisotropic one) we refer to [EOR].*

Example 8 *Let* $1 \leq p < \infty$, $\Omega \in C^{0,1}$, $\beta \in \mathbb{R}$ *and let* $d(x) = \operatorname{dist}(x, \partial\Omega)$ *for* $x \in \Omega$. *Set* $X(\Omega) = Y(\Omega) = L^p(\Omega; d^\beta)$. *Then* $W(X,Y) = W^{1,p}(\Omega; d^\beta, d^\beta)$. *Suppose that* $F \in W^{1,p}(\Omega; d^\beta, d^\beta)^*$ *is such that* $F(\chi_\Omega) \neq 0$. *From Theorem 28 we obtain:*

(i) If $\beta \in (-1, \infty)$, *then there is a constant* K *such that*

$$\int_\Omega |u(x)|^p d^\beta(x)\,\mathrm{d}x \leq K \left(|F(u)|^p + \sum_{i=1}^N \int_\Omega \left| \frac{\partial u}{\partial x_i}(x) \right|^p d^\beta(x)\,\mathrm{d}x \right)$$

for all $u \in W^{1,p}(\Omega; d^\beta, d^\beta)$.

(ii) If $\beta \in (-\infty, -1]$, *then there is a constant* K *such that*

$$\int_\Omega |u(x)|^p d^\beta(x)\,\mathrm{d}x \leq K \sum_{i=1}^N \int_\Omega \left| \frac{\partial u}{\partial x_i}(x) \right|^p d^\beta(x)\,\mathrm{d}x.$$

Example 9 *Let* $\Omega \in C^{0,1}$, $N > 1$ *and*

$$\Phi(x) = \begin{cases} \left(\frac{e}{\alpha+1}\right)^{\alpha+1} x^{\alpha+1}, & 0 < x < \alpha+1, \\ e^x, & x \geq \alpha+1, \end{cases}$$

where $0 < \alpha < N-1$. *Set* $X(\Omega) = Y(\Omega) = L_\Phi(\Omega)$. *Then* $W(X,Y) = W^1 L_\Phi(\Omega)$ *and, using Theorem 28, we obtain:*

There exists a constant K *such that for all* $u \in W^1 L_\Phi(\Omega)$,

$$\|u\|_{\Phi,\Omega} \leq K \left(\left| \int_\Omega u\,\mathrm{d}x \right| + \|\nabla u\|_{\Phi,\Omega} \right).$$

5.2 FRIEDRICHS INEQUALITY

Definition 9 *Suppose that* $X = X(\Omega)$ *and* $Y = Y(\Omega)$ *are Banach function spaces,* $W(X,Y)$ *is the corresponding abstract Sobolev space. We say that* $[X,Y]$ supports the Friedrichs inequality *if there is a constant* K *such that for all* $u \in C_0^\infty(\Omega)$,

$$\|u\|_{X(\Omega)} \leq K\, \|\nabla u\|_{Y(\Omega)}.$$

Define

$$A^0 = \limsup_{n\to\infty} \left\{ \|u\chi_{\Omega^n}\|_{X(\Omega)} : u \in C_0^\infty(\Omega), \|u\|_{W(X,Y)} \leq 1 \right\}.$$

Note that this limit always exists and that $A^0 \in [0,1]$.

Theorem 31 *Assume that* (56) *and the following condition hold:*

$$\text{If } \chi_\Omega \in W(X,Y) \text{ and } \{u_j\} \subset C_0^\infty(\Omega),\ u_j \to c \text{ in } W(X,Y), \text{ where } c \text{ is a constant function on } \Omega, \text{ then } c = 0. \tag{60}$$

Then $[X,Y]$ *supports the Friedrichs inequality if, and only if,* $A^0 < 1$.

There is another characterization of the Friedrichs inequality:

Theorem 32 *Let $Z(\Omega)$ be a Banach function space such that*

$$W_0(X,Y) \hookrightarrow Z(\Omega) \hookrightarrow X(\Omega).$$

Suppose that condition (60) is fulfilled. Then the following statements are equivalent:

(i) *$[X,Y]$ supports the Friedrichs inequality.*

(ii) *There exists $C > 0$ such that for all $u \in W_0(X,Y)$,*

$$\|u\|_{Z(\Omega)} \leq C\|\nabla u\|_{Y(\Omega)}.$$

(iii) *The operator $T\, W_0(X,Y) \to [Y(\Omega)]^N$ given by $Tu = \nabla u$ has closed range.*

Example 10 *Let us make a remark on Example 8. Using Theorem 31, one can show that if $\Omega \in C^{0,1}$ and $\beta < p-1$, then there exists a constant K such that the inequality*

$$\int\limits_\Omega |u(x)|^p d^\beta(x)\,\mathrm{d}x \leq K \sum_{i=1}^N \int\limits_\Omega \left|\frac{\partial u}{\partial x_i}(x)\right|^p d^\beta(x)\,\mathrm{d}x$$

holds for all $u \in C_0^\infty(\Omega)$.

Example 11 *Let $1 \leq p < \infty$, $\alpha, \beta \in \mathbb{R}$, $\Omega = \mathbb{R}^N \setminus \overline{G}$ where G is a bounded domain in $\mathbb{R}^N$ such that $0 \in G$. Then there is a constant K such that the inequality*

$$\int\limits_\Omega |u(x)|^p e^{\alpha|x|}\,\mathrm{d}x \leq K \sum_{i=1}^N \int\limits_\Omega \left|\frac{\partial u}{\partial x_i}(x)\right|^p e^{\beta|x|}\,\mathrm{d}x$$

holds for all $u \in C_0^\infty(\Omega)$ if, and only if, $\alpha \leq \beta$, $\beta \neq 0$.

REFERENCES

[L] W. A. J. Luxemburg: Banach function spaces (Thesis: Technische Hogeschool te Delft, 1955).

[HA] P. R. Halmos: Measure theory. Graduate Texts in Math. **18**. Springer–Verlag, New York–Heidelberg–Berlin, 1974.

[BS] C. Bennett and R. Sharpley: Interpolation of operators. Pure Appl. Math. **129**, Academic Press, Inc., 1988.

[BR] J.S. Bradley: Hardy inequalities with mixed norms. Canad. Math. Bull. **21** (1978), no. 4, 405–408.

[M] V.G. Maz'ya: Sobolev Spaces. Springer–Verlag, Berlin–Heidelberg–New York, 1985. (An extended and improved version of the books Einbettungssätze für Sobolevsche Räume 1, 2; Zur Theorie der Sobolevsche Räume. Teubner–Texte zur Mathematik, Vols. **21**, **28**, **38**. Teubner, Leipzig, 1979, 1980, 1981.)

[SI] G. J. Sinnamon: Weighted Hardy and Opial-type inequalities. J. Math. Anal. Appl. **160** (1991), 434–445.

[P1] L. Pick: Weighted modular estimates for the Hardy operator. Preprint.

[BK2] S. Bloom and R. Kerman: Weighted L_Φ integral inequalities for operators of Hardy type. Preprint, Brock University.

[S] E. Sawyer: Weighted Lebesgue and Lorentz norm inequalities for the Hardy operator. Trans. Amer. Math. Soc. **281** (1984), no. 1, 329–337.

[B] E. I. Berezhnoi: Inequality with weights for Hardy operators in function spaces. Function spaces, Teubner–Texte zur Mathematik, Band **120**. B.G. Teubner Verlagsgesellschaft, Stuttgart–Leipzig, 1991, 75–79.

[KN] K. Knopp: Über Reihen mit positiven Gliedern. J. London Math. Soc. **3** (1928), 205–211.

[H] H. P. Heinig: Weighted inequalities in Fourier analysis. Nonlinear Analysis, Function Spaces and Applications, **4**. Teubner–Texte zur Mathematik, Band **119**, Leipzig, 1990, 42–85.

[HKK] H. P. Heinig, R. Kerman, M. Krbec: Weighted exponential inequalities. Preprint no. **79**, Math. Inst. Czech. Acad. Sci., Prague, 1992, 30 pp.

[OG] B. Opic and P. Gurka: Weighted inequalities for geometric means. Preprint no. **75**, Math. Inst. Czech. Acad. Sci., Prague, 1992, 11 pp. (To appear in Proc. Amer. Math. Soc.)

[PO] L. Pick and B. Opic: On the geometric mean operator. Preprint no. **78**, Math. Inst. Czech. Acad. Sci., Prague, 1992, 9 pp. (An extended version to appear in J. Math. Anal. Appl.)

[LP] Q. Lai and L. Pick: The Hardy operator, L_∞, and BMO. (To appear in J. London Math. Soc.)

[OK] B. Opic and A. Kufner: Hardy-type inequalities. Pitman Research Notes in Mathematics Series, **211**. Longman Scientific and Technical, Harlow, 1990.

[O] B. Opic: On the distance of the Riemann–Liouville operator from compact operators. Preprint no. **80**, Math. Inst. Czech. Acad. Sci., Prague, 1992, 10 pp. (To appear in Proc. Amer. Math. Soc.)

[MS] F. J. Martin–Reyes and E. Sawyer: Weighted inequalities for Riemann–Liouville fractional integrals of order one and greater. Proc. Amer. Math. Soc. **106** (1989), 727–733.

[ST1] V. D. Stepanov: Weighted inequalities for a class of Volterra convolution operators. J. London Math. Soc. (2) **45** (1992), 232–242.

[ST2] V. D. Stepanov: Two-weighted estimates for Riemann–Liouville integrals. Preprint no. **39**, Math. Inst. Czech. Acad. Sci., Prague, 1988, 28 pp.

[BK1] S. Bloom and R. Kerman: Weighted norm inequalities for operators of Hardy type. Proc. Amer. Math. Soc. **113** (1991), 135–141.

[J] R. K. Juberg: The measure of non-compactness in L^p for a class of integral operators. Indiana Univ. Math. J. 23 (1974), 925–936.

[K] V. P. Kabaila: On the inclusion of $L_p(\nu)$ into $L_r(\mu)$ (Russian). Lit. Mat. Sb. **21** (1981), 143–148.

[AV] A. Avantaggiati: On compact embedding theorems in weighted Sobolev spaces. Czechoslovak Math. J. **29** (104) (1979), no. 4, 635–648.

[KP] M. Krbec and L. Pick: On imbeddings between weighted Orlicz spaces. Z. Anal. Anwend. **10** (1991), no. 1, 105–116.

[P2] L. Pick: A remark on continuous imbeddings between Banach function spaces. Colloquia Mathematica Societatis János Bolyai, Vol. **58**. Approximation theory, Kecskemét (Hungary), 1990, 571–581.

[CG–F] J. A. Cavanati and F. Galaz–Fontes: Compactness of imbeddings between Banach spaces and applications to Sobolev spaces. J. London Math. Soc. (2) **41** (1990), 511–525.

[GO] P. Gurka and B. Opic: Continuous and compact imbeddings of weighted Sobolev spaces I, II, III. Czechoslovak Math. J. **38** (133) (1988), 730–744; ibid. **39** (114) (1989), 78–94; ibid. **41** (116) (1991), 317–341.

[OR] B. Opic and J. Rákosník: Estimates for mixed derivatives of functions from anisotropic Sobolev–Slobodeckij spaces with weights. Quart. J. Math. Oxford (2) **42** (1991), 347–363.

[KOP] M. Krbec, B. Opic and L. Pick: Imbedding theorems for weighted Orlicz–Sobolev spaces. J. London Math. Soc. **46** (3) (1992), 543–556.

[EO] D. E. Edmunds and B. Opic: Weighted Poincaré and Friedrichs inequalities. (To appear in J. London Math. Soc.)

[EOP] D. E. Edmunds, B. Opic and L. Pick: Poincaré and Friedrichs inequalities in abstract Sobolev spaces. (To appear in Math. Proc. Camb. Phil. Soc.)

[EOR] D. E. Edmunds, B. Opic and J. Rákosník: Poincaré and Friedrichs inequalities in abstract Sobolev spaces II. (To appear in Math. Proc. Camb. Phil. Soc.)

[AM] C. J. Amick: Some remarks on Rellichs theorem and the Poincaré inequality. J. London Math. Soc. (2) **18** (1978), 81–93.

[EE1] D. E. Edmunds and W. D. Evans: Spectral theory and embeddings of Sobolev spaces. Quart. J. Math. Oxford (2) **30** (1979), 431–453.

[EE2] D. E. Edmunds and W. D. Evans: Spectral theory and differential operators. Oxford Mathematical Monographs. Clarendon Press, Oxford, 1987.

[EH] W. D. Evans and D. J. Harris: Sobolev embeddings for generalized ridged domains. Proc. London Math. Soc. (3) **54** (1987), 141–175.

[HLP] G. H. Hardy, J. E. Littlewood and G. Pólya: Inequalities. University Press, Cambridge, 1952.

[KJF] A. Kufner, O. John and S. Fučík: Function spaces. Academia, Prague, 1977.

[KO] V. M. Kokilashvili: On Hardy's inequalities in weighted spaces (Russian). Soobshch. Akad. Nauk Gruzin. SSR **96** (1979), no.1, 37–40.

[MU] B. Muckenhoupt: Hardy's inequality with weights. Studia Math. **44** (1972), 31–38.

[T] G. Tomaselli: A class of inequalities. Boll. Un. Mat. Ital. (4) **2** (1969), 622–631.

Mathematical Institute
Academy of Sciences of the Czech Republic
Žitná 25 115 67 Praha 1
Czech Republic
e-mail: `krbecm@csearn.bitnet`
`opic@csearn.bitnet`
`rakosnik@csearn.bitnet`
Current address of L. Pick
University of Wales
College of Cardiff
Senghennydd Road
Cardiff CF2 4AG
Wales
e-mail: `pickl@taff.cardiff.ac.uk`

Approximations of Functions on Manifolds

Pjotr I. Lizorkin, Sergej M. Nikol'skij

Introduction

The article presents a review of investigations by the authors and some other Soviet mathematicians with respect to the approximation of functions on a sphere and on more general manifolds. Of special interest in this context are functions with differential properties, say functions belonging to H_p^r and $B_{p,q}^r$.
The classical theory of approximation on the 1-dimensional torus $\mathbb{T}^1$ or on the real axis $\mathbb{R}^1$ is essentially based on the action of the translation group, which influences the technical procedure of approximation and the structure of functions to be approximated. Evidently, this carries over to $\mathbb{T}^n$ and $\mathbb{R}^n$.
Otherwise, for approximation on an interval $[a, b]$, or, more generally, on a subset $\Omega \subset \mathbb{R}^n$ the group properties are lost. In some simple cases the problem can be regarded as a periodical one. In general it is necessary to extend the function defined on Ω to a function defined on $\mathbb{R}^n$ and to ask how to approximate the trace of this function by traces of the usual approximating functions, say polynomials or functions of exponential type.
So there are two ways for approximating functions on a manifold. The first one is using manifolds with a transitive group action. This is closely related to Lie groups and homogeneous spaces. The second one is approximation of functions defined on a subset of the euclidean n-space by traces of polynomials or functions of exponential type.
The second method is described in chapter 2 of this review containing recent results of S. M. Nikol'skij.
Take the unit sphere $\sigma \subset \mathbb{R}^n$ as an example. On the one hand it is a symmetric manifold on which the group of rotations acts transitively and isomorphically. Here the spectral theory of the Laplace- Beltrami operator leads to an approximation procedure via spherical harmonics. On the other hand the unit sphere is a simple geometric object and the spherical harmonics turn out to be the traces of certain algebraic polynomials on $\mathbb{R}^n$.
The paper is organized as follows. Chapter 1 deals with the approximation of functions on special manifolds imbedded in $\mathbb{R}^n$, i.e. spheres. Chapter 2 generalizes these results to more general manifolds. Chapter 3 is dedicated to the group theoretical approach, the manifolds in question are hyperbolic spaces regarded as an hyperboloid with Poincarè metric, the Heisenberg group and Riemannian manifolds with non positive curvature, i.e. Cartan-Hadamard manifolds. Additionals remarks and the discussion of further questions can be found in chapter 4.

1. Approximation of functions on the sphere

1.1. Special features of our investigations on approximation of functions on the sphere

$\sigma \subset \mathbb{R}^n$ (cf.[2-7]) and adjacent investigations of A. P. Terechin are:
1. the use of averaged differences of higher order and construction of corresponding moduli of smoothness
2. the explicit form of approximating polynomials in the proof of a Jackson-type theorem.
At the beginning of our investigations the moduli we used caused some difficulties (cf. [10]). These difficulties were overcome partly by us ([2]), but mainly by G. A. Kaljabin ([11]) and H. P. Rustamov ([12]). Now there is a rich theory of approximation based on moduli of smoothness.
In our works we pay great attention to cases where the modulus of smoothness is close to power one. These cases are connected with the approximative description of the so-called Besov-type spaces (when direct and inverse theorems of approximation are reversed), see 1.3..
1.2. Let $\sigma = \sigma^{n-1}$ be the unit sphere in $\mathbb{R}^n$:

$$\sigma = \{x \in \mathbb{R}^n, |x|^2 = \sum_{i=1}^{n} x_i^2 = 1\}.$$

For a function $f(x)$ on σ we introduce the averaging operator S_t:

$$S_t : f(x) \longrightarrow S_t f(x),$$

$$S_t f(x) = \frac{1}{|\tilde{\sigma}| \sin t^{n-2}} \int\limits_{xy=\cos t} f(y)dy. \tag{1.1}$$

Here $|\tilde{\sigma}|$ is the full measure of the unit sphere $\tilde{\sigma} = \sigma^{n-2} \subset \mathbb{R}^{n-1}$, dy is the Lebesgue measure on $\tilde{\sigma}$ induced from $\mathbb{R}^{n-1}$. For fixed $x \in \sigma$ and $t > 0$ the expression

$$S_t f(x) - f(x) = (S_t - I) f(x)$$

is the averaged increment of the function $f(x)$ under the displacement along geodesics with the step t. Thus the operator S_t represents some analogon of the shift operator in the classical theory of approximation, although the group properties are lost: $S_t \cdot S_u \neq S_{t+u}$.
Following classical traditions many authors considered moduli of smoothness on the sphere which are constructed vià the operator $(S_t - I)^k$:

$$\omega_{k,p}(f,\delta) = \sup_{0<t\leq\delta} ||(S_t - I)^k f||_{L_p(\sigma)}. \tag{1.2}$$

So $\omega_{p,k}(f,\delta)$ is the modulus of classical type.
Instead of (1.2) we have suggested a new construction - the so-called star modulus:

$$\omega^*_{k,p}(f,\delta) = \sup_{0<t\leq\delta} ||(\Delta^k_{t,\tau} S_t f)|_{t=0}||_{L_p(\sigma)}. \tag{1.3}$$

Here the expression in brackets denotes the difference of order k relative to t with the step τ calculated for $t = 0$. Formula (1.3) describes the averaging of the k-th difference of the function f along different geodesics, cf. (1.6).

1.3. Now we introduce the notion of Besov spaces $B^r_{p,q}(\sigma)$ on the sphere σ and give their characterizations in terms of the best approximation.

Definition The space $B^r_{p,q}(\sigma)$, $r > 0$, $1 \le p, q \le \infty$ is the collection of functions $f \in L_p(\sigma)$ such that for $k > r$

$$b^r_{p,q}(f) := \left(\int_0^\pi \frac{\omega^*_{k,p}(f,\tau)^q}{\tau^{1+rq}} d\tau \right)^{1/q} < \infty \tag{1.4}$$

with the usual modification for $q = \infty$.
It is easy to see that $B^r_{p,q}(\sigma)$ is a Banach space with the norm

$$\|f|B^r_{p,q}(\sigma)\| = \|f\|_{L_p(\sigma)} + b^r_{p,q}(f).$$

Note that the dependence on k of this definition is not essential.
Denote by $T_N(x)$, $x \in \sigma$, the trace of an algebraic polynomial of power N. This trace is represented as a polynomial in the spherical harmonics $Y_m(x)$ of order m:

$$T_N(x) = \sum_{m=0}^{N} Y_m(x). \tag{1.5}$$

Let $E_N(f)_p$ be the best approximation of f by the spherical harmonics of order $\le N$ in the metric of $L_p(\sigma)$:

$$E_N(f)_p = \inf_{k \le N} \|f - T_k\|_{L_p(\sigma)}.$$

Theorem 1 *A Function $f \in L_p(\sigma)$ belongs to the space $B^r_{p,q}(\sigma)$ if and only if the expression*

$$\left(\sum_{m=0}^{\infty} 2^{mrq} E^q_{2^m}(f)_p \right)^{1/q}$$

is finite. This one is majorized by the seminorm $b^r_{p,q}(f)$ and vice versa:

$$b^r_{p,q}(f) \le c \left(\|f\|_{L_p(\sigma)} + \left(\sum_{m=0}^{\infty} 2^{mrq} E^q_{2^m}(f)_p \right)^{1/q} \right).$$

Hereby the expression in the outer brackets is equivalent to the norm $\|f|B^r_{p,q}(\sigma)\|$.
For the spaces $B^r_{p,\infty}(\sigma)$, $1 \le p < \infty$, this theorem was proved in the paper [5], the case of Zygmund spaces $B^r_{\infty,\infty}(\sigma)$ was recently investigated in [3]. There a Jackson-type theorem on σ was obtained for the first time. Concerning the proofs, we used different forms of Bernstein inequalities. For example, the following inequality holds:

$$\|(\Delta^k_{t,\tau} S_t T_N)|_{t=0}\|_{L_p(\sigma)} \le (N\tau)^k \|T_N\|_{L_p(\sigma)}.$$

Note that there are many equivalent norms for the spaces $B^r_{p,q}(\sigma)$. So, in (1.4) one can replace the star modulus $\omega^*_{k,p}(f,t)$ by a difference norm:

$$\|(\Delta^k_{t,\tau} S_t f)|_{t=0}\|_{L_p(\sigma)}$$

cf. also [13].

1.4. As already mentioned in 1.2., we have the following identity:

$$\Delta^k_{t,\tau} S_t f(x)\big|_{t=0} = \frac{1}{|\tilde{\sigma}|} \int_{\tilde{\sigma}} \Delta^k_{t,\tau} f(x\cos t + y\sin t)\big|_{t=0} dy. \tag{1.6}$$

Here $y \in \sigma$ is a vector orthogonal to x, $\tilde{\sigma} = \sigma^{n-2}(x)$ and dy is the corresponding measure. Thus (1.6) is the "true" mean of the indicated difference. In [7] we considered the θ-moduli which were constructed on the basis of averaging to the power θ, $1 \le \theta \le p$:

$$\omega^\theta_{k,p}(f,\delta) = \sup_{0<\tau\le\delta} \left\| \left(\frac{1}{|\tilde{\sigma}|} \int_{\tilde{\sigma}} |\Delta^k_{t,\tau} f(x\cos t + y\sin t)\big|_{t=0}|^\theta dy \right)^{1/\theta} \right\|_{L_p(\sigma)}. \tag{1.7}$$

It is important to note that for $p = \theta$ the p-moduli can be expressed in the form

$$\omega^p_{k,p}(f,\delta) = \sup_{0<t\le\delta} \left(\int_{SO(n)} |\Delta^k_{t,\tau} f(ux\cos t + uy\sin t)\big|_{t=0}|^p d\mu(u) \right)^{1/p}, \tag{1.7'}$$

where x, y are fixed and orthogonal points of σ and integration is carried out with respect to the Haar measure μ on $SO(n)$. With the help of (1.7) the following crucial estimate can easily be proved:

$$\omega^p_{k,p}(f,\lambda\delta) \le (\lambda+1)^k \omega^p_{k,p}(f,\delta), \tag{1.8}$$

where $\lambda > 0$. We remark that the moduli (1.7) were implicitly used in our starting paper [3], where we studied the approximation on σ in the $C(\sigma)$ metric (identified with $L_\infty(\sigma)$) for the Zygmund spaces $B^r_{\infty,\infty}(\sigma) = C^r(\sigma)$:

$$C^r(\sigma) = \{f \in C(\sigma) : |\Delta^k_{t,h} f(x\cos t + y\sin t)\big|_{t=0}| \le c|h|^r \text{ for } k > r\}.$$

Note that the δ-moduli are well adjusted for the approximation procedures via polynomials for differences like (1.6), cf. [3] and [8]. This fact and estimates like (1.8) make it possible to prove the following Jackson-type theorem.

Theorem 2 *Let $f \in L_p(\sigma)$, $1 \le p \le \infty$. Then the following inequality holds for $m = 0, 1, \ldots$*

$$E_m(f)_p \le c\, \omega^p_{k,p}(f, \frac{\pi}{m+1}).$$

The constant c does not depend on f and m.

For $p = \infty$ this theorem was proved in [3], for the other p the prove is similar to this one.

1.5. Another strengthening of the construction (1.3) was proposed by A. P. Terechin. Let us consider the averaging operator

$$S_t \;:\; L_p(\sigma) \ni f(x) \longrightarrow S_t f(x) =: u_f(x,t) \tag{1.9}$$

as a map of $L_p(\sigma)$ onto a closed subspace U of the space $L_{p,\infty}(\sigma \times (o, 2\pi]) =: L_{p,\infty}$ of functions $u(x,t)$ with the mixed norm:

$$\|u|L_{p,\infty}\| = \sup_t \|u(\cdot,t)\|_{L_p(\sigma)}.$$

The map (1.9) establishes an isometric isomorphism between $L_p(\sigma)$ and U which reduces the question of approximation of elements from $L_p(\sigma)$ to those of U.
But in the spaces $L_{p,\infty}$ and U the group of translations A_τ acts isometrically:

$$A_\tau \; : \; u(x,t) \longrightarrow u(x,t+\tau).$$

Therefore one can turn to the theory of approximation in the Banach space U, cf. [14]. Then, instead of the modulus (1.3), the "tight modulus" arises:

$$\omega^T_{k,p}(f,\delta) = \sup_{0<\tau\leq\delta} \sup_t \|\Delta^k_{t,\tau} S_t f(\cdot)\|_{L_p(\sigma)}. \tag{1.10}$$

With its help A. P. Terechin obtained direct and inverse theorems of approximation in U which admit the interpretation in terms of L_p- approximation on σ by means of polynomials in spherical harmonics.
We must say that for the sphere this way seems to be rather hard despite of its elegance. The main difficulty is the interpretation of the apparatus of approximation. In case of a sphere this interpretation leads to polynomials in eigenfunctions of the Laplace-Beltrami operator, i.e. to polynomials in spherical harmonics. For more general manifolds the matter is much more complicated, see 3.2. and 3.3. .

1.6. Now we concentrate upon the results connected with the classical moduli (1.2). Here standard constructions of approximation via the operator $(I - S_t)^k$, $k \geq 1$, do not give raise for spherical harmonics. Here one has to use procedures which are not well adapted to the moduli of type (1.2) - Cesaro means, Riesz-Bochner means .... Therefore a proof of the direct theorem turned out to be difficult. The first step was made in [2] for $p = 2$. Using the multiplier-theorem on the sphere and specializing estimates of the Gegenbauer polynomials, G. A. Kalyabin obtained a Jackson-type theorem in terms of the moduli (1.2) for $1 \leq p \leq \infty$, cf. [11]. Finally H. P. Rustamov completed this work by proving the following estimate:

$$E_m(f)_p \leq c\, \omega_{k,p}\left(f, \frac{\pi}{m+1}\right)$$

for $f \in L_p(\sigma)$, $1 \leq p \leq \infty$, $k \geq 1$ and $m = 0,1,2,\ldots$ He made use of the Cesaro means. The results are to be published. In [15] H. P. Rustamov announced a proof for the equivalence of different moduli on the sphere.

2. Approximation by entire functions of exponential type and algebraic polynomials on general manifolds

The m-dimensional manifold $\Gamma = \Gamma_m \subset \mathbb{R}^n$, $1 \leq m \leq n$, is defined as follows.
Every point $x^0 \in \Gamma$ can be covered by a rectangle Δ that looks like:

$$\Delta = \{x \in \mathbb{R}^n \; : \; |x_i - x_i^0| < \alpha \quad \text{for} \quad i = 1,\ldots,m,$$

$$|x_j - x_j^0| < \beta \quad \text{for} \quad j = m+1,\ldots,n\}$$

after a suitable denumeration of the coordinates. On the projection

$$\Delta' = \{x' \in R^m \; : \; |x_i - x_i^0| < \alpha \quad \text{for} \quad i = 1,\ldots,m\}$$

the continuously differentiable functions

$$x_j = \psi_j(x') = \psi_j(x_1, \ldots, x_m) \tag{2.1}$$

are defined for $j = 1, \ldots, m$, describing the intersection of Γ and Δ

$$\gamma = \Gamma \cap \Delta \tag{2.2}$$

We assume Γ to be compact, i.e. Γ is a finite union of such intersections γ:

$$\Gamma = \bigcup_{j=1}^{N} \gamma_j \tag{2.3}$$

We write $\Gamma \in C^s$ if the functions ψ_j are s times continuously differentiable. Let us mention that Γ can be given by differentiable functions of parameters

$$x_j = \phi_j(t_1, \ldots, t_m) = \phi_j(t)$$

for $t \in \omega$ and $j = 1, \ldots, n$.
We define (see [19] - [22] for the details):

$$\|f\|_{L_p(\Gamma)} = \left(\int_\Gamma |f(x)|^p d\Gamma \right)^{1/p} \quad , \quad 1 \leq p \leq \infty ,$$

where $d\Gamma$ is the volume element of Γ. Furthermore, for functions f on $\mathbb{R}^n$ we define the norm

$$|\!|\!| f(x) |\!|\!|_{L_p(\Gamma)} = \sup_{y \in \mathbb{R}^n} \|f(x+y)\|_{L_p(\Gamma)}$$

Evidently we have for $z \in \mathbb{R}^n$:

$$|\!|\!| f(x+z) |\!|\!|_{L_p(\Gamma)} = |\!|\!| f(x) |\!|\!|_{L_p(\Gamma)}$$

For a domain $\Omega \in \mathbb{R}^n$ we define the L_p-norm as usual:

$$\|f\|_{L_p(\Omega)} = \left(\int_\Omega |f(x)|^p dx \right)^{1/p} \qquad (= \|f\|_p \quad \text{for} \quad \Omega = \mathbb{R}^n)$$

Let $g_\nu(x)$ be an entire function of the exponential type $\nu > 0$. Then the following estimates hold:

$$|\!|\!| g_\nu |\!|\!|_{L_p(\Gamma)} \leq c\nu^{\frac{n-m}{p}} \|g_\nu\|_p \tag{2.4}$$

for $1 \leq m \leq n$, $\Gamma \in C^1$ and

$$|\!|\!| \Delta^k_{t_i,h} g_\nu |\!|\!|_{L_p(\Gamma)} \leq c(\nu|h|)^k |\!|\!| g_\nu |\!|\!|_{L_p(\Gamma)} \tag{2.5}$$

for $\nu \geq 1$, $\Gamma \in C^k$. Here c is a constant that does not depend on the succeeding factor, and $\Delta^k_{t_i,h}$ denotes the difference of order k along t_i with step h. By definition for $r > 0$,

$1 \le p \le \infty$ and for a domain $\Omega \subset \mathbb{R}^n$ it holds $f \in H_p^r(\Omega)$ if and only if the following norms are finite:

$$\|f\|_{H_p^r(\Omega)} := \|f\|_{L_p(\Omega)} + \|f\|_{h_p^r(\Omega)},$$

$$\|f\|_{h_p^r(\Omega)} := \sup_{y \in \mathbb{R}^n} \frac{\|\Delta_y^k f(x)\|_{L_p(\Omega_y)}}{|y|^r}, \qquad k > r,$$

cf. also [23]; we write H_p^r if $\Omega = \mathbb{R}^n$. For a function f given on Γ we define (see also (2.1)-(2.3)):

$$\|f\|_{H_p^r(\gamma)} := \|F\|_{H_p^r(\Delta')},$$

where $F(x') = f(x_1, \ldots, x_m, \psi_{m+1}(x'), \ldots, \psi_n(x'))$ and $x \in \Delta' = \Delta' \cap \gamma$. Finally, put

$$\|f\|_{H_p^r(\Gamma)} := \sum_{\gamma} \|f\|_{H_p^r(\gamma)}.$$

It is well known that a function $f \in H_p^{r+\frac{n-m}{p}}$ on $\mathbb{R}^n$ admits a trace on Γ which belongs to $H_p^r(\Gamma)$ and, vice versa, a function $f \in H_p^r(\Gamma)$ can be extended to a function on $\mathbb{R}^n$ belonging to $H_p^{r+\frac{n-m}{p}}$, cf. [23] for the details.
Now we shall define two linear operations which make it possible to approximate a function $f \in H_p^{r+\frac{n-m}{p}}$ on Γ. The first operation is the following one:

$$g_\nu(x) - f(x) = \frac{1}{|\sigma|} \int_{\mathbb{R}^n} g(|y|) \Delta_{y/\mu}^k f(x) dy. \tag{2.6}$$

Here

$$g(u) = \kappa \left(\frac{\sin u}{u} \right)^{[n+r+\frac{n-m}{p}]+2}$$

is a Jackson-type kernel. The constant κ normalizes g:

$$\frac{1}{|\sigma|} \int_{\mathbb{R}^n} g(|\sigma|) dy = \int_0^\infty g(u) u^{n-1} du = 1.$$

and $|\sigma|$ is the $(n-1)$-dimensional measure of the unit sphere $\sigma \subset \mathbb{R}^n$, $\nu = (n + [r] + 2)\mu$, $[r]$ denotes the largest integer $< r$.
By definition g_ν is an entire function of exponential type ν.

Theorem 1 *Let $f \in H_p^{r+\frac{n-m}{p}}$, $\Gamma \in C^k$, $k > r$. Then the entire functions g_ν approximate f as follows:*

$$\|f - g_\nu\|_{L_p(\Gamma)} \le \frac{cK}{\nu^r}, \qquad K = \|f\|_{H_p^{r+\frac{n-m}{p}}}, \tag{2.7}$$

$$\|g_\nu\|_p \le c\|f\|_p, \qquad \nu \ge 1, \tag{2.8}$$

$$\|g_\nu\|_{H_p^r(\Gamma)} \le cT, \qquad T = \|f\|_p + K, \tag{2.9}$$

$$\|f\|_{H_p^r(\Gamma)} \leq cT. \tag{2.9'}$$

We give an outline of the proof. For $f \in H_p^{r+\frac{n-m}{p}}$ from (2.6) it follows that

$$\|g_\nu - f\|_p \leq c\nu^{-(r+\frac{n-m}{p})s}. \tag{2.10}$$

We put

$$f = \sum_{s=0}^{\infty} q_s \quad , \quad q_s = g_{2^s} - g_{2^s-1} \quad , \quad q_0 = g_1 \quad , \quad s = 1, 2, \ldots \tag{2.11}$$

Here the q_s are entire functions of the type 2^s which can be estimated from above:

$$\|q_s\|_p \leq c2^{-(r-\frac{n-m}{p})s}$$

Then according to (2.4) we get

$$|\!|\!|q_s|\!|\!|_{L_p(\Gamma)} \leq c2^{-sr} \quad , \quad s = 0, 1, 2, \ldots$$

Therefore the series (2.11) converges to f in the metric $|\!|\!|\cdot|\!|\!|_{L_p(\Gamma)}$. But in this metric Bernstein-type inequalities are valid. This gives us the possibility to estimate the values of

$$\frac{|\!|\!|\Delta^k_{t_s,h} f|\!|\!|_{L_p(\gamma)}}{|h|^r} \quad \text{and} \quad \frac{|\!|\!|\Delta^k_{t_s,h} g_\nu|\!|\!|_{L_p(\Gamma)}}{|h|^r}.$$

Therefore the $H_p^r(\Gamma)$-norms of f and g_ν can be estimated. This leads to the inequalities (2.9) and (2.9'), because $|\!|\!|\cdot|\!|\!|_{L_p(\Gamma)} \geq \|\cdot\|_{L_p(\Gamma)}$. So (2.5) leads to a new proof of the trace theorem $H_p^{r+\frac{n-m}{p}} \longrightarrow H_p^r(\Gamma)$. The theorem is proved.

Let us remark that (2.7) and (2.9) directly involve (2.9'). This fact is usually called the "converse part" of approximation.

The above theory has its origins in the classical theory of approximation by trigonometric polynomials going back to Jackson, Bernstein, de la Valee-Poussen and Zygmund.

In the second step we approximate the entire functions g_ν of exponential type ν. Here the inequality (2.8) will become important. We conclude

$$|g_\nu| \leq c_1\nu^{n/p}\|g_\nu\|_p \leq c\nu^{n/p}\|f\|_p \tag{2.12}$$

with the help of (2.4). This second linear operation gives us the possibility to approximate the functions $g_\nu(x)$ by algebraic polynomials $P_N(x)$ of order N which are the partial sums of the Chebychev series of $g_\nu(x)$ in the cube

$$V_\rho = \{x \,:\, |x_j| \leq \rho\,,\, j = 1, \ldots, n\} \supset \Gamma.$$

Here

$$P_N(x) = \sum_{|k|<4\rho\nu n} a_k \cos\left(k_1 \arccos\frac{x_1}{\rho}\right)\cdots\cos\left(k_n \arccos\frac{x_n}{\rho}\right), \tag{2.13}$$

$$k = (k_1, \ldots, k_n) \quad , \quad k_j \geq 0 \quad , \quad |k| = \sum_{j=1}^{n} k_j.$$

Theorem 2 *Let $f \in H_p^{r+\frac{n-m}{p}}$ and suppose g_ν to be the entire function from Theorem 1 and let P_N be defined as in (2.13). Then the following estimates hold:*

$$\|f - P_N\|_{L_p(\Gamma)} \leq cKN^{-r}, \tag{2.14}$$

$$\|P_N\|_{H_p^r(\Gamma)} \leq cK \quad \nu \geq 1, \tag{2.15}$$

$$K = \|f\|_{H^{r+\frac{n-m}{p}}} \quad , \quad \Gamma \in C^k \quad , \quad k > r.$$

Conversely, (2.14) *and* (2.15) *involve $f \in H_p^r(\Gamma)$ and after extension of f on $\mathbb{R}^n$ this finally results in $f \in H^{r+\frac{n-m}{p}}$.*

The proof makes essential use of the following estimates:

$$|a_s| \leq Me^{\nu n \rho}e^{-H\sum_{j=1}^n |s_j|},$$

$$|g_\nu(x)| \leq M \quad , \quad \sinh H = 1,$$

which involve:

$$\|g_\nu - P_N\|_{L_p(\Gamma)} \leq cKq^N,$$

$$\|g_\nu - P_N\|_{H_p^r(\Gamma)} \leq cKq^N$$

for some q, $0 < q < 1$. Now, using Theorem 1, we obtain:

$$\|f - P_N\|_{L_p(\Gamma)} \leq \mathcal{O}(N^{-r}) + \mathcal{O}(q^N) = \mathcal{O}(N^{-r}),$$

$$\|P_N\|_{H_p^r(\Gamma)} \leq \|g_\nu\|_{H_p^r(\Gamma)} + \|g_\nu - P_N\|_{H_p^r(\Gamma)} \leq c_1 + \mathcal{O}(q^N) \leq c.$$

The converse part of Theorem 2 is almost evident. Note that from the simpler inequalities

$$\|f - g_\nu\|_{L_p(\Gamma)} = \mathcal{O}(N^{-r}), \tag{2.16}$$

$$\|P_N\|_{L_p(\Gamma)} = \mathcal{O}(1) \tag{2.17}$$

the converse conclusion is true only for some relatively simple manifolds such as a sphere or an algebraic surface isomorphic to a sphere.

By the proposed method of approximation we obtain polynomials which are bounded not only with respect to their $L_p(\Gamma)$-norm but with respect to their $H_p^r(\Gamma)$-norm, too.

From (2.6) one obtains the estimate

$$\|f - g_\nu\|_{L_p(\Gamma)} \leq \int_0^\infty u^{n-1} g(u) \omega^k(f, u)_{L_p(\Gamma)} du, \tag{2.18}$$

where $\omega^k(f, u)_{L_p(\Gamma)}$ is the modulus of continuity of f of order k with respect to the metric $\|\cdot\|_{L_p(\Gamma)}$.

The inequality (2.15) could be generalized to other classes of functions, e.g. Besov spaces or Triebel-Lizorkin spaces ...

Furthermore it is possible to investigate the corresponding periodic problems by changing the operator (2.6) slightly, cf. [5], [6].

3. Approximation on homogeneous and Riemannian manifolds

3.1. First we talk a little about approximation on compact symmetric spaces $\mathcal{M}$ of rank 1. Denote by G the group of isometries of $\mathcal{M}$. Then there is a distinguished G-invariant differential operator - the Laplace-Beltrami operator $\Delta_{\mathcal{M}}$. Its spectrum is discrete and the corresponding invariant eigenspaces are finite dimensional and span $L_2(\mathcal{M}, \tilde{\mu})$ as direct summands. Here $\tilde{\mu}$ is the invariant measure on $\mathcal{M}$ corresponding to the Riemannian metric. The system of polynomials constructed via the projections on these subspaces can be used as the apparatus of approximation on $\mathcal{M}$ in the norm of $L_2(\mathcal{M}, \tilde{\mu}) = L_2(\mathcal{M})$.
Recall that $\mathcal{M}$ can be identified with the homogeneous space G/H, where H is a maximal closed subgroup of G. Fix a point $x \in \mathcal{M}$ and select an arbitrary unit vector X from the Lie algebra of G. Then the point $e^{tX} \cdot x$, $t \geq 0$, moves along a geodesic line through x. For a function $f(x)$, $x \in \mathcal{M}$, we define the k-difference relative to t with the step τ calculated for $t = 0$ by equality

$$\Delta_\tau^k f(e^{tX} \cdot x)|_{t=0} = \sum_{j=0}^{k} (-1)^{(k-j)} \binom{k}{j} f(e^{j\tau X} \cdot x). \tag{3.1}$$

Let μ be the Haar measure on G and let normalize the measures on G and H as follows:

$$\int\limits_G F(u) d\mu(u) = \int\limits_{\mathcal{M}} \left(\int\limits_H F(uh) dh \right) d\tilde{\mu}.$$

The modulus of smoothness $\omega_{k,p}^p(f, \delta)$ on $\mathcal{M}$ is defined by the following expression:

$$\omega_{k,p}^p(f, \delta) = \sup_{0<\tau\leq\delta} \left(\int\limits_G |\Delta_\tau^k f(u \cdot e^{tX} \cdot x)|_{t=0}|^p d\mu(u) \right)^{1/p}. \tag{3.2}$$

This expression can be generalized by means of differentiation along the geodesics. On the base of the modulus (3.2) one can construct the theory of approximation and can consider Besov spaces on $\mathcal{M}$ similar as in Chapter 1 (using eigenfunctions of the operator $\Delta_{\mathcal{M}}$ instead of spherical harmonics). Such an approach is realized in [25] for $p = \infty$ including interesting results associated with inequalities of Bernstein-, Favard-, Nikol'skij-type. It would be important to complete these investigations.

3.2. A different situation appears in the case of noncompact symmetric spaces. In the papers [26] and [27] the questions of approximation were considered for the Lobachevskii space Λ^n realized as a hyperboloid $\mathcal{H}$ in the pseudo-euclidean space $E^{n,1}$ with Poincaré metric. Here the isometry group is $SO(n,1)$ and the manifold can be represented in the form $\mathcal{H} = SO(n,1)/SO(n)$. Often Λ^n is called the 'hyperbolic space' and the pseudo-rotations are called hyperbolic rotations $SH(n+1)$, cf. [28]. As it was said, $\mathcal{H}$ is the hyperboloid on E^{n+1}:

$$\mathcal{H} = \{x = (x_1, \ldots, x_{n+1}) \ : \ x_1^2 + \cdots + x_n^2 - x_{n+1}^2 = -1 \ , \ x_{n+1} > 0\}$$

The pseudo-euclidean metric in $E^{n,1}$

$$dx_1^2 + \cdots + dx_n^2 - dx_{n+1}^2$$

induces the Riemannian metric on $\mathcal{H}$ which can be written in polar coordinates (θ, ϕ) as:

$$d\varrho^2 = d\theta^2 + \sinh^2\theta (d\varrho_\phi)^2 \tag{3.3}$$

The relation between x-coordinates and the polar coordinates is given by formulas similar to the transformation of euclidean to spherical coordinates in $\mathbb{R}^{n+1}$, but with hyperbolic instead of trigonometric functions of θ. Note that θ measures the global Riemannian distance of an $x \in \mathcal{H}$ from the pole $(0, \ldots, 0, 1)$ while $\phi(\varphi_{n-1}, \ldots, \varphi_1)$ is a "direction" and ϕ "travels" over the sphere $\sigma^{n-1} = \tilde{\sigma} \in \mathbb{R}^{n-1}$. Furthermore $d\varrho_\phi$ in (3.3) denotes the infinitesimal distance on $\tilde{\sigma}$. The distance $\varrho(x, y)$ between the points $x, y \in \mathcal{H}$ is given by the formula

$$\cosh \varrho(x, y) = [x, y] = -x_1 y_1 - \ldots - x_n y_n + x_{n+1} y_{n+1}.$$

Let $\Sigma_r(x^0)$ be the Riemannian sphere on $\mathcal{H}$ with radius r and center x^0. Recall the expression for the invariant measure on $\mathcal{H}$:

$$d\mu = \frac{dx_1 \ldots dx_n}{x_{n+1}} = \sinh^{n-1}\theta \, d\theta \, d\tilde{\sigma}(\phi).$$

The Laplace-Beltrami operator $\mathcal{L}$ on $\mathcal{H}$ has a continuous spectrum. The spectral decomposition with respect to $\mathcal{L}$ can be expressed by the following formula (cf.[26]):

$$f(x) = \int_0^\infty Y_\alpha(f, \alpha) \, d\nu(\alpha). \tag{3.4}$$

It resembles the Laplace series - the functions $Y_\alpha(f, \alpha)$ play the role of projections on the eigenspaces of $\mathcal{L}$:

$$Y_\alpha(f, x) = \int_{\mathcal{H}} f(y) P_\alpha([x, y]) \, dy.$$

Here $P_\alpha(\cosh\theta)$ is the associated Legendre function and

$$d\nu(\alpha) = \frac{1}{(2\pi)^{n/2}} \left| \frac{\Gamma(\frac{n-2}{2} + i\alpha)}{\Gamma(i\alpha)} \right|^2 .$$

For a locally summable function $f(x)$ on $\mathcal{H}$ we define the generalized shift by means of averaging (cf.(1.1)):

$$S_t^\# f(x) = \frac{1}{|\Sigma_r(x)|} \int_{\Sigma_r(x)} f(y) \, d_\Sigma y = \frac{1}{|\tilde{\sigma}| \sinh^{n-1} r} \int_{[x,y]=\cosh r} f(y) \, d_\Sigma y, \tag{3.5}$$

where $d_\Sigma y$ denotes the measure on Σ_r . With its help one can define different moduli of smoothness.
In [26] the author used the classical modulus

$$\omega_{k,p}^\#(f, \delta) = \sup_{0<t\leq\delta} \|(I - S_t^\#)^k f\|_{L_p(\mathcal{H})}$$

and developed a theory of approximation on $\mathcal{H}$ in the metric of $L_2(\mathcal{H})$ by means of partial segments of the integral (3.1). There a Jackson-type theorem and a Bernstein-type inequality were proved. Furthermore Besov spaces $B_{2,q}^\alpha$ were introduced and their connections to approximation were established.

However, soon it was found out that the generalization of these results to the case $1 \leq p < 2$ is not possible, cf. [27]. In fact, if $f \in L_p(\mathcal{H})$, $1 \leq p < 2$, the following function (cf. (3.3))

$$\int_0^\infty \eta(\alpha) Y(f,\alpha)\, d\nu(\alpha)$$

doesn't belong to $L_p(\mathcal{H})$ while $\eta(\alpha)$ is a bounded C^∞- function. So this method of approximation fails.
In the paper [27] the author started with the "tight" modulus (cf. (1.8)):

$$\omega_{k,p}^{T\,\#}(f,\delta) = \sup_{0<\tau\leq\delta} \sup_t \|\Delta_{t,\tau}^k S_t^\# f\|_{L_p(\mathcal{H})}$$

and acted via the scheme of 1.5. . This theory of approximation on $\mathcal{H}$ and the theory of Besov spaces behaves as usual for all p, $1 \leq p \leq \infty$. The apparatus of approximation - "entire vectors of finite power" - appears quite naturally here. However, its deeper investigation remains an open and interesting problem - in contrast to the sphere.

3.3. For sake of shortness we only mention the papers [29] and [30] where the results of [27] are generalized. In [29] our results were transferred to the case of Cartan-Hadamard manifolds $\mathcal{M}$ - i.e. complete simply connected Riemannian manifolds of non-positive sectional curvature. It is well known that such $\mathcal{M}$ are homeomorphic to the euclidean space. Moreover, the geodesics starting from some $x \in \mathcal{M}$ cover $\mathcal{M}$ singlefold. This makes it possible to supply the methods of [27] in the new situation. Thus the lack of homogeneity is not an obstacle for the construction of global approximation procedures on $\mathcal{M}$. Of course the comparison of local situations in different points turns out to be impossible.
In [30] the problem of approximation on the Heisenberg group *Heis*^{n} is considered. As a model for this group $\mathbb{R}^{2n+1}$ equipped with the standard group operation but without metric and affine structures is taken. Recall that on *Heis*^{n} automorphisms (dilatations) act which resemble a little bit the homothetic transformations of the euclidean space (however, with anisotropic character). In *Heis*^{n} a metric can be introduced which is naturally connected with the group operations and dilatations. It is an interesting fact that this metric does not arise from any infinitesimal Riemannian metric. The spaces $B_{p,q}^\alpha(Heis^n)$ considered in [30] are isotropic in some sense - according to the above mentioned facts. They are characterized by means of the best approximations with the help of "entire vectors of finite power".

4. Conclusion

Unfortunately we have no possibility to speak about the results of A. Kamzolov close to ours (see for example [31]), about the investigations of the Georgian school of mathematicians (see survey [32]) and about series of other important works (for instance G. Kušnirenko [33], Ar. Djafarov [34], etc.).
We should like to give our credit for the publications of German mathematicians, which exerted influence on our investigations (H. Berens, P. Butzer, H. Johnen, S. Pawelke, see [10], [35], [36]).
Finaly let us mention that the recent papers [37] and [38], having some connections to our topics.

References

[1] Nikol'skij, S. M. Approximation of functions of several variables and imbedding theorems. (Russian) Nauka, Moscow, 1969

[2] Nikol'skij, S. M. A theorem concerning approximation on the sphere. Analysis Mathematica, 9, 1983, 207-221

[3] Nikol'skij, S. M., Lizorkin, P. I., Approximation with spherical polynomials. (Russian) Trudy Mat. Inst. Steklov., 166, 1984, 186-200

[4] Nikol'skij, S. M., Lizorkin, P. I., Inequalities for harmonical, spherical and algebraical polynomials. (Russian) Dokl. Akad. Nauk SSSR, 289(3), 1986, 541-545

[5] Nikol'skij, S. M., Lizorkin, P. I., Function spaces on a sphere and the theory of approximations. (Russian) Mat. Zam., 41(4), 1987, 509-516

[6] Nikol'skij, S. M., Lizorkin, P. I., A new way to the theory of funftion spaces on the sphere. (Russian) Trudy Mat. Inst. Steklov., 181, 1988, 213-221

[7] Nikol'skij, S. M., Lizorkin, P. I., Approximation on a sphere - a survey. Banach center publications, Warsaw, 22, 1989, 281-292

[8] Terechin, A. P., Uniform approximation with algebraic polynomials on a sphere of a noncountably dimensional space. (Russian) Mat. Zam., 41(3), 1987, 333-341

[9] Terechin, A. P., Approximations with spherical harmonics and classes of differentiable functions on a sphere. (Russian)

[10] Butzer, P., Johnen, H., Lipschitz spaces on compact manifolds. J. Funct. Anal., 7, 1971, 242-266

[11] Kaljabin, G. A., On moduli of smoothness of functions on a sphere. (Russian) Dokl. Akad. Nauk SSSR, 294(5), 1987, 1051-1054

[12] Rustamov, H. P., On approximation of functions on a sphere. (Russian) Dokl. Akad. Nauk SSSR, 320(6), 1991, 1319-1325

[13] Lizorkin, P. I., Rustamov, H. P., Nikol'skij-Besov spaces on a sphere and the theory of approximation. (Russian) Trudy Mat. Inst. Steklov., 204, 1992

[14] Terechin, A. P., The bounded group of operators and best possible approximation. (Russian) Diff. uravnenija i vyčeslitelnaja mat., Saratov. Univ., 2, 1975, 3-28

[15] Rustamov, H. P., On the equivalency of several moduli of smothness on a sphere. (Russian) Dokl. Akad. Nauk SSSR, 321(1), 1991, 23-27

[16] Šanjkov, V. Ju., Coercitive properties of degenerate equations. (Russian) Dokl. Akad. Nauk SSSR, 281(3), 1985, 1320-1322

[17] Nikol'skij, S. M., Approximation of functions with trigonometric polynomials on manifolds. (Russian) Dokl. Akad. Nauk SSSR, part 1: 319(6), 1991, part 2: 320(1), 1991, 40-44

[18] Nikol'skij, S. M., Approximation of functions from the Besov class on manifolds with entire functions of exponential type. (Russian) Dokl. Akad. Nauk SSSR, 319(1), 1991

[19] Nikol'skij, S. M., Approximation with entire functions of exponential type and with polynomials. (Russian) Trudy Mat. Inst. Steklov., 201, 1992

[20] Nikol'skij, S. M., Again about the approximation with entire functions of exponential type and with polynomials. (Russian) Trudy Mat. Inst. Steklov., 204, 1993 (to appear)

[21] Nikol'skij, S. M., Approximation with entire functions of exponential type in a Banach space metric. (Russian) Dokl. Akad. Nauk Rossii, 324(2), 1992

[22] Nikol'skij, S. M., On a approximation method for functions with polynomials on a manifold. (Russian) Dokl. Akad. Nauk Rossii, 324(3), 1992, 529-532

[23] Nikol'skij, S. M., Properties of several classes of functions on differentiable manifolds. (Russian) Mat. Sb., 33(75), no. 2, 1953, 261-325

[24] Helgason, S., Differential geometry, Lie groups and symmetric spaces. Academic Press, New York, 1978

[25] Ivanov, V. A., Approximation of functions on homogeneous spaces. (Russian) PhD thesis, Moskov. Gos. Univ., 1984

[26] Petrova, I. V., Approximation on the hyperboloid in the L_2-metric. (Russian) Trudy Mat. Inst. Steklov., 194, 1992, 215-228

[27] Lizorkin, P. I., Direct and inverse theorems in the theory of approximation of functions on the Lobačevskii space. (Russian) Trudy Mat. Inst. Steklov., 194, 1992, 120-147

[28] Vilenkin, H. Ja., Special functions and representation theory of groups. (Russian) Nauka, Moscow, 1965

[29] Lizorkin, P. I., Approximation on Riemannian manifolds. (Russian) Trudy Mat. Inst. Steklov., 200, 1991, 222-235

[30] Lizorkin, P. I., Guliev, V. S., Function spaces and problems of approximation on the Heisenberg group. (Russian) Trudy Mat. Inst. Steklov., 201, 1992, 245-272

[31] Kamzolov, A. I., On the best approximation in $W_p^\alpha(S^n)$ with polynomials and spherical harmonics. (Russian) Mat. Zam., 32(3), 1982, 285-293

[32] Žižiašvili, L. V., Topurija, S. B., Fourier-Laplace series on the sphere. (Russian) Itogi Nauki i Techn., Ser. Mat. Anal., 15, 1977, 83-130

[33] Kušnirenko, G. G., Problems of approximation of continuous functions on the sphere with finite spherical sums. (Russian) Trudy Charkov. Polytechn. Inst., Ser. Ing. -Phys., 25(3), 1959, 3-22

[34] Džafarov, Ar. S., On spherical analoga of the classical theorems of J. Jackson and S. N. Bernstein. (Russian) Dokl. Akad. Nauk SSSR, 203(2), 1972, 278-281

[35] Berens, H., Butzer, P. L., Pawelke, S., Limitierungsverfahren von Reihen mehrdimensionaler Kugelfunktionen. Publ. RIMS, Kyoto Univ., Ser. A, 4(2), 1968, 201-268

[36] Pawelke, S., Über die Approximationsordnung bei Kugelfunktionen und algebraischen Polynomen. Tôhoku Math. J., 24, 1972,475-486

[37] Li Luoqing, Approximation for spherical functions by Bochner-Riesz means. Approx. Theory and its Appl., 7(4), 1991, 93-120

[38] Bondioli, Cr., Function spaces of Nikol'skij type on compact manifolds. Rendiconti Lincei, Mat. e appl. ser. IX., 14(3), 1992, 185-194

Pjotr I. Lizorkin
Sergej M. Nikol'skij
Steklov Mathematical Institute RAN
Vavilov 42
117 966 Moscow GSP-1, Russia
Fax: (095)-135-05-55

The Sharp Apriori Estimates for Some Superlinear Degenerate Elliptic Problems

Stanislav I. Pohožaev

Keywords and phrases: superlinear elliptic problem, the smoothness barrier, integral identity, sharp apriori estimates.

ABSTRACT

The principal objective of this paper (talk) is the systematic development of the general integral identities and their applications to nonlinear elliptic problems.

Here we shall restrict our consideration to specific nonlinear elliptic problems of the "simple" form

$$\begin{cases} \Delta u + f(x,u) = h(x) & \text{in} \quad \Omega \subset \mathbb{R}^N, \qquad N \geq 3 \\ u = 0 & \text{on} \quad \partial\Omega, \end{cases}$$

containing such equations as

$$\Delta u + \frac{1}{|x|^\gamma}|u|^{p-2}u = h(x)$$

with $2 < p < \frac{2N}{N-2}$ and $2\gamma > 2N - (N-2)p$.

We shall establish the existence of the smoothness barrier for these solutions, we shall deduce the sharp estimates for these solutions, for the gradients of solutions and for the second order derivatives of solutions. These a priori estimates are based on new integral identities.

We construct examples from which it follows that these a priori estimates are unimprovable ones.

Introduction

We consider the following equation

$$\Delta u + f(x,u) = h(x)$$

in bounded domain $\Omega \subset \mathbb{R}^N$ ($N \geq 3$) with smooth boundary $\partial\Omega$ under the Dirichlet condition

$$u|_{\partial\Omega} = 0.$$

First we shall establish the existence of the "smoothness barrier" for solutions of superlinear elliptic equations. This barrier is explained by the existence of the inherent singularity of solutions of such equations.

The "smoothness barrier" separates out the solutions for which the regularity theory is valid, and the solutions, for which the regularity theory is not valid in general.

At first we give the example (1.4), which is a special case of the example (2.1) where $\gamma = 0$. This will be done in order to show that the existence of the "smoothness barrier" is not connected with the singularity of the coefficients of the equation.

Furthermore we shall prove sharp a priori estimates. Unfortunately the wellknown, classical methods [1, 2] are not applicable to obtain the global a priori estimates for superlinear elliptic problems. Here the principal objective is the method of deducing these a priori estimates. This method is based on the new integral identity, in its turn which is the "weighted" version of the identity [4].

As we mentioned above, we have restricted ourselves to the simplest case of variational identities for nonlinear elliptic problems. More general identities can be found in the papers [5–8].

The contents of the present paper (talk) is as follows.

In Section 1 we shall prove the existence of the smoothness barrier for solutions of regular superlinear elliptic problems.

In Section 2 we shall prove the existence of the smoothness barrier for solutions of singular superlinear elliptic problems.

In Section 3 we shall deduce the basic integral identity.

In Section 4 we shall obtain the sharp (i.e. unimprovable) a priori estimates for both the solution and its gradient in corresponding spaces.

In Section 5 we shall give the example showing that the above a priori estimates cannot be improved.

In Section 6 we shall obtain sharp a priori estimates for the second derivatives of the solutions in corresponding spaces.

In Section 7 we shall discuss the example showing that the obtained a priori estimates are not improvable under certain conditions.

Section 8 is devoted to a special superlinear singular elliptic problem.

Remark on notations. Constants from the right-hand sides of inequalities don't depend on evaluated functions from corresponding spaces. In a given context, the same Const will (generally) be used to denote different constants.

1 The smoothness barrier for regular superlinear elliptic equations

Let us consider a problem of the form

$$\begin{cases} \Delta u + f(x,u) = h(x) & \text{in} \quad \Omega \subset \mathbb{R}^N, \qquad N \geq 3, \\ u = 0 & \text{on} \quad \partial\Omega, \end{cases} \tag{1.1}$$

where Ω is a bounded domain in $\mathbb{R}^N$ ($N \geq 3$) with smooth boundary $\partial\Omega$.

The regularity theory for weak solutions of such equations is well known, see [1, 2], if the Caratheodory (in particular continuous) function $f : \Omega \times \mathbb{R} \to \mathbb{R}$ possesses the coercive property, i.e.

$$\limsup_{|u|\to\infty} \frac{uf(x,u)}{|u|^2} \leq 0.$$

Here the increasing smoothness of h implies the increasing smoothness of the solution of the corresponding equation.

The situation is different for anticoercive superlinear equations, i.e.

$$\liminf_{|u|\to\infty} \frac{uf(x,u)}{|u|^2} = +\infty.$$

It turns out that there exists a wide class of such superlinear elliptic equations for which the solutions possess inhere singularity independent of the smoothness of the function h.

Now we give the following

Definition 1 *The (primary) L_{p_0}-solution ($p_0 > 1$) of the problem*

$$\Delta v = g(x) \quad in \quad \Omega, \qquad v = 0 \quad on \quad \partial\Omega$$

with $g \in L_{q_0}(\Omega)$, $q_0 > 1$, is called L_{p_0}-strong limit of the smooth solutions of

$$\Delta v_m = g_m(x) \quad in \quad \Omega, \qquad v_m = 0 \quad on \quad \partial\Omega,$$

if $g_m \to g$ strongly in L_{q_0} as $m \to \infty$.

Theorem 1 *Let the Caratheodory (in particular continuous) function $f : \Omega \times \mathbb{R} \to \mathbb{R}$ satisfy*

$$|f(x,y)| \leq \text{Const.}(1 + |u|^{p-1}) \tag{1.2}$$

with $p > p_* = \frac{2N-2}{N-2}$ ($N \geq 3$).

Let the solution u of problem (1.1) belong to L_{p_0}.

Let

$$p_0 > (p-2)\frac{N}{2}. \tag{1.3}$$

Then $u \in W_q^2(\Omega)$, if $h \in L_q(\Omega)$ with $q > 1$.

The proof is based on the well-known estimates for the solutions of linear elliptic problems and on the Sobolev embedding theorems for the spaces $W_t^2(\Omega)$ with $t > 1$.

These results together with the "bootstrap" arguments provide the proof of Theorem 1.

Corollary 1 *Let $p_0 = p$. Then "increasing of smoothness" occur if*

$$p < p_c = \frac{2N}{N-2} \qquad (N \geq 3 \quad and \quad p > p_*).$$

Corollary 2 *Let the a priori estimate for $\|u\|_{\overset{\circ}{W}{}_2^1(\Omega)}$ be known. Then "increasing of smoothness" occur if*

$$p < p_c \qquad (N \geq 3 \quad and \quad p > p_*)$$

Remark 1 *Corollary 2 is in general not valid if $p > p_0$ (see Remark 2 below).*

The inequality (1.3) is unimprovable. We show that the inequality (1.3) cannot be substituted by

$$p_0 = (p-2)\frac{N}{2} - \varepsilon \tag{1.3'}$$

for sufficiently small $\varepsilon > 0$. We consider the following example

$$\begin{cases} \Delta u + |u|^{p-2}u = h_\lambda(x) & \text{in } \Omega_0 = \{x \in \mathbb{R}^N \,|\, |x| < 1\}, \\ u = 0 & \text{on } \partial\Omega_0 \end{cases} \tag{1.4}$$

with $p > p_*$, $N \geq 3$, where

$$h_\lambda(x) = A^{p-1}|x|^{\beta-2}[(1-|x|^{\lambda-\beta})^{p-1} - 1] - A\lambda(\lambda+N-2)|x|^{\lambda-2},$$

$$\beta = -\frac{2}{p-2}, \qquad A = \left\{\frac{2}{(p-2)^2}[(N-2)p - 2N + 2]\right\}^{\frac{1}{p-2}}$$

and $\lambda > \beta$.

The solution of (1.4) is

$$u_\lambda(x) = A(|x|^\beta - |x|^\lambda).$$

For $\lambda = 2$ we have $h_2 \in L_\infty(\Omega)$ and

$$u_2 \in L_{p_0}(\Omega) \quad \text{for} \quad p_0 = (p-2)\frac{N}{2} - \varepsilon,$$

where ε is a sufficiently small positive number. Nevertheless the solution u_2 does not belong to $W_q^2(\Omega)$ with $q > 1$. Moreover, this solution even does not belong to $L_{p_1}(\Omega)$ with $p_1 = (p-2)\frac{N}{2}$.

Remark 2 *We note that* $u_2 \in \overset{\circ}{W}{}_2^1(\Omega)$ *for* $p > p_c = \frac{2N}{N-2}$.

"Inner" singularity of solutions of some superlinear elliptic equations.

Let us consider the family of problems (1.4) with parameter $\lambda > \beta$.

The smoothness of the right-hand side increases with the growth of λ. Nevertheless the smoothness of the solution u_λ does not increase. This is explained by the structure of the solution

$$u_\lambda(x) = v_0(x) + v_\lambda(x),$$

where $v_0(x) = A|x|^\beta$ and $v_\lambda(x) = -A|x|^\lambda$.

Thus the solution u_λ has its singular component $v_0(x)$ independent of the smoothness of the right-hand side. Actually, this component $v_0(x)$ determines the class of solutions for $\lambda > \beta$.

This fact occures for more general superlinear elliptic equations.

2 The smoothness barrier for singular superlinear elliptic equations

We consider a problem of the form

$$\begin{cases} x^\gamma \Delta u + |u|^{p-2}u = g(x) & \text{in } \Omega_0 = \{x \in \mathbb{R}^N \,|\, |x| < 1\}, \qquad N \geq 3, \\ u = 0 & \text{on } \partial\Omega_0, \end{cases}$$

or the corresponding problem

$$\begin{cases} \Delta u + \frac{1}{|x|^\gamma}|u|^{p-2}u = h_{\lambda,\gamma}(x) & \text{in } \Omega_0 \\ u = 0 & \text{on } \partial\Omega_0 \end{cases} \tag{2.1}$$

with $p > 2$ and $2 > \gamma > 2N - 2 - (N-2)p$.

We note that the admissible interval contains the positive values associated to the degenerated operator $-|x|^\gamma \Delta u$.

When $p > p_*$ this interval contains also nonpositive values γ.

Let in the problem (2.1)

$$h_{\lambda,\gamma}(x) = A_\gamma^{p-1}|x|^{\beta_\gamma - 2}[(1 - |x|^{\lambda - \beta_\gamma})^{p-1} - 1] - A_\gamma \lambda(\lambda + N - 2)|x|^{\lambda - 2}, \tag{2.2}$$

$$\beta_\gamma = \frac{\gamma - 2}{\beta - 2}, \qquad A_\gamma = \left\{\frac{(2-\gamma)[\gamma + (N-2)p - 2N + 2]}{(p-2)^2}\right\}^{\frac{1}{p-2}}, \qquad \lambda > \beta_\gamma.$$

The problem (2.1) for the function $h_{\lambda,\gamma}$, determined by the formula (2.2), has the solution

$$u_{\lambda,\gamma}(x) = v_{0,\gamma}(x) + v_{\lambda,\gamma}(x),$$

where

$$v_{0,\gamma}(x) = A_\gamma |x|^{\beta_\gamma}, \qquad v_{\lambda,\gamma}(x) = -A_\gamma |x|^\lambda.$$

We note that the smoothness of the function $h_{\lambda,\gamma}$ increases with the growth of λ. Moreover, the order of zero of this function at $x_0 = 0$ increases also

$$h_{\lambda,\gamma}(x) = -A_\gamma[(p-1)A_\gamma^{p-2} + \lambda(\lambda + N - 2)]|x|^{\lambda-2}(1 + o(1))$$

as $x \to 0$.

Nevertheless the smoothness of the solution $u_{\lambda,\gamma}$ does not increase because there exists the singular component $-v_{0,\gamma}(x)$.

Remark 3 *We note, that under the conditions of the problem* (2.1)–(2.2) *and* $\gamma > \frac{2N-(N-2)p}{2}$ *we have* $u_{\lambda,\gamma} \in \overset{0}{W}{}_2^1(\Omega)$.

Now consider the general problem (1.1).

Theorem 2 *Let the Caratheodory function* $f : \Omega \times \mathbb{R} \to \mathbb{R}$ *satisfy*

$$|f(x,u)| \leq \frac{\text{Const}}{|x|^\gamma}(1 + |u|^{p-1}) \tag{2.3}$$

with $2 > \gamma \geq 0$ *and* $p > p_\gamma = \frac{2N-2-\gamma}{N-2}$ $(N \geq 3)$.

Let the solution of (1.1) *belong to the space* $L_{p_0}(\Omega)$.

Then, if

$$p_0 > (p-2)\frac{N}{2-\gamma}, \tag{2.4}$$

from $h \in L_q(\Omega)$, $q > 1$, *it follows that* $u \in W_q^2(\Omega)$.

The proof of this theorem is analogous to the proof of Theorem 1.

Remark 4 *The inequality* (2.4) *cannot be replaced by*

$$p_0 = p_0(\varepsilon) \equiv (p-2)\frac{N}{2-\gamma} - \varepsilon \tag{2.4'}$$

with sufficiently small positive ε.

This fact follows from the problem (2.1)–(2.2) with $2 > \gamma \geq 0$. In this case the solution $u_{\lambda,\gamma}$ of this problem (for $\lambda \geq 2$) only belongs to the space $L_{p_0(\varepsilon)}(\Omega)$ with sufficiently small positive ε, although $h_\lambda \in L_\infty(\Omega)$ for $\lambda \geq 2$.

3 The integral identity

Now consider the problem (1.1) under following assumptions.

Let Ω be a bounded domain with sufficiently smooth boundary $\partial\Omega$. The point $x_0 = 0$ is contained in interior of Ω and the domain Ω is starshaped with respect to the point $x_0 = 0$. That means that the scalar product of the outward normal $\vec{\nu}$ to the boundary $\partial\Omega$ at the point $x \in \partial\Omega$ and the radius-vector $\vec{x}$ is nonnegative

$$(\vec{x}, \vec{\nu}) \geq 0 \quad \text{on} \quad \partial\Omega. \tag{3.1}$$

Condition (f_1). Let the function $f : \Omega \times \mathbb{R} \to \mathbb{R}$ be a Caratheodory function, i.e. $f(x, \cdot)$ is continuous on $\mathbb{R}$ for almost all $x \in \Omega$ and $f(\cdot, u)$ is measurable on Ω for all $u \in \mathbb{R}$.

Let the function f satisfy the inequality

$$|f(x,u)| \leq \frac{\text{Const}}{|x|^\gamma}(1 + |u|^{p-1}) \tag{3.2}$$

with

$$2 < p < \frac{2N}{N-2} \qquad (N \geq 3) \tag{3.3}$$

and

$$\frac{2N-(N-2)p}{2} < \gamma < 2. \tag{3.4}$$

Furthermore, let the function $F(x,u) \equiv \displaystyle\int_0^u f(x,t)\,dt$ have the partial derivatives $\frac{\partial F}{\partial x_i}$ $(i = 1, 2, \ldots, N)$ such that $\frac{\partial F}{\partial x_i} : \Omega \times \mathbb{R} \to \mathbb{R}$ are the Caratheodory functions satisfying the inequality

$$\left| \sum_{i=1}^N x_i \frac{\partial F(x,u)}{\partial x_i} \right| \leq \frac{\text{Const}}{|x|^\gamma}(|u| + |u|^p).$$

We introduce the following abbreviation

$$\mu_* = \frac{2(N-\gamma) - (N-2)p}{p-2}.$$

Note that (3.4) implies $\mu_* < 0$.

We suppose that the function h satisfies the following

Condition (*h*). Let

$$h \in L_{q_0}(\Omega) \quad \text{with} \quad q_0 > \frac{2N}{N+1}. \tag{3.5}$$

We introduce the following spaces

$$L_{p,\mu-\gamma}(\Omega) = \left\{ u(x) | \int_\Omega |u|^p |x|^{\mu-\gamma}\, dx < \infty \right\},$$

$$\overset{\circ}{W}{}^1_{2,\mu}(\Omega) = \left\{ u(x) | \int_\Omega |Du|^2 |x|^\mu\, dx < \infty,\ u|_{\partial\Omega} = 0 \right\}.$$

Now let the possible solution u of the problem (1.1) belong to the space

$$\overset{\circ}{W}{}^1_{2,\mu}(\Omega) \bigcap L_{p,\mu-\gamma}(\Omega) \quad \text{with} \quad \mu > \mu_*.$$

Consider this solution in the domain $\Omega_\rho = \Omega \setminus \overline{B}_\rho$, where B_ρ is the ball with centre at the point $x_0 = 0$ and sufficiently small radius $\rho > 0$ such that the closure $\overline{B}_\rho \subset \Omega$.

Remember that the well-known theory of local a priori estimates (see [1–3]) implies that the above-mentioned possible solution belongs to the Sobolev space $W^2_{q_0}(\Omega_\rho)$.

Multiplying the equation from (1.1) by the "integrating factor"

$$\left(\sum_{i=1}^N x_i \frac{\partial u}{\partial x_i} + \alpha u \right) |x|^\mu,$$

where α and μ are real parameters, integrating the obtained product over Ω_ρ and using the boundary condition from (1.1), we get

$$\frac{1}{2}\int_{\partial\Omega} |Du|^2 \cdot (x,\nu) \cdot |x|^\mu\, ds + \rho^{\mu+1} \int_{|x|=\rho} \left[\frac{1}{2}|Du|^2 - \left(\frac{\partial u}{\partial r}\right)^2 - F(x,u) - \alpha \frac{u}{\rho}\frac{\partial u}{\partial r} + \frac{\alpha\mu}{2}\frac{u^2}{\rho^2} \right] ds +$$

$$+ \left(\frac{N-\mu}{2} - 1 - \alpha \right) \int_{\Omega_\rho} \left(\frac{\partial u}{\partial r}\right)^2 |x|^\mu\, dx + \left(\frac{N+\mu}{2} - 1 - \alpha \right) \int_{\Omega_\rho} \left| \frac{1}{r} D_\omega u \right|^2 |x|^\mu\, dx +$$

$$+ \int_{\Omega_\rho} \left[\alpha u f(x,u) - (N+\mu)F(x,u) - \sum_{i=1}^N x_i \frac{\partial F(x,u)}{\partial x_i} \right] |x|^\mu\, dx +$$

$$+ \frac{1}{2}\alpha\mu(\mu+N-2) \int_{\Omega_\rho} u^2 |x|^{\mu-2}\, dx = \int_{\Omega_\rho} h(x) \left(\sum_{i=1}^N x_i \frac{\partial u}{\partial x_i} + \alpha u \right) |x|^\mu\, dx, \tag{3.6}$$

where $u = u(x)$.

Here $\vec{\nu}$ is a unit vector of the outward normal to the boundary $\partial\Omega$, (r,ω) are spherical coordinates in $\mathbb{R}^N$ with $r = |x|$,

$$|Du|^2 = \left(\frac{\partial u}{\partial r}\right)^2 + \left| \frac{1}{r} D_\omega u \right|^2 = \left(\frac{\partial u}{\partial x_1}\right)^2 + \left(\frac{\partial u}{\partial x_2}\right)^2 + \ldots + \left(\frac{\partial u}{\partial x_N}\right)^2,$$

$$F(x,u) = \int_0^u f(x,t)\, dt,$$

$\frac{\partial F(x,u)}{\partial x_i}$ is a partial derivative of $F(x,u)$ with respect to x_i $(i = 1, 2, \ldots, N)$.

Lemma 1 *Let the condition* (f_1) *be satified. Let* $u \in \overset{\circ}{W}{}^1_{2,\mu}(\Omega) \cap L_{p,\mu-\gamma}(\Omega)$ *with* $\mu > \mu_*$. *Then*

$$\liminf_{\rho\to 0} \rho^{\mu+1} \int_{|x|=\rho} \left[|Du|^2 + |F(x,u)| + \frac{|u|}{\rho}\left|\frac{\partial u}{\partial r}\right| + \frac{u^2}{\rho^2}\right] ds = 0. \tag{3.7}$$

Proof. Let $B_{\rho_1} = \{x \,|\, |x| < \rho_1\}$ be a ball of sufficiently small radius ρ_1 such that $\overline{B}_{\rho_1} \subset \Omega$. Then by assumption we have

$$\int_{B_{\rho_1}} \left(|Du|^2 + \frac{|u|^p}{|x|^\gamma}\right)|x|^\mu \, dx < \infty. \tag{3.8}$$

Furthermore, for $\mu > \mu_*$ we have

$$\int_{B_{\rho_1}} \left(\frac{|u| \cdot |Du|}{|x|} + \frac{u^2}{|x|^2}\right)|x|^\mu \, dx \le \text{Const} \cdot \int_{B_{\rho_1}} \left(|Du|^2 + \frac{|u|^p}{|x|^\gamma}\right)|x|^\mu \, dx$$

where the Const does not depend on the function u. Then by (3.8) we obtain

$$\int_{B_{\rho_1}} \left(|Du|^2 + \frac{|u|^p}{|x|^\gamma} + \frac{|u|}{|x|}\left|\frac{\partial u}{\partial r}\right| + \frac{1}{|x|^2}u^2\right)|x|^\mu \, dx < \infty. \tag{3.9}$$

Now we use arguments from a paper by M. Berestycki and P. Lions (see [9, p. 321]). We have

$$\int_{B_{\rho_1}} \left(|Du|^2 + \frac{|u|^p}{|x|^\gamma} + \frac{|u|}{|x|}\left|\frac{\partial u}{\partial r}\right| + \frac{1}{|x|^2}u^2\right)|x|^\mu \, dx =$$

$$= \int_0^{\rho_1} \left[\int_{|x|=\rho} \left(|Du|^2 + \frac{|u|^p}{|x|^\gamma} + \frac{|u|}{|x|}\left|\frac{\partial u}{\partial r}\right| + \frac{1}{|x|^2}u^2\right)|x|^\mu \, ds\right] d\rho < \infty.$$

Hence there exists a sequence $\rho_n \to 0$ such that

$$\rho_n^{\mu+1} \int_{|x|=\rho_n} \left(|Du|^2 + \frac{|u|^p}{\rho_n^\gamma} + \frac{|u|}{\rho_n}\left|\frac{\partial u}{\partial r}\right| + \frac{1}{\rho_n^2}u^2\right) ds \to 0, \tag{3.10}$$

as $n \to \infty$.

Indeed, if

$$\liminf_{\rho\to 0} \rho^{\mu+1} \int_{|x|=\rho} \left(|Du|^2 + \frac{|u|^p}{\rho^\gamma} + \frac{|u|}{\rho}\left|\frac{\partial u}{\partial r}\right| + \frac{1}{\rho^2}u^2\right) ds > 0,$$

then the integral of the function

$$\rho^\mu \int_{|x|=\rho} \left(|Du|^2 + \frac{|u|^p}{\rho^\gamma} + \frac{|u|}{\rho}\left|\frac{\partial u}{\partial r}\right| + \frac{u^2}{\rho^2}\right) ds$$

does not converge in $L_1[0, \rho_1]$. This fact contradicts to the inequality (3.9).

We further note that by (f_1) it follows that

$$|F(x,u)| \le \frac{\text{Const}}{|x|^\gamma}(|u| + |u|^p).$$

Then

$$\rho^{\mu+1}\int_{|x|=\rho}|F(x,u)|\,ds \le \text{Const}\cdot\rho^{\mu+1}\int_{|x|=\rho}\left(\frac{|u|^p}{\rho^\gamma}+\frac{|u|}{\rho^\gamma}\right)ds \le$$
$$\le \text{Const}_1\cdot\rho^{\mu+1}\int_{|x|=\rho}\frac{|u|^p}{\rho^\gamma}\,ds+\text{Const}_2\cdot\rho^{\mu+1}\int_{|x|=\rho}\frac{1}{\rho^\gamma}\,ds \le$$
$$\le \text{Const}_1\cdot\rho^{\mu+1}\int_{|x|=\rho}\frac{|u|^p}{\rho^\gamma}\,ds+\text{Const}_3\cdot\rho^{\mu+N-\gamma}.$$

Hence for $\mu > \mu_*$ we have

$$\rho_n^{\mu+N-\gamma}\to 0 \quad \text{as} \quad n\to\infty.$$

Then from (3.10) it follows that

$$\rho_n^{\mu+1}\int_{|x|=\rho_n}\left(|Du|^2+|F(x,u)|+\frac{|u|}{\rho_n}\left|\frac{\partial u}{\partial r}\right|+\frac{1}{\rho_n^2}u^2\right)ds\to 0, \tag{3.11}$$

as $n\to\infty$.

Thus the lemma is proved.

Theorem 3 *Let all the conditions from Section* 3 *be satisfied. Let the solution of the problem* (1.1) *belong to the space*

$$\overset{\circ}{W}{}^1_{2,\mu}(\Omega)\bigcap L_{p,\mu-\gamma}(\Omega) \quad \textit{with} \quad \mu>\mu_* \quad \textit{and} \quad \int_\Omega h^2(x)|x|^{\mu+2}\,dx<\infty.$$

Then this solution u satisfies the following integral representation

$$\frac{1}{2}\int_{\partial\Omega}|Du|^2(\vec{x},\vec{\nu})|x|^\mu\,ds+\left(\frac{N-\mu}{2}-1-\alpha\right)\int_\Omega\left(\frac{\partial u}{\partial r}\right)^2|x|^\mu\,dx+$$
$$+\left(\frac{N+\mu}{2}-1-\alpha\right)\int_\Omega\left|\frac{1}{r}D_\omega u\right|^2|x|^\mu\,dx+$$
$$+\int_\Omega\left[\alpha uf(x,u)-(N+\mu)F(x,u)-\sum_{i=1}^N x_i\frac{\partial F}{\partial x_i}(x,u)\right]|x|^\mu\,dx+$$
$$+\frac{1}{2}\alpha\mu(\mu+N-2)\int_\Omega u^2|x|^{\mu-2}\,dx=\int_\Omega h(x)\left(\sum_{i=1}^N x_i\frac{\partial u}{\partial x_i}+\alpha u\right)|x|^\mu\,dx. \tag{3.12}$$

Proof. Consider the integral representation (3.6) for $\rho=\rho_n$, where $\{\rho_n\}$ is the sequence from Lemma 1.

We note that for $\mu>\mu_*$ the inequality

$$\int_\Omega u^2|x|^{\mu-2}\,dx\le\text{Const}\cdot\int_\Omega\frac{|u|^p}{|x|^\gamma}|x|^\mu\,dx,$$

holds, where Const doesn't depend on the function u.

Using the assumptions of the theorem the corresponding integrals over domains Ω_{ρ_n} converge to the corresponding integrals over the domain Ω as $n\to\infty$.

Using the limit (3.11) the corresponding integrals over the spheres $|x|=\rho$ converge to zero as $n\to\infty$.

Then taking the limit of the integral (3.6) for $\rho=\rho_n$ as $n\to\infty$, we obtain the representation (3.12).

4 A priori estimates for solutions and their derivatives of first order

Let us introduce an additional condition on the function f.

Let now

$$\mu^* = \frac{(N-2)p - 2(N-\gamma)}{p+2}$$

Then (3.4) implies $\mu^* > 0$.

Condition (f_2). Let $\mu \in \mathbb{R}$ and $\alpha \in \mathbb{R}$ be such parameters that

$$\mu_* < \mu < \mu^*, \tag{4.1}$$

$$K_1 := \frac{N - |\mu| - 2}{2} - \alpha > 0, \tag{4.2}$$

and

$$\alpha u f(x,u) - (N+\mu)F(x,u) - \sum_{i=1}^{N} x_i \frac{\partial F}{\partial x_i}(x,u) \geq \frac{K_2}{|x|^\gamma}|u|^p - \frac{K_3}{|x|^\gamma} \tag{4.3}$$

with $K_2 > 0$ and $K_3 > 0$ for almost all $x \in \Omega$ and for all $u \in \mathbb{R}$.

Let us give an example of the function $f(x,u)$ satisfying both Condition (f_1) and Condition (f_2).

Example 1 *Consider the function* $f(x,u) = \frac{|u|^{p-2}u}{|x|^\gamma}$ *with* p *and* γ *satisfying the inequalities* (3.3) *and* (3.4), *respectively. Then we have*

$$\alpha u f(x,u) - (N+\mu)F(x,u) - \sum_{i=1}^{N} x_i \frac{\partial F}{\partial x_i}(x,u) = \left(\alpha - \frac{N+\mu-\gamma}{p}\right)\frac{|u|^p}{|x|^\gamma}$$

The inequality (4.3) *is satisfied if*

$$\alpha > \frac{N+\mu-\gamma}{p} \tag{4.4}$$

Thus in this case Condition (f_2) *is reduced to compatibility of the inequalities* (4.1), (4.2) *and* (4.4). *The compatibility of this system of inequalities follows from* (3.3) *and* (3.4).

Theorem 4 *Let all conditions from Section* 3 *and Condition* (f_2) *be satisfied. Let the solution* u *of* (1.1) *belong to the space*

$$\overset{\circ}{W}{}^1_{2,\mu}(\Omega) \bigcap L_{p,\mu-\gamma}(\Omega) \quad \textit{for} \quad \mu : \mu_* < \mu < \mu^*$$

and

$$\int_\Omega h^2(x)|x|^{\mu+2}\,dx < \infty.$$

Then we have the following a priori estimate for u

$$\int_{\partial\Omega} |Du|^2(\vec{x},\vec{\nu})|x|^\mu\,ds + \int_\Omega |Du|^2|x|^\mu\,dx +$$

$$+ \int_\Omega |u|^p|x|^{\mu-\gamma}\,dx \leq C_3 \int_\Omega h^2(x)|x|^{\mu+2}\,dx + C_4, \tag{4.5}$$

where $C_3 > 0$ *and* $C_4 > 0$ *do not depend neither on* u *nor on* h *from the indicated classes.*

Proof. We have

$$\left|\frac{1}{2}\alpha\mu(\mu+N-2)\int_\Omega u^2|x|^{\mu-2}\,dx\right| \le \frac{1}{4}K_2\int_\Omega \frac{|u|^p}{|x|^\gamma}|x|^\mu\,dx + C_1,$$

where

$$C_1 = \frac{p-2}{p}\left(\frac{8}{pK_2}\right)^{\frac{2}{p-2}}\left|\frac{1}{2}\alpha\mu(\mu+N-2)\right|^{\frac{p}{p-2}}\cdot\int_\Omega |x|^{\frac{(p-2)\mu+2(\gamma-p)}{p-2}}\,dx < \infty$$

for $\mu > \mu_*$. Furthermore, we have

$$\left|\int_\Omega h(x)\sum_{i=1}^N x_i\frac{\partial u}{\partial x_i}|x|^\mu\,dx\right| \le \frac{K_1}{2}\int_\Omega |Du|^2|x|^\mu\,dx + \frac{1}{2K_1}\int_\Omega h^2(x)|x|^{\mu+2}\,dx$$

and

$$\left|\alpha\int_\Omega h(x)u|x|^\mu\,dx\right| \le \frac{K_2}{4}\int_\Omega\frac{|u|^p}{|x|^\gamma}|x|^\mu\,dx + \frac{\alpha^2}{4}\int_\Omega h^2(x)|x|^{\mu+2}\,dx + C_2,$$

where

$$C_2 = \frac{p-2}{p}\left(\frac{8}{pK_2}\right)^{\frac{p}{p-2}}\int_\Omega |x|^{\frac{(p-2)\mu+2(\gamma-p)}{p-2}}\,dx < \infty$$

for $\mu > \mu_*$.

We also have for $\mu > \mu_*$

$$\int_\Omega \frac{|x|^\mu}{|x|^\gamma}\,dx < \infty.$$

Now from these inequalities and from the integral representation (3.12) we obtain the a priori estimate (4.5). Hence Theorem 4 is proved.

5 Unimprovability of the a priori estimate (4.5)

Consider the following problem

$$\begin{cases} \Delta u + \frac{1}{|x|^\gamma}|u|^{p-2}u = h_0(x) & \text{in } \Omega_0 = \{x\in\mathbb{R}^N \mid |x|<1\}, \quad N\ge 3,\\ u = 0 & \text{on } \partial\Omega_0, \end{cases} \tag{5.1}$$

where p satisfies the condition (3.3) and γ satisfies the condition (3.4). Here

$$h_0(x) = A_0^{p-1}|x|^{\beta-2}[(1-|x|^{2-\beta})^{p-1}-1] - 2A_0N,$$

$$\beta = \frac{\gamma-2}{p-2} < 0, \qquad A_0 = |\beta(\beta+N-2)|^{\frac{1}{p-2}}.$$

The function h_0 satisfies the condition

$$\int_{\Omega_0} h_0^2(x)|x|^{\mu+2}\,dx < \infty, \qquad \forall\mu > \mu_*,$$

and $h_0 \in L_{q_0}(\Omega_0)$ with $q_0 > \frac{2N}{N+1}$.

Thus all conditions of Theorem 4 are satisfied for the problem (5.1).

This problem has the solution

$$u_0(x) = A_0(|x|^\beta - |x|^2), \tag{5.2}$$

which belongs to $\overset{\circ}{W}{}^1_{2,\mu}(\Omega_0) \cap L_{p,\mu-\gamma}(\Omega_0)$ for any $\mu > \mu_*$ and does not belong neither to $\overset{\circ}{W}{}^1_{2,\mu_*}(\Omega_0)$ nor to $L_{p,\mu_*-\gamma}$.

Thus the weight exponent $\mu > \mu_*$ and hence the a priori estimate (4.5) is unimprovable in the considered class of degenerated quasilinear elliptic problems (1.1).

6 A priori estimates for derivatives of second order

In this section we obtain the sharp a priori estimates for derivatives of second order to the solution of the problem (1.1).

At first we shall establish the preliminary estimates for the function

$$f^*(x) = f(x, u(x)),$$

where f satisfies Condition (f_1), i.e. satisfies all inequalities from (f_1). Let $N = 3$ and assume that

$$\gamma p' < 3 \tag{6.1}$$

holds.

Lemma 2 *Let $u \in L_{p,\mu-\gamma}(\Omega)$ and*

$$\nu_0 = (p'-1)\gamma + \mu \le 0, \qquad \frac{1}{p} + \frac{1}{p'} = 1.$$

Then $f^* \in L_{p'}(\Omega)$ and

$$\int_\Omega |f^*(x)|^{p'}\, dx \le \text{Const} \int_\Omega |u(x)|^p |x|^{\mu-\gamma}\, dx + \text{Const}, \tag{6.2}$$

where constants are independent of the function u.

Proof. The proof is evident. Indeed, we have

$$\int_\Omega |f^*(x)|^{p'}\, dx \le \text{Const}_1 \int_\Omega |u(x)|^p \cdot |x|^{-\gamma p'}\, dx + \text{Const}_2 \int_\Omega |x|^{-\gamma p'}\, dx =$$

$$= \text{Const}_1 \int_\Omega |u(x)|^p \cdot |x|^{\mu-\gamma} |x|^{-\nu_0}\, dx + \text{Const}_2 \int_\Omega |x|^{-\gamma p'}\, dx \le$$

$$\le \text{Const}_3 \int_\Omega |u(x)|^p |x|^{\mu-\gamma}\, dx + \text{Const}_4$$

under the above conditions.

Lemma 3 *Let* $u \in L_{p,\mu-\gamma}(\Omega)$ *and*

$$(p'-1)\gamma + \mu > 0.$$

Then $f^* \in L_q(\Omega)$ *and*

$$\int_\Omega |f^*(x)|^q \, dx \leq \mathrm{Const}(q)\left(\int_\Omega |u(x)|^p \cdot |x|^{\mu-\gamma} \, dx\right)^{\frac{q}{p'}} + \mathrm{Const}(q) \tag{6.3}$$

for any positive $q < \frac{Np'}{(p'-1)\gamma+\mu+N}$, *where the constants* $\mathrm{Const}(q) > 0$ *are independent of the function* u.

Proof. We have

$$\int_\Omega |f^*(x)|^q \, dx \leq \mathrm{Const}_1(q) \int_\Omega \frac{|u(x)|^{(p-1)q}}{|x|^{\gamma q-\nu}} \cdot \frac{1}{|x|^\nu} \, dx + \mathrm{Const}_2(q) \int_\Omega \frac{1}{|x|^{\gamma q}} \, dx. \tag{6.4}$$

Putting

$$\nu = \frac{q}{p'}[(p'-1)\gamma + \mu], \qquad t = \frac{p'}{q} > 1 \quad \text{and} \quad t' = \frac{t}{t-1},$$

we obtain

$$\int_\Omega \frac{|u(x)|^{(p-1)q}}{|x|^{\gamma q-\nu}} \frac{1}{|x|^\nu} \, dx \leq \left(\int_\Omega \frac{|u(x)|^{(p-1)qt}}{|x|^{(\gamma q-\nu)t}} \, dx\right)^{1/t} \times$$

$$\times \left(\int_\Omega \frac{1}{|x|^{\nu t'}} \, dx\right)^{1/t'} = \mathrm{Const}_3(q) \cdot \left(\int_\Omega \frac{|u(x)|^p}{|x|^\gamma} |x|^\mu \, dx\right)^{q/p'}.$$

Furthermore, we have

$$\int_\Omega \frac{1}{|x|^{\gamma q}} \, dx \leq \mathrm{Const}_4(q)$$

under the above conditions.

Substituting the obtained inequalities in (6.4) we get the assertion of Lemma 3.

Theorem 5 *Let all the conditions from Section 3, Condition* (f_2) *and*

$$\gamma > \frac{2N - (N-2)p}{p'} \tag{6.5}$$

holds.

Suppose that (6.1) *and* $p > 3$ *are satisfied for* $N = 3$.

Let the solution u *of* (1.1) *belong to the space*

$$\overset{\circ}{W}{}^1_{2,\mu_0}(\Omega) \cap L_{p,\mu_0-\gamma} \quad \textit{for} \quad \mu_0 = -(p'-1)\gamma, \qquad \mu_* < \mu_0 < \mu^*,$$

and let

$$\int_\Omega h^2(x)|x|^{\mu_0+2} \, dx < \infty.$$

Then this solution u *belongs to* $W^2_{p'}(\Omega)$ *and*

$$\|u\|_{W^2_{p'}(\Omega)} \leq \mathrm{Const}\left[\left(\int_\Omega h^2(x)|x|^{\mu_0+2} \, dx\right)^{1/p'} + \left(\int_\Omega h^2(x)|x|^{\mu_0+2} \, dx]\right)^{1/2} + 1\right],$$

where the positive constant Const *does'nt depend neither on* u *nor* h *from these classes.*

Remark 5 *Evidently it follows from the inequality* (6.5) *that* γ *is greater than the left-hand side of* (3.4).

Proof. Theorem 4 implies for $\mu = \mu_0$

$$\int_\Omega |u|^p |x|^{\mu_0-\gamma}\, dx \le \text{Const} \int_\Omega h^2(x)|x|^{\mu_0+2}\, dx + \text{Const}.$$

Then by Lemma 2 for $\mu = \mu_0$ we obtain

$$\int_\Omega |f^*(x)|^{p'}\, dx \le \text{Const} \int_\Omega h^2(x)|x|^{\mu_0+2}\, dx + \text{Const}, \tag{6.6}$$

where the positive constants Const are independent of u and h.

Now we get by the conditions of the theorem $h \in L_{p'}(\Omega)$ and

$$\|h\|_{L_{p'}(\Omega)} \le \text{Const}\left(\int_\Omega h^2(x)|x|^{\mu_0+2}\, dx\right)^{1/2}.$$

Furthermore, applying the well-known theory of the linear problem

$$\begin{cases} \Delta u = -f^*(x) + h(x) & \text{in } \Omega, \\ u = 0 & \text{on } \partial\Omega, \end{cases}$$

we have

$$\|u\|_{W^2_{p'}(\Omega)} \le \text{Const}(\|f^*\|_{L_{p'}(\Omega)} + \|h\|_{L_{p'}(\Omega)}),$$

where the positive constant Const doesn't depend neither on f nor on h from the above classes.

Then the statement of Theorem 5 follows from the estimate (6.6).

Remark 6 *Once more we notice that the different constants are denoted by the same letter.*

Now we consider the case

$$\gamma \le \frac{2N - (N-2)p}{p'} \tag{6.7}$$

under the condition (3.4). Let

$$q(\mu) = \frac{Np'}{(p'-1)\gamma + \mu + N}$$

where $(p'-1)\gamma + \mu \ge 0$.

Then $q(\mu) > 1$, if $\mu < \frac{N-\gamma}{p-1}$.

Theorem 6 *Let all the conditions from Section 3,* (f_2) *and* (6.7) *be satisfied.*

Moreover, suppose that (6.1) *holds for* $N = 3$. *Take* μ *such that*

$$\mu_* < \mu < \min\left(\mu^*, \frac{N-\gamma}{p-1}\right). \tag{6.8}$$

Let the solution of (1.1) *belong to the space*

$$\overset{\circ}{W}{}^{1}_{2,\mu}(\Omega)\bigcap L_{p,\mu-\gamma}(\Omega).$$

Let $\int_\Omega h^2(x)|x|^{\mu+2}\,dx<\infty$ *and* $h\in L_q(\Omega)$ *for every* q *with*

$$1<q<q(\mu). \tag{6.9}$$

Then for every such q*, the solution* u *belongs to* $W^2_q(\Omega)$ *and*

$$\|u\|_{W^2_q(\Omega)}\le \text{Const}\left[\|h\|_{L_q(\Omega)}+\left(\int_\Omega h^2(x)|x|^{\mu+2}\,dx\right)^{1/p'}+1\right], \tag{6.10}$$

where the positive constant Const *doesn't depend neither on* u *nor on* h *from the above classes.*

Proof. From Theorem 4 we get

$$\int_\Omega |u|^p|x|^{\mu-\gamma}\,dx\le C_3\int_\Omega h^2(x)|x|^{\mu+2}\,dx+C_4,$$

where the corresponding constants are determined in Section 4 with μ satisfying (6.8).
Then by Lemma 2 we have for the function f^*

$$\|f^*\|_{L_q(\Omega)}\le \text{Const}\left(\int_\Omega h^2(x)|x|^{\mu+2}\,dx+\text{Const}\right)^{1/p'} \tag{6.11}$$

for every q satisfying (6.9), where the constants do not depend neither on u nor on h.
Furthermore, by the well-known theory of the linear problem

$$\begin{cases}\Delta u=-f^*(x)+h(x) & \text{in }\ \Omega,\\ u=0 & \text{on }\ \partial\Omega,\end{cases}$$

we have

$$\|u\|_{W^2_q(\Omega)}\le \text{Const}(\|f^*\|_{L_q(\Omega)}+\|h\|_{L_q(\Omega)}),$$

where the constant Const >0 doesn't depend neither on f^* nor on h.

Then we obtain the statement of Theorem 6 from (6.11). Now on the basis of this theorem we establish the "limit" a priori estimate for the derivatives of second order of the solution to the problem (1.1).

Calculate the limit

$$q^*=q(\mu_*)=\frac{N(p-2)}{2p-2-\gamma}$$

for $(p'-1)\gamma+\mu_*\ge 0$, if γ satisfies the inequality (6.7). Notice that it follows from (3.4) that $q^*>1$ for $N\ge 3$ and $p>2$.

Theorem 7 *Let all the conditions from Section 3 and* (f_2), (6.7) *be satisfied. Suppose that* (6.1) *is satisfied for* $N=3$. *Let* $\int_\Omega h^2(x)|x|^{\mu_*+2}\,dx<\infty$ *and* $h\in L_{q^*}(\Omega)$. *Let the solution* u *of* (1.1) *belong to* $\overset{\circ}{W}{}^{1}_{2,\mu}(\Omega)\cap L_{p,\mu-\gamma}(\Omega)$ *for any* $\mu>\mu_*$.

Then for every q

$$1<q<q^* \tag{6.12}$$

this solution belongs to $W_q^2(\Omega)$ *and*

$$\|u\|_{W_q^2(\Omega)} \leq \text{Const}\left[\|h\|_{L_{q^*}(\Omega)} + \left(\int_\Omega h^2(x)|x|^{\mu_*+2}\,dx\right)^{1/p'} + 1\right], \tag{6.13}$$

where the positive constant Const *doesn't depend neither on u nor on h from the above classes.*

Proof. For every $q:\ 1 < q < q^*$ we define

$$\varepsilon = q^* - q > 0 \quad \text{and} \quad \mu_\varepsilon : q(\mu_\varepsilon) = q^* - \varepsilon/2.$$

Then for $\mu = \min(\mu_\varepsilon, 0)$ with $q(\mu) \geq q^* - \varepsilon/2$ the conditions of Theorem 6 are satisfied. Hence we get the desired result for for $q:\ 1 < q = q^* - \varepsilon < q(\mu)$

7 Unimprovability of the a priori estimate for the second derivatives

In this section we show that the a priori estimate (6.13) is sharp one, i.e. it is unimprovable in the class of the problem (1.1), satisfying the conditions of Theorem 7. We show that under these conditions the exponent q cannot be replaced by $q = q^*$ without additional assumptions.

We consider the problem (5.1) from Section 5 under conditions and (6.1), and (6.7) for γ again. Concerning this problem, as noticed in Section 5, all the conditions from Section 3 and (f_2) are satisfied. Moreover, we have

$$\int_{\Omega_0} h_0(x)|x|^{\mu_*+2}\,dx < \infty, \qquad \int_{\Omega_0} |h_0(x)|^{q^*}\,dx < \infty.$$

The solution of this problem determined by (5.2), as noticed in Section 5, belongs to $\overset{\circ}{W}{}^1_{2,\mu}(\Omega_0)$ for any $\mu > \mu_*$. Thus all the conditions of Theorem 7 are satisfied for the problem (5.1).

Concerning the solution $u_0(x)$ with $u_0|_{\partial\Omega_0} = 0$, we have

$$\|u_0\|_{W_q^2(\Omega_0)} \asymp \left(\int_{\Omega_0} |\Delta u|^q\,dx\right)^{1/q} = \text{Const}\left[\int_0^1 (r^{\beta-2} + \text{Positive Const})^q r^{N-1}\,dr\right]^{1/q},$$

where $\beta - 2 = \frac{\gamma-2}{p-2} - 2 < 0$. Here p and γ satisfy the following conditions

$$2 < p < \frac{2N}{N-2} \qquad (N \geq 3),$$

$$\frac{2N-(N-2)p}{2} < \gamma < \min\left(2, \frac{2N-(N-2)p}{p'}\right)$$

and additionally $\gamma p' < 3$, if $N = 3$; the sign $\asymp$ denotes the equivalence.

Then we obtain

$$\|u_0\|_{W_q^2(\Omega_0)} < \infty \quad \text{for} \quad (\beta-2)q + N > 0,$$

i.e. for $q < \frac{N(p-2)}{2p-2-\gamma} = q^*$ and $\int_{\Omega_0}|\Delta u_0|^{q^*}\,dx = \infty$. Hence u_0 does not belong to the space $W_{q^*}^2(\Omega_0)$.

Thus, under the conditions of Theorem 7 the exponent $q:\ 1 < q < q^*$ cannot be replaced by $q = q^*$.

8 Special problem (1.1)

Consider the following boundary value problem

$$\begin{cases} \Delta u + \frac{1}{|x|^\gamma}|u|^{p-2}u = h(x) & \text{in } \Omega \subset \mathbb{R}^N, \quad N \geq 3, \\ u = 0 & \text{on } \partial\Omega, \end{cases} \tag{8.1}$$

in the domain $\Omega \subset \mathbb{R}^N$ with smooth boundary $\partial\Omega$ satisfying the condition (3.1), where p and γ satisfy the inequalities (3.3) and (3.4), respectively.

Remark 7 *We do not suppose here the additional condition* (6.1) *for* $N = 3$.

Then all theorems given above are valid for this problem. For example, we present the theorem on the sharp a priori estimates for the second derivatives.

Theorem 8 *Let the conditions* (3.1), (3.3), (3.4), (3.5) *and* (6.7) *hold. Let*

$$\int_\Omega h^2(x)|x|^{\mu_*+2}\,dx < \infty \quad \textit{and} \quad h \in L_{q^*}(\Omega).$$

Let the solution u *of the problem* (1.1) *belong to* $\overset{\circ}{W}{}^1_{2,\mu}(\Omega) \cap L_{p,\mu-\gamma}(\Omega)$ *with any* $\mu > \mu_*$.

Then for every q *satisfying* (6.12) *this solution* u *belongs to* $W_q^2(\Omega)$ *and satisfies the inequality* (6.13).

Remark 8 *In Theorem 8 the condition* (3.4) *may be replaced by the inequality*

$$2\gamma > 2N - (N-2)p.$$

References

[1] O. A. Ladyzhenskaya and N. N. Ural'tseva. Linear and Quasilinear Elliptic Equations. Moscow: Nauka, 1973.

[2] D. Gilbarg, N. S. Trudinger. Elliptic Partial Differential Equations of Second Order. Berlin–New-York, 1977.

[3] S. Agmon, A. Douglis, L. Nirenberg. Estimates near the Boundary for Solutions of Elliptic Partial Differential Equations Satisfying General Boundary Conditions. *Comm. Pure Appl. Math.* **12** (1959) 623–727.

[4] S. I. Pohožaev. Eigenfunctions of the Equation $\Delta u + \lambda f(u) = 0$. *Soviet Math. Dokl.* **6** (1965) 1408–1411. (English transl. *Dokl. Akad. Nauk SSSR.* **165** (1965) 33–36.)

[5] S. I. Pohožaev. On the Eigenfunctions of Quasilinear Elliptic Problems. *Math. USSR Sbornik* **11** (1970) No. 2, 171–188. (English transl. *Mat. Sbornik* **82** (1970) No. 2, 192–212.)

[6] P. Pucci and J. Serrin. A General Variational Identity, *Indiana Univ. Math. J.* **35** (1986) No. 3, 681–703.

[7] N. Kawano, W.-M. Ni, Sh. Youtsutani. A Generalized Pohožaev Identity and Its Applications, *J. Math. Soc. Japan* **42** (1990) No. 3, 541–564.

[8] R. C. A. M. van der Vorst. Variational Identities and Applications to Differential Systems, *Arch. Rat. Mech. Anal.* **116** (1991) 375–398.

[9] H. Berestycki, P. L. Lions. Nonlinear Scalar Field Equations, I, Existence of a Ground State, *Arch. Rat. Mech. Anal.* **82** (1983) No. 4, 313–345.

Stanislav I. Pohožaev
Steklov Mathematical Institute RAN
Vavilov 42
117 966 Moscow GSP-1, Russia
Fax: 0007-095-135-05-55
E-Mail: POHOZAEV@LOG.MIAN.SU

Singularity Theory in Function Spaces

Thomas Runst

1 INTRODUCTION

The aim of the survey is to describe how one can use results of the singularity theory (in the sense of H. Whitney [40] and R. Thom [38]) in quasi- Banach and Banach spaces, respectively, for the investigation of the solution structure of the Dirichlet problem for semilinear elliptic equations.
At first we consider the following nonlinear elliptic equation

$$\begin{aligned} -\Delta\, u(x) - f(u(x)) &= h(x), \quad x \in \Omega \\ u(x) &= 0, \qquad x \in \partial\Omega, \end{aligned} \tag{1}$$

where $\Omega \subset \mathbb{R}^n$ is a bounded and smooth domain with boundary $\partial\Omega$, h is a given data function and f is a sufficiently smooth real-valued function on $\mathbb{R}^1$ which generates a nonlinear superposition operator.
We are interested in existence and multiplicity results of solutions of equation (1) in dependence on the kind of the nonlinearity f and on the given function h.
Let $0 < \lambda_1 < \lambda_2 \leq \cdots \leq \lambda_k \leq \cdots$ denote the sequence of eigenvalues of the equation

$$\begin{aligned} -\Delta v &= \lambda v \text{ in } \Omega \\ v &= 0 \quad \text{on } \partial\Omega. \end{aligned} \tag{2}$$

It is known that our problem (1) admits multiplicity results in dependence on the behaviour between the values of the first derivative f' and the spectrum of $-\Delta$ under Dirichlet conditions if h belongs to the classical Hölder space $C^\alpha(\Omega), 0 < \alpha < 1$ fixed. First, note that if the range of f' contains no eigenvalue λ_k,i.e.

$$f'(\mathbb{R}^1) \cap \sigma(-\Delta) = \emptyset,$$

where $\sigma(-\Delta) = \{\lambda_k, k \in \mathbb{N}\}$ denotes the spectrum of $-\Delta$ under Dirichlet condition, then (1) is uniquely solvable in $C^{2+\alpha}(\Omega)$. We refer to A.Hammerstein [17] and C.L.Dolph [10].
More interesting are the cases when

$$f'(\mathbb{R}^1) \cap \sigma(-\Delta) \neq \emptyset.$$

The first result, where the range of f' contains only the first eigenvalue λ_1 was proved by A. Ambrosetti, G. Prodi [2] in 1972. In their paper "On the inversion of some differentiable mappings with singularities between Banach spaces" they showed that under the following conditions for $f \in C^2(\mathbb{R}^1)$:

$$f(0) = 0,\ 0 < f'(-\infty) < \lambda_1 < f'(+\infty) < \lambda_2 \text{ and } f'' > 0 \text{ on } \mathbb{R}^1$$

the solution structure of equation (1) can be completely described.

THEOREM 1 (A. Ambrosetti, G. Prodi) . *Assume that f satisfies the above assumptions. Then there exists a smooth hypersurface $M_1 \subset C^\alpha(\Omega)$ of codimension 1 which disconnects $C^\alpha(\Omega)$ in two components M_0 and M_2 with the following property:*

if $h \in M_0$, then (1) has no solution,
if $h \in M_1$, then (1) has exactly one solution ,
if $h \in M_2$, then (1) has exactly two solutions in $C^{2+\alpha}(\Omega)$.

For the proof they used the local behaviour of the corresponding Fredholm mapping $\Phi : C_0^{2+\alpha}(\Omega) \longrightarrow C^\alpha(\Omega)$ defined by $\Phi(u) = -\Delta u - f(u)$. In that case the origin is a singular point of $\Phi'(u)[v] = -\Delta v - f'(u)v, \Phi' : C_0^{2+\alpha}(\Omega) \longrightarrow C^\alpha(\Omega)$. Using tools from singularity theory the behaviour of Φ can be completely described by means of the so-called normal form corresponding to such a singularity. The above assumptions with respect to f imply the existence of a global fold point at 0. A corresponding result within the framework of Besov and Triebel–Lizorkin spaces was proved by M. Geisler, T. Runst [15]. A. Manes, A. M. Micheletti [24] replaced the condition $0 < f'(-\infty) < \lambda_1$ by $-\infty < f'(-\infty) < \lambda_1$. The next important result was obtained by J. L. Kazdan, F. W. Warner [19] for more general nonlinear functions f. For example, they showed if one decomposes $h = h_1 + t\varphi_1$ (φ_1: normalized eigenfunction to λ_1, h_1: smooth function which satisfies $\int_\Omega h_1(x)\varphi_1(x)dx = 0$), then there exists a $t_0 = t_0(h_1)$ such that (1) has no solution for $t > t_0$ and at least one solution for $t < t_0$. Simultaneously, E. N. Dancer [9] and H. Amann, P. Hess [1] showed that if f satisfies $f'(-\infty) < \lambda_1 < f'(\infty)$ and f' is bounded on $[0, \infty)$, then (1) has at least two solutions if $h = h_1 + t\varphi_1$ and $t < t_0(h_1)$ and at least one solution if $t = t_0$, see also M. S. Berger, E. Podolak [5] and E.Podolak [28]. In [21], A. C. Lazer, P. J. McKenna showed if $f'(-\infty) < \lambda_1$ and $\lambda_{2k} < \lambda_{2k+1}, k \geq 1$ then there exists $t_1 \leq t_0$ such that (1) has at least three solutions if $h = h_1 + t\varphi_1$ and $t < t_1$. Further results in this direction may be found in P. Hess, B. Ruf [18], B.Ruf [30] and S. Solimini [37]. Furthermore, A. C. Lazer, P. J. McKenna [22] obtained if $f'(-\infty) < \lambda_1$ and $\lambda_2 < f'(\infty) \leq \lambda_3$, then (1) has at least four solutions if $h = h_1 + t\varphi_1$ and t is sufficiently small. Results with respect to function spaces of Besov–Triebel– Lizorkin spaces may be found in P. Drábek, T. Runst [11], L. Päivärinta, T. Runst [27] and T. Runst [35].

Here we consider problem (1) using tools from singularity theory. The aim of the following survey is to describe how one can apply results of the singularity theory in quasi- and Banach spaces, respectively, to obtain a good describtion of the solution structure. As above mentioned that approach was already taken by A. Ambrosetti, G. Prodi [2] in the case of Banach spaces. In recent years, the connection between singularity theory in Banach spaces and multiplicity results has been developped further by M. S. Berger, P. T. Church, J. G. Timourian [3], [4], P. T. Church, J. G. Timourian [8], F. Lazzeri, A. M. Micheletti [23], F. Giannoni, A. M. Micheletti [16], A. M. Micheletti [25], V. Cafagna, G. Tarantello [7], B. Ruf [31], [32], [33], ...

The singularity theory was elaborated for Banach spaces. For general quasi- Banach spaces some striking tools fail: not locally convex, the quasi-norm is not necessarily continuous, the dual space may be trivial, ... However, if there are " enough " continuous functionals over the quasi- Banach spaces some results of that theory can be carried over. For detailed proofs of the following results we refer to M. Geisler, T. Runst [15] and T. Runst [36].

2 DIFFERENTIABLE MAPPINGS BETWEEN QUASI-BANACH SPACES

Let X be a (real or complex) linear vector space. $\| \cdot \mid X \|$ is said to be a quasi-norm if $\| \cdot \mid X \|$ satisfies the usual conditions of a norm with the exception of the triangle inequality which is replaced by

$$\| x+y \mid X \| \leq K(\| x \mid X \| + \| y \mid X \|), \tag{3}$$

i.e. there exists a positive constant $K > 0$ such that (3) holds for all $x \in X$ and all $y \in X$. Of course, we have $K \geq 1$. (If $K = 1$ is admissible, then X is a normed space.) A quasi-normed space is said to be a quasi-Banach space if it is complete.
Let X be a quasi-Banach space. By a theorem due to S. Rolewicz [29] (see also G. Köthe [20]) we may assume without loss of generality that

$$\| x+y \mid X \|^{\lambda} \leq \| x \mid X \|^{\lambda} + \| y \mid X \|^{\lambda}$$

holds for suitable $\lambda, 0 < \lambda \leq 1$. This converts X into a linear metric space with translation invariant metric

$$d_X(x,y) := \| x-y \mid X \|^{\lambda} .$$

Let X, Y be two quasi-Banach spaces with quasi-norm $\| \cdot \mid X \|$ and $\| \cdot \mid Y \|$, respectively. A linear mapping $T : X \to Y$ is continuous if

$$\| T \mid X \to Y \| := \sup_{x \neq 0} \frac{\| Tx \mid Y \|}{\| x \mid X \|} < \infty.$$

Clearly $\| \cdot \mid X \to Y \|$ is a quasi-norm. For continuous $R, T : X \to X$ one has

$$\| R \cdot T \mid X \to X \| \leq \| R \mid X \to X \| \cdot \| T \mid X \to X \| .$$

LEMMA 1 (M. Geisler, T. Runst) . *Let X be a quasi-Banach space and T be an isomorphism of X onto X. Then for all linear mappings $S : X \to X$ with $\| S \mid X \to X \|$ small enough T-S is a an isomorphism, too.*

DEFINITION 1 *Let X and Y be quasi-Banach spaces. A map $\Phi : U \to Y$, $U \subset X$ open, is of class $C^k(U)$: $k = 0,1,2,\dots$ if Φ has a Taylor expansion with continuous multilinear forms*

$$\begin{aligned}\Phi(x+h) &= \Phi(x) + \Phi'(x)[h] + \frac{1}{2}\Phi''(x)[h,h] + \cdots \\ &+ \frac{1}{k!}\Phi^{(k)}(x)[h,\dots,h] + r_k(x,h)\end{aligned}$$

for all $x, x+h \in U$ *such that*
(i) $\Phi^{(j)}(x)$ *depends continuously on* $x, j = 0, 1, \ldots, k$,
(ii) $\lim_{\| h | X \| \to 0} \frac{\| r_k(x,h) | Y \|}{\| h | X \|^k} = 0$.

As in the case of Banach spaces one gets an implicit function theorem.

LEMMA 2 (M. Geisler, T. Runst) . *Let X, Y, Z be quasi-Banach spaces, U be a neighbourhood of the origin (0;0) in* $X \times Y$ *and* $\Phi : U \to Z$ *be of class* $C^k(U)$ *such that*
(i) $\Phi(0,0) = 0$
(ii) $\Phi'_y(0,0)$ *is invertible as a map of* $U \cap Y$ *in Z.*
Then there exist a neighbourhood $V \subset X, 0 \in V$, *and a unique map* $\varphi : V \to Y$ *of class* $C^k(V)$ *which solves the equation* $\Phi(x,y) = 0$*; that is,* $\Phi(x, \varphi(x)) = 0$.

Remark 1 : Let a mapping Φ of class C^1 have an invertible (Fréchet-)derivative $\Phi'(x)$ at the point x. Then Φ maps a neighbourhood of x C^1-diffeomorphically onto a neighbourhood of $\Phi(x)$. We say that Φ is locally invertible at x.

DEFINITION 2 *Let* $\Phi : X \to Y$ *be a* C^1 *-map. A point* $x \in X$ *is called singular if* $\Phi'(x)$ *is not invertible; that is,* $\Phi'(x)[u] = 0$ *has a nontrivial solution* $v \in X$. *Otherwise, x is a regular point of* Φ. *The collection of all singular points of* Φ *we call the singular set and denote it by* S_Φ. *A point* $y \in Y$ *is said to be a critical point if* $y \in \Phi(S_\Phi)$.

The importance of the singular set S_Φ for understanding the solution structure becomes clear from the following theorem.

THEOREM 2 (M. Geisler, T. Runst) . *Let X, Y be quasi-Banach spaces, and let* $\Phi : X \to Y$ *be a proper* $C^1(X)$*-map (i.e. for every compact* $K \subset Y$, *the preimage* $\Phi^{-1}(K) \subset X$ *is compact). Then the following assertions hold.*
(i) For $S_\Phi = \emptyset$, Φ *is a global homeomorphism of X onto Y.*
(ii) For $y \in Y \setminus \Phi(S_\Phi)$ *the cardinality of* $\Phi^{-1}(y)$ *is finite and constant on every connected component of* $Y \setminus \Phi(S_\Phi)$.

An essential tool for further investigation is the use of bifurcation systems (splitting of an equation via projections). Therefore one should assume that a factor space can be identified with a closed subspace of the original linear space. For Banach spaces this can be done easily by virtue of the Hahn–Banach extension theorem. For quasi-Banach spaces the situation is more complicated.

DEFINITION 3 *Let A be a quasi-Banach space and* $B \subset A$ *be a closed subspace. The factor space* A/B *is the collection of all sets* $x + B = \{x + y \mid y \in B\}$ *equipped with a quasi-norm*

$$\| x + B \mid A/B \| := \inf_{y \in B} \| x - y \mid A \| .$$

Remark 2 : It follows that $\| \cdot \mid A/B \|$ is a quasi-norm and that A/B is complete with respect to this quasi-norm.
To describe the factorization in quasi-Banach spaces we introduce the concept of dual-rich quasi-Banach spaces. We explain " how many " continuous functionals a space should have to become a " good " space in our considerations.

DEFINITION 4 *A quasi-Banach space X is said to be dual-rich if for all $0 \neq x \in X$ there exists a continuous linear functional $x^\star$ with $x^\star(x) = 1$.*

Remark 3 : Clearly, every Banach space and also the quasi-Banach spaces $l_p, 0 < p < 1$, is dual-rich; on the other hand, $L_p(\mathbb{R}^1)$ with $0 < p < 1$ is not dual-rich because its dual is trivial.
Let A be a dual-rich quasi-Banach space and B be a finite-dimensional subspace spanned by $x_1, \ldots, x_n$. Then there exists a dual system $x_1^\star, \ldots, x_n^\star$ such that $x_i^\star(x_k) = \delta_{ik}$ (Kronecker's symbol). Put for $x \in A$

$$\pi_B x = \sum_{i=1}^{n} x_i^\star(x) x_i, \pi_B^\perp x = x - \pi_B x.$$

Clearly, π_B and $\pi_B^\perp$ are continuous linear operators with

$$\pi_B x = x, \pi_B^\perp x = 0 \text{ if and only if } x \in B,$$

and

$$\pi_B^2 = \pi_B, \pi_B^{\perp^2} = \pi_B^\perp.$$

So $\pi_B^\perp$ is well-defined on A/B, that is, $\pi_B^\perp(x+B) = \pi_B^\perp x$, and creates an isomorphism between A/B and $\pi_B^\perp(A)$. On the other hand, let E be a closed subspace of finite codimension, that is, there exist linearly independent $y_1, \ldots, y_n \notin E$ such that $x \in A$ can be written as $x = \lambda_1 y_1 + \cdots + y_n + y$ with $y \in E$. In the same way as before we define

$$\pi_E^\perp x = \sum_{i=1}^{n} y_i^\star(x) y_i, \pi_E x = x - \pi_E^\perp$$

and obtain

$$\pi_E^\perp x = 0 \text{ if and only if } x \in E,$$

and

$$\pi_E^2 = \pi_E, \pi_E^{\perp^2} = \pi_E^\perp.$$

So $\pi_E^\perp$ is well-defined on A/E and creates an isomorphism between A/E and $\pi_E^\perp(A)$, also.

DEFINITION 5 *Let X be a quasi-Banach space. A subset $M \subset X$ is called a C^k- manifold of codimension 1 if for all $m \in M$ one can find a neighbourhood U of m and a C^k-functional $F : U \to \mathbb{R}^1$ such that*

$$(i)\ F'(m) \neq 0\,,\ (ii)\ M \cap U = \{x \in U\,,\ F(x) = 0\}$$

Remark 4 : It is not hard to see that $a^\star(x) = 0$ defines a closed linear subspace of codimension 1 if $a^\star$ is a non- vanishing continuous linear functional.

LEMMA 3 *Let M be a closed connected C^1-manifold of codimension 1 in a dual-rich quasi-Banach space X. Then $X \setminus M$ has at most two components.*

3 SINGULARITY THEORY IN DUAL-RICH QUASI-BANACH SPACES

We give a short survey on singularity theory in dual-rich quasi-Banach spaces.

DEFINITION 6 *Let E, F be dual-rich quasi-Banach spaces. A mapping $\Phi \in C^1(E,F)$ is called a (nonlinear) Fredholm operator if the Fréchet derivative of Φ at u, $\Phi'(u)$, is a linear Fredholm map belonging to $\mathcal{L}(\mathcal{E},\mathcal{F})$ for each $u \in E$. In this case the index of Φ, ind Φ, is defined by*

$$ind\,\Phi(u) := ind\,\Phi'(u) = dim\,ker\,\Phi'(u) - dim\,coker\,\Phi'(u) \text{ for all } u \in E.$$

It is known that the index of Φ is independent of $u \in E$. In order to describe the behaviour of Φ in a neighbourhood of a singular point we need a classification in the sense of Thom's list of elementary catastrophes. In what follows, we assume that E and F are dual-rich quasi-Banach spaces, and that

$$\Phi : E \to F$$

is a (nonlinear) Fredholm operator of index 0, with

$$dim\,ker\,\Phi'(u) \leq 1 \text{ for all } u \in E.$$

The basic tools for the classification of the singularities are the following two lemmas. We remark that the singularities investigated here belong to infinite-dimensional quasi-Banach spaces. But by our assumptions with respect to Φ these mappings can be handled like their finite-dimensional counterpart. The first result is called the local representation lemma, here in dual-rich quasi-Banach spaces. For the proof we refer to T .Runst [36].

LEMMA 4 *Let E, F be dual-rich quasi-Banach spaces, $\Phi : E \to F$ as above. Let $\bar{u} \in E$ be a singular point, i.e. by our assumptions*

$$dim\,ker\,\Phi'(\bar{u}) = dim\,coker\,\Phi'(\bar{u}) = 1 \text{ at } \bar{u} \in E.$$

Then there exist neighbourhoods $U(\bar{u}) \subset E$, $V(\Phi(\bar{u}) \subset F$, a quasi-Banach space X and smooth diffeomorphisms

$$\alpha \ : \ U \to \alpha(U) \subset \mathbb{R}^1 \times X$$
$$\beta \ : \ V \to \beta(V) \subset \mathbb{R}^1 \times X$$

such that

$$\Psi : \beta \circ \Phi \circ \alpha^{-1} : \alpha(U) \to \beta(V)$$

has the form

$$\Psi(t,x) = (f(t,x), x)\,.$$

Remark 5 : This lemma is a generalization of results obtained by B. Ruf [33], see also V. Cafagna, F. Donati [6], within the framework of dual-rich quasi-Banach spaces.
Furthermore, we apply the Malgrange preparation theorem with respect to dual- rich quasi-Banach spaces. A proof may be found in T. Runst [36].

LEMMA 5 *Let X be a dual-rich quasi-Banach space, U an open neighbourhood of the origin (0 ; 0) of $\mathbb{R}^1 \times X$ and $f : U \to \mathbb{R}^1$ a smooth function such that*

$$\frac{\partial^l}{\partial t^l} f(0,0) = \begin{cases} 0: & l = 0, 1, \dots, k-1 \\ \neq 0: & l = k \end{cases}$$

Then there exist smooth functions

$$a_j : U \to \mathbb{R}^1, j = 0, 1, \dots, k-1$$

vanishing at (0 ; 0) such that

$$t^k = \sum_{j=0}^{k-1} a_j\,(f(t,z), z) \cdot t^j.$$

Remark 6 : The infinite-dimensional version of the Malgrange preparation theorem can be found in P. Michor [26] and V. Cafagna, G. Tarantello [7]. It is possible to carry over that result to dual-rich quasi- Banach spaces.

4 SPACES OF BESOV–TRIEBEL–LIZORKIN TYPE

For our purpose Besov spaces $B^s_{p,q}(\mathbb{R}^n)$ and Triebel-Lizorkin spaces $F^s_{p,q}(\mathbb{R}^n)$ are of special interest. We start by listing some basic material. Let $\mathcal{S} = \mathcal{S}(\mathbb{R}^n)$ denote the Schwartz space of rapidly decreasing functions on the euclidean n-space $\mathbb{R}^n$ and let $\mathcal{S}' = \mathcal{S}'(\mathbb{R}^n)$ be its topological dual. Then one can find functions ϕ_0, $\phi \in \mathcal{S}(\mathbb{R}^n)$ such that

$$\text{supp}\ \phi_0 \subset \{x \mid |x| \leq 2\}\ ,\ \text{supp}\ \phi \subset \{x \mid 1 \leq |x| \leq 4\}$$

$$0 \leq \phi_0(x), \phi(x) \leq 1,\ \phi_j(x) := \phi(2^{-j+1}x),\ \sum_{j=0}^{\infty} \phi_j(x) = 1.$$

We call these a dyadic resolution of unity.

DEFINITION 7 *For $-\infty < s < \infty, 0 < p, q \leq \infty$ we define*
$B^s_{p,q}(\mathbb{R}^n) = \{f \in \mathcal{S}'(\mathbb{R}^n) :$

$$\|f|B^s_{p,q}(\mathbb{R}^n)\| = \left(\sum_{j=0}^{\infty} 2^{jsq} (\int_{\mathbb{R}^n} \mid \mathcal{F}^{-1}\phi_j \mathcal{F} f(x) \mid^p dx)^{\frac{q}{p}} \right)^{\frac{1}{q}} < \infty\}$$

and
$F^s_{p,q}(\mathbb{R}^n) = \{f \in \mathcal{S}'(\mathbb{R}^n) :$

$$\|f|F^s_{p,q}(\mathbb{R}^n)\| = \left(\int_{\mathbb{R}^n} (\sum_{j=0}^{\infty} 2^{jsq} \mid \mathcal{F}^{-1}\phi_j \mathcal{F} f(x) \mid^q)^{\frac{p}{q}} dx \right)^{\frac{1}{p}} < \infty\}$$

(with obvious modification for $p = \infty$ and /or $q = \infty$).

For $F^s_{p,q}(\mathbb{R}^n)$ we always assume that $p<\infty$. Here $\mathcal{F}$ stands for Fourier transform in $\mathcal{S}'$ and $\mathcal{F}^{-1}$ for its inverse; $F^s_{p,q}(\mathbb{R}^n)$ and $B^s_{p,q}(\mathbb{R}^n)$ are Banach spaces for $p,q\geq 1$ and quasi-Banach spaces otherwise.
We have the following continuous embeddings:

$$\mathcal{S}\subset B^s_{p,q}(\mathbb{R}^n)\subset\mathcal{S}' \text{ and } \mathcal{S}\subset F^s_{p,q}(\mathbb{R}^n)\subset\mathcal{S}'.$$

LEMMA 6 *For $-\infty<s<\infty$, $0<p,q\leq\infty$, $F^s_{p,q}(\mathbb{R}^n)$ and $B^s_{p,q}(\mathbb{R}^n)$ are dual-rich.*

Proof: We know that $\mathcal{S}\subset B^s_{p,q}(\mathbb{R}^n)\subset\mathcal{S}'$ and $\mathcal{S}\subset F^s_{p,q}(\mathbb{R}^n)\subset\mathcal{S}'$, where the embeddings are continuous. If $\psi\in\mathcal{S}$, this means that

$$|\, f(\psi)\,|\leq \|f\mid B^s_{p,q}(\mathbb{R}^n)\|\, P(\psi).$$

Thereby c does not depend on f and ψ, and P stands for a proper seminorm in the $\mathcal{S}$-topology. From this we conclude that $\mathcal{S}\subset(B^s_{p,q}(\mathbb{R}^n))'$, where the prime denotes duality. If $f(\psi)=0$ for all $\psi\in\mathcal{S}$ then $f=0$ in $B^s_{p,q}(\mathbb{R}^n)$. Consequently, for all $f\neq 0$ we find $\psi\in\mathcal{S}$ with $f(\psi)=1$. The same arguments for $F^s_{p,q}(\mathbb{R}^n)$ finish the proof.

Let us remark that the two scales of spaces $B^s_{p,q}(\mathbb{R}^n)$ and $F^s_{p,q}(\mathbb{R}^n)$ cover many well- known function spaces. We give some examples; for details we refer to H. Triebel [39].
Let $s>0$; then $B^s_{\infty,\infty}(\mathbb{R}^n)=\mathcal{C}^s(\mathbb{R}^n)$, a Zygmund space. If s is not an integer then $B^s_{\infty,\infty}(\mathbb{R}^n)=C^s(\mathbb{R}^n)$, a Hölder space. Let $1<p<\infty$, $-\infty<s<\infty$, then $F^s_{p,2}(\mathbb{R}^n)=H^s_p(\mathbb{R}^n)$, a Bessel-potential space. Let $1\leq p<\infty, 1\leq q\leq\infty, s>0$, then $B^s_{p,q}(\mathbb{R}^n)=\Lambda^s_{p,q}(\mathbb{R}^n)$, the classical Besov space. Furthermore, it holds

$$B^s_{2,2}(\mathbb{R}^n)=F^s_{2,2}(\mathbb{R}^n)=H^s_2(\mathbb{R}^n).$$

For dealing with boundary value problems it is useful to define Besov and Triebel–Lizorkin spaces on domains. Let Ω be a bounded C^∞- domain in $\mathbb{R}^n$ with boundary $\partial\Omega$. Then one can introduce the spaces $B^s_{p,q}(\partial\Omega)$ and $F^s_{p,q}(\partial\Omega)$ by standard procedure via local charts, see H. Triebel [39]. The spaces $B^s_{p,q}(\Omega)$ and $F^s_{p,q}(\Omega)$ are defined as usually by the restriction method, see next definition. If $g\in\mathcal{S}'$ then the restriction of g to Ω is an element of $\mathcal{D}'(\Omega)$ which will be denoted by $g_{|\Omega}$. We recall that $\mathcal{D}'(\Omega)$ is the collection of all complex-valued distributions on Ω.

DEFINITION 8 *(i) Let $0<p\leq\infty$, $0<q\leq\infty$ and $-\infty<s<\infty$. Then*

$$B^s_{p,q}(\Omega)=\{f\in\mathcal{D}'(\Omega),\ \exists g\in B^s_{p,q}(\mathbb{R}^n)\ with\ g_{|\Omega}=f\}$$

$$\|\,f\mid B^s_{p,q}(\Omega)\,\|=\inf\|\,g\mid B^s_{p,q}(\mathbb{R}^n)\,\|,$$

where the infimum is taken over all $g\in B^s_{p,q}(\mathbb{R}^n)$ with $g_{|\Omega}=f$.
(ii) Let $0<p<\infty$, $0<q\leq\infty$ and $-\infty<s<\infty$. Then

$$F^s_{p,q}(\Omega)=\{f\in\mathcal{D}'(\Omega),\ \exists g\in F^s_{p,q}(\mathbb{R}^n)\ with\ g_{|\Omega}=f\}$$

$$\|\,f\mid F^s_{p,q}(\Omega)\,\|=\inf\|g\mid F^s_{p,q}(\mathbb{R}^n)\,\|,$$

where the infimum is taken over all $g\in F^s_{p,q}(\mathbb{R}^n)$ in the above sense.

As above the following embeddings are continuous:

$$\mathcal{D}(\Omega) \subset B^s_{p,q}(\Omega) \subset \mathcal{D}'(\Omega)\,,$$

$$\mathcal{D}(\Omega) \subset F^s_{p,q}(\Omega) \subset \mathcal{D}'(\Omega)\,,$$

This yields an analogue to the last lemma.

LEMMA 7 *Let $-\infty < s < \infty$, $0 < p,q \leq \infty$, and let Ω be a bounded domain with smooth boundary. Then the spaces $F^s_{p,q}(\Omega)$ and $B^s_{p,q}(\Omega)$ are dual-rich.*

It is well-known that for

$$0 < p,q \leq \infty, s > s_0 := \frac{1}{p} + \max\left(0, (n-1)(\frac{1}{p} - 1)\right)$$

the spaces $B^s_{p,q}(\Omega)$ and $F^s_{p,q}(\Omega)$ (here $p \neq \infty$) admits traces on the boundary $\partial\Omega$. These boundary values can be characterized in a similar fashion, namely as $B^{s-\frac{1}{p}}_{p,q}(\partial\Omega)$ and $F^{s-\frac{1}{p}}_{p,p}(\partial\Omega)$, respectively. For the subspaces with vanishing trace in the above sense we write $B^s_{p,q,0}(\Omega)$ and $F^s_{p,q,0}(\Omega)$, respectively.
Let Δ denote the Laplacian. If $0 < p,q \leq \infty$ and $s > s_0$ then Δ yields an isomorphic map from $B^s_{p,q,0}(\Omega)$ onto $B^{s-2}_{p,q}(\Omega)$. If $0 < p < \infty$ a corresponding result holds in $F^s_{p,q,0}(\Omega)$.
For details we refer to H. Triebel [39]. In the following, the real part of the spaces $B^s_{p,q}(\mathbb{R}^n)$, etc., is denoted by $\tilde{B}^s_{p,q}(\mathbb{R}^n), \ldots$ (for exact definitions see J. Franke, T. Runst [12]). Now we can formulate the generalization of the above mentioned Ambrosetti–Prodi result (see M. Geisler, T. Runst [15]. Let $0 < \lambda_1 < \lambda_2 \leq \lambda_3 \leq \ldots$ denote the sequence of eigenvalues of the homogeneous Dirichlet problem $-\Delta\;|\overset{o}{B}{}^1_{2,2}(\Omega)$ ($\overset{o}{B}{}^s_{p,q}(\Omega)$ denotes the completion of $D(\Omega)$ in $B^s_{p,q}(\Omega)$).

THEOREM 3 *Let $\Omega \subset \mathbb{R}^n$ be a bounded connected domain with smooth boundary $\partial\Omega$. Let $0 < p \leq \infty$, $0 < q \leq \infty$ and $s > \dfrac{n}{p}$. Assume that the real-valued smooth function f fulfils the following conditions*

$$f(0) = 0, 0 < f'(-\infty) < \lambda_1 < f'(+\infty) < \lambda_2, f'' > 0$$

where λ_1, λ_2 have the above meaning. Then for the solutions of the boundary value problem

$$\begin{aligned} -\Delta u(x) - f(u(x)) &= h(x) \text{ in } \Omega \\ u(x) &= 0 \quad \text{on } \partial\Omega \end{aligned} \tag{4}$$

the following assertions are valid:
(i) There exists a closed connected C^1-manifold M_1 of codimension 1 in $\tilde{B}^{s-2}_{p,q}(\Omega)$ such that $\tilde{B}^{s-2}_{p,q}(\Omega)\backslash M_1$ consists of two connected components M_0 and M_2 with the following properties:
(a) if $h \in M_0$ then (4) has no solution in $\tilde{B}^s_{p,q,0}(\Omega)$,
(b) if $h \in M_1$ then (4) has exactly one solution in $\tilde{B}^s_{p,q,0}(\Omega)$,
(c) if $h \in M_2$ then (4) has exactly two solutions in $\tilde{B}^s_{p,q,0}(\Omega)$.
(ii) Let $p < \infty$ then a corresponding result holds for $\tilde{F}^s_{p,q,0}(\Omega)$, too.

We remark that $B^s_{p,q}(\Omega) \subset C(\bar{\Omega})$ if $s > \frac{n}{p}$ or $s = \frac{n}{p}$ and $0 < q \leq 1$. A corresponding result holds within the framework of Triebel–Lizorkin spaces: We have $F^s_{p,q}(\Omega) \subset C(\bar{\Omega})$ if $s > \frac{n}{p}$ or $s = \frac{n}{p}$ and $0 < p \leq 1$.

5 CLASSIFICATION OF SINGULARITIES

In what follows, we consider the nonlinear operator

$$\Phi_\lambda : E \to F\,,$$

generated by the semilinear elliptic boundary value problem with Dirichlet condition

$$\begin{aligned} -\Delta u - \lambda u + f(u) &= h \text{ in } \Omega \\ u &= 0 \text{ on } \partial\Omega\,, \end{aligned} \tag{5}$$

where $\Omega \subset \mathbb{R}^n$ is a bounded and smooth domain, $\lambda \in \mathbb{R}^1$ is a parameter, f is a smooth real-valued function on $\mathbb{R}^1$ with the property that

$$g(t) := f(t) - \lambda t \to \pm\infty \text{ when } t \to \pm\infty.$$

Furthermore, h is a given data function which belongs to $\tilde{F}^{s-2}_{p,q}(\Omega)$ where $s > \frac{n}{p}+2, 0 < p < \infty$ and $0 < q \leq \infty$. Notice that corresponding results hold for the spaces of Besov type, too (Here $p = \infty$ is also admissible.). Now we put

$$E := \tilde{F}^s_{p,q,0}(\Omega)\ ,\ \ F := \tilde{F}^{s-2}_{p,q}(\Omega).$$

We remark that by our assumptions $F \subset \tilde{C}(\bar{\Omega})$, see 4. By S. Fučík [14] we obtain the following result. If λ_1 denotes the smallest eigenvalue of the homogeneous Dirichlet problem in the above sense then it holds $\lambda_1 > 0$, and λ_1 is simple. Furthermore, there exists a unique normed positive eigenfunction $\varphi_1 \in \tilde{C}^\infty(\bar{\Omega})$ to λ_1 with $\varphi_1(x) > 0$ in Ω and $\int_\Omega \varphi_1^2(x)dx = 1$. By [34], [36] we obtain that

$$\Phi_\lambda(u) := -\Delta u - \lambda u + f(u)\,,\ \Phi_\lambda : E \to F$$

is a proper mapping and the linearization of Φ_λ, given by

$$\Phi'_\lambda(u)[v] = -\Delta v - \lambda v + f'(u)v,$$

maps E in F for all $u \in E$. By the linear theory of the Laplacian we obtain that $\Phi_\lambda{}'(u)$ is a Fredholm operator of index 0. Furthermore, let $\mu_1^\lambda(u) < \mu_2^\lambda(u) \leq \mu_3^\lambda(u) \leq \cdots$ denote the eigenvalues of $\Phi_\lambda{}'(u)$:

$$\Phi'_\lambda(u)[v_i^\lambda(u)] = (-\Delta - \lambda + f'(u))\, v_i^\lambda(u),$$

where $\{v_i^\lambda(u)\}_{i=1}^\infty \subset E$ denotes a corresponding orthonormal system of eigenfunctions (with respect to the Hilbert space $L_2(\Omega)$). We remark that the first eigenvalue is simple and the following results are known :

(i) Equation (5) has a unique solution in E if h belongs to F and $\lambda < f'(t)+\lambda_1$ for all $t \in \mathbb{R}^1$ (The linearization of Φ_λ, i.e.,

$$\Phi'_\lambda(u)[v] = -\Delta v - \lambda v + f'(u)v$$

is regular at every $u \in E$, see Theorem 2).
(ii) If $\lambda > f'(t)+\lambda_1$ for some $t \in \mathbb{R}^1$ then Φ_λ admits singular points, i.e., there exist nontrivial functions $v \in E$ with

$$\begin{aligned} -\Delta v - \lambda v + f'(u)v &= 0 \text{ in } \Omega \\ v &= 0 \text{ on } \partial\Omega. \end{aligned} \tag{6}$$

From now on, we suppose that $f'(t) \geq f'(0) = 0$. Examples are $f(t) = t^{2n+1}, n = 1, 2, \ldots$ and smooth perturbations. In that case we obtain a nice describtion of S_Φ^λ.

THEOREM 4 *For $\lambda_1 < \lambda < \lambda_2$ one has*

$$S_\Phi^\lambda = \{u \in E \,,\, \mu_1^\lambda(u) = 0\} \tag{7}$$

and S_Φ^λ is a smooth manifold of codimension 1 *in E. Furthermore, we have*

$$\mu_2^\lambda \geq \lambda_2 - \lambda \textit{ for every } u \in E. \tag{8}$$

Proof: Assertion (7) follows from (8) since for $\lambda < \lambda_2$ we get $\mu_2^\lambda > 0$ for all $u \in E$. The inequality in (8) one obtains by the variational characterization of the eigenvalues and the properties of f.

Remark 7 : Furthermore, one can consider the case $\lambda_{k-1} < \lambda < \lambda_k$, where λ_{k-1} is a simple eigenvalue, too. Then $S_\Phi^\lambda = \{u \in E \,,\, \mu_{k-1}^\lambda(u) = 0\}$. By the hypothesis on λ one gets $dim\, ker\, \Phi'_\lambda(u) = 1$ for $u \in S_\Phi^\lambda$.
Now we begin with the classification of the singularities using the methods going back to B. Ruf [33]. He showed that the kinds of singularities can completely described by derivatives of the function $\mu^\lambda(u)$. As above we consider the case $\lambda_1 < \lambda < \lambda_2$. According to the last theorem we put for $\lambda_1 < \lambda < \lambda_2$ fixed:

$$S := S_\Phi^\lambda,\ \mu(u) := \mu_1^\lambda(u),\ v(u) := v_1^\lambda(u) \text{ and } \Phi := \Phi_\lambda.$$

DEFINITION 9 *(a) A point $u \in S$ is 1-transverse, if there exists a $w \in E$ such that $\mu'(u)[w] \neq 0$. We define $S_1 = \{u \in S,\ u \text{ is 1- transverse}\}$.*
(b) $u \in S_1$ is a fold point if

$$\mu'(u)[v(u)] \neq 0. \tag{9}$$

This is equivalent to say that $v(u) \notin T_uS$, where T_uS denotes the tangent space to S at the point u.

Remark 8 : Because of $\lambda_1 < \lambda < \lambda_2$ we get that $S = S_1$ holds.
By J. Franke, T. Runst [13] we obtain that the function $f(t,x)$ from Lemma 5 has the following form in our case

$$f(t,x) = \int_\Omega \Phi\left(tv + y(t,x)\right) v d\omega,$$

where $0 \neq v \in E$ with $\Phi'(u)[v] = 0$. Applying Lemma 4 and Lemma 5 we obtain the normal form for folds.

THEOREM 5 *If u is a fold point of Φ, then there exist neighbourhoods $U(u) \subset E$, $V(\Phi(u)) \subset F$, a quasi-Banach space X and diffeomorphisms $\alpha : U \to \alpha(U) \subset \mathbb{R}^1 \times X$, $\beta : V \to \beta(V) \subset \mathbb{R}^1 \times X$ such that $\beta \circ \Phi \circ \alpha^{-1}(t,x) = \Psi_F(t,x)$ where $\Psi_F : \mathbb{R}^1 \times X \to \mathbb{R}^1 \times X$ is given by $(t;x) \to (t^2;x)$.*

Examples:

(i) If $\Phi = \Phi_\lambda = -\Delta - \lambda + f$ then condition $\mu'(u)[v(u)] \neq 0$ is equivalent to $\int_\Omega f''(u)v^3(u)d\omega \neq 0$.

(ii) In the above mentioned paper by A. Ambrosetti, G. Prodi [2] the conditions of the function f imply that the origin is a global fold point.

(iii) Further results concerning fold points may be found in the paper by V. Cafagna, G. Tarantello [7]: If $u \in E$ is a fold point then there exists a smooth hypersurface $\Sigma \subset V \subset F$ which disconnects V into two components A and B such

if $h \in A$, then $\Phi^{-1}(h)$ has no preimages in U
if $h \in B$, then $\Phi^{-1}(h)$ has exactly two regular preimages in U,
if $h \in \Sigma$, then $\Phi^{-1}(h)$ has exactly one preimages in U.

In a similar way, one can introduce higher singularities.

DEFINITION 10 *(a) Let $u \in S_1$ be not a fold point, i.e. $\mu'(u)[v(u)] = 0$. We say that u is 1-1-transverse, if there exists a $w \in T_uS_1$ such that*

$$\mu_{vw} = (\mu'(u)[v(u)])'[w] \neq 0. \tag{10}$$

Let $S_{1,1} = \{u \in S_1 ,\ u \text{ is 1-1-transverse } \}$.
(b) A point $u \in S_{1,1}$ is said to be a cusp point, if

$$\mu_{vv}(u) = (\mu'(u)[v(u)])'[v(u)] \neq 0. \tag{11}$$

This is equivalent to say that $v(u)$ is in T_uS_1 and does not belong to the tangent space of $S_{1,1}$. Applying the methods due to B. Ruf [33], see also V. Cafagna, F. Donati [6], the following result concerning the normal form for cusps can be proved by application of Lemma 4 and Lemma 5.

THEOREM 6 *If $u \in S$ is a cusp point for Φ, then there are neighbourhoods $U(u) \subset E$, $V(\Phi(u)) \subset F$, a quasi-Banach space X, and diffeomorphisms $\alpha : U \to \alpha(U) \subset \mathbb{R}^2 \times X$, $\beta : V \to \beta(V) \subset \mathbb{R}^2 \times X$ such that $\beta \circ \Phi \circ \alpha^{-1}(s,t,x) = \Psi_C(s,t,x)$ where $\Psi_C : \mathbb{R}^2 \times X \to \mathbb{R}^2 \times X$ is given by $(s,t;x) \to (s^3 - st, t; x)$.*

Examples:

(i) If $\Phi = \Phi_\lambda = -\Delta - \lambda + f$ then the cusp condition (11) is equivalent to

$$C_\lambda(u) := \int_\Omega f'''(u)v^4(u)d\omega + 3\int_\Omega f''(u)v^2(u)v_v(u)d\omega \neq 0, \tag{12}$$

where

$$v_v(u) = (-\Delta - \lambda + f'(u))^{-1}\left(-f''(u)v^2(u)\right) \text{ and } u = 0 \text{ on } \partial\Omega. \tag{13}$$

We remark that $f''(u)v^2(u)$ is normal to $v(u)$ because u is not a fold point.

(ii) B. Ruf [33] considered the case $f(u) = u^3$.

(iii) Further results concerning cusp singularities may be found in the paper by V. Cafagna, G. Tarantello [7]:
If u is a cusp point then there exists a smooth hypersurface $\Sigma \subset V \subset F$ which disconnects V into two components A and B such that

$$\text{if } h \in A, \text{ then } \Phi^{-1}(h) \text{ has exactly one regular preimage in } U \in E,$$
$$\text{if } h \in B, \text{ then } \Phi^{-1}(h) \text{ has exactly three regular preimages in } U.$$

Moreover, there is a smooth hypersurface $\tilde{\Sigma} \subset \Sigma$ such that

$$\text{if } h \in \tilde{\Sigma}, \text{ then } \Phi^{-1}(h) \text{ has exactly one preimage in } U,$$
$$\text{if } h \in \Sigma \setminus \tilde{\Sigma}, \text{ then } \Phi^{-1}(h) \text{ has exactly two preimages in } U.$$

In the same way one can introduce higher singularities.

DEFINITION 11 *(a) Let $u \in S_{1,1}$ be not a cusp point,i.e. $\mu_{vv}(u) = 0$. We say that u is 1-1-1-transverse, if there exists a $w \in T_u S_{1,1}$ such that*

$$\mu_{vvw}(u) \neq 0. \tag{14}$$

Let $S_{1,1,1} = \{u \in S_{1,1}\ ,\ u \text{ is 1-1-1-transverse}\}$.
(b) A point $u \in S_{1,1,1}$ is a swallow tail singularity, if

$$\mu_{vvv}(u) \neq 0. \tag{15}$$

Applying Lemma 4 and Lemma 5 one obtain the normal form for swallow tails, see also B. Ruf [33].

THEOREM 7 *If $u \in S$ is a swallow tail singularity for Φ, then there exist neighbourhoods $U(u) \subset E$, $V(\Phi(u)) \subset F$, a quasi-Banach space X, and local diffeomorphisms $\alpha : U \to \alpha(U) \subset \mathbb{R}^3 \times X$ and $\beta : V \to \beta(V) \subset \mathbb{R}^3 \times X$ such that $\beta \circ \Phi \circ \alpha^{-1}(r,s,t,x) = \Psi_S(r,s,t,x)$ where $\Psi_S : \mathbb{R}^3 \times X \to \mathbb{R}^3 \times X$ is given by $(r,s,t;x) \to (t^4 + st^2 + rt, s, r; x)$.*

DEFINITION 12 *(a) Let $u \in S_{1,1,1}$ be not a swallow tail,i.e. $\mu_{vvv}(u) = 0$. We say that u is 1-1-1-1-transverse, if there exists a $w \in T_u S_{1,1,1}$ such that*

$$\mu_{vvvw}(u) \neq 0. \tag{16}$$

Let $S_{1,1,1,1} = \{u \in S_{1,1,1}\ ,\ u \text{ is 1-1-1-1-transverse}\}$.
(b) A point $u \in S_{1,1,1,1}$ is a butterfly singularity, if

$$\mu_{vvvv}(u) \neq 0. \tag{17}$$

THEOREM 8 *If $u \in S$ is a butterfly singularity, then there exist neighbourhoods $U(u) \subset E$, $V(\Phi(u)) \subset F$, a quasi-Banach space X, and local diffeomorpisms $\alpha : U \to \alpha(U) \subset \mathbb{R}^4 \times X$, $\beta : V \to \beta(V) \subset \mathbb{R}^4 \times X$ such that $\beta \circ \Phi \circ \alpha^{-1}(q,r,s,t,x) = \Psi_B(q,r,s,t,x)$, where $\Psi_B : \mathbb{R}^4 \times X \to \mathbb{R}^4 \times X$ is given by $(q,r,s,t;x) \to (t^5 + st^3 + rt^2 + qt, s, r, q; x)$.*

6 CONCLUDING REMARKS

As above showed (see Theorem 4) the singular set S_{Φ}^{λ} of our semilinear boundary value problem

$$\begin{aligned} -\Delta u - \lambda u + f(u) &= h \text{ in } \Omega \\ u &= 0 \text{ on } \partial\Omega \end{aligned} \qquad (18)$$

is defined by

$$S_{\Phi}^{\lambda} = \{u \in E,\ \mu_1^{\lambda}(u) = 0\} \qquad (19)$$

if $\lambda_1 < \lambda < \lambda_2$.
In the case when $E = C_0^{2+\alpha}(\Omega)$, $f(u) = u^3$, $\lambda_2 > 2.5\lambda_1$ and $\lambda_1 < \lambda < \frac{\lambda_2 + 4.5\lambda_1}{7}$ holds B.Ruf [32] proved that $C_\lambda(u)$ is positive for all $u \in S_{\Phi}^{\lambda}$ (see (12)). It means that the singular set contains only fold and cusp points. A similar result (for a different range of λ) holds also if $f(t) = t^3$ is replaced by $f(t) = h(t)t^3$, where h is a bounded $C^3(\mathbb{R}^1)$-function. Up to now it is a open problem whether corresponding results hold also for more general nonlinearities f. Furthermore, B. Ruf [33] proved that for certain domains (for example $\Omega = (0,1)$) the singular set contains also higher singularities (swallow tail and butterfly points) when $\lambda < \lambda_2$ is large enough. To prove it B. Ruf [33] used step functions $z(x) \in B_{p,p}^{\frac{1}{p}}(0,1) \cap L_\infty(0,1)$ which does not belong to the classical spaces $C^{2+\alpha}(0,1)$ and $W_2^1(0,1)$, respectively. This shows that it is useful to try to deal with problems of the above type not only, say, in Hölder and Sobolev spaces, but in more general function spaces. It was the aim of this survey to describe such an approach with respect to the two scales of function spaces of Besov and Triebel- Lizorkin type. For further results and open questions in Hölder spaces we refer also to B. Ruf [33] .

References

[1] Amann, H., and Hess, P., *A multiplicity result for a class of elliptic boundary value problems* . Proc. Roy. Soc. Edinb. Sect.A **84**(1979), 145-151.

[2] Ambrosetti, A., and Prodi, G., *On the inversion of some differentiable mappings with singularities between Banach spaces* . Ann. Mat. Pura Appl. **93**(1972), 231-246.

[3] Berger, M. S., Church, P. T., and Timourian, J.G., *Folds and cusps in Banach spaces with application to nonlinear partial differential equations. I* . Indiana Univ. Math. J. **34** (1985), 1-19.

[4] Berger, M. S., Church, P. T., and Timourian, J.G., *Folds and cusps in Banach spaces. II* . Trans. Amer. Math. Soc. **307** (1988), 225-244.

[5] Berger, M. S., and Podolak, E.,*On the solutions of a nonlinear Dirichlet problem* . Indiana Univ. Math. J. **24** (1975), 837-849.

[6] Cafagna, V., and Donati, F., *Singularity theory of Fredholm maps and the number of solutions to some nonlinear first order differential equations* . preprint.

[7] Cafagna, V., and Tarantello, G., *Multiple solutions for some semilinear elliptic equations* Math. Ann. **276** (1987), 643-656.

[8] Church, P. T., and Timourian, J.G., *The singular set of a nonlinear elliptic operator* . Michigan Math. J. **35** (1988), 197-213.

[9] Dancer, E. N., *On the range of certain weakly nonlinear elliptic partial differential equations* . J. Math. Pure Appl. **57** (1978) , 351-366.

[10] Dolph, C. L, *Nonlinear integral equations of the Hammerstein– type* . Trans. AMS **60**, 289-307.

[11] Drábek, P., and Runst, T., *On the existence of solutions of a semilinear elliptic boundary value problem with superlinear nonlinearities* . Z. Anal. Anwendungen **9** (1990), 105-112.

[12] Franke, J., and Runst, T., *On the admissibility of function spaces of type* $B^s_{p,q}$ and $F^S_{p,q}$, *and boundary value problems for non-linear partial differential equations* . Analysis Math. **13** (1987), 3-27.

[13] Franke, J., and Runst, T., *Non-linear perturbations of linear non-invertible boundary value problems in function spaces of type* $B^s_{p,q}$ and $F^s_{p,q}$ .Czechoslovak Math. J. **38** (1988), 623-641.

[14] Fučík, S., *Solvability of Nonlinear Equations and Boundary Value Problems* . D. Reidel Publ. Company, Holland 1980.

[15] Geisler, M., and Runst, T., *On a superlinear Ambrosetti– Prodi problem in Besov and Triebel–Lizorkin spaces* .,J. London Math. Soc. **43** (1991), 324-336.

[16] Giannoni, F., and Micheletti, A.M., *Some remarks about multiplicity results for some semilinear elliptic problems by singularity theory* . Rend. Mat. Roma, Serie VII, **8** (1988), 367-384.

[17] Hammerstein, A., *Nichtlineare Integralgleichungen nebst Anwendungen* . Acta Math. **54** (1929), 117-176.

[18] Hess, P., and Ruf, B., *On a superlinear elliptic boundary value problem* . Math. Z. **164** (1978), 9-14.

[19] Kazdan, J.L., and Warner, F.W., *Remarks on some quasi-linear elliptic equations* . Comm. Pure Appl. Math. **28** (1981), 282-294.

[20] Köthe, G.,*Topologische lineare Räume. I* . Springer, Berlin, 1960.

[21] Lazer, A.C., and McKenna, P.J., *On the number of solutions of a nonlinear Dirichlet problem* . J. Math. Anal. Appl. **84** (1981), 282- 294.,

[22] Lazer, A.C., and McKenna, P.J., *Multiplicity of solutions of nonlinear boundary value problems with nonlinearities crossing several higher eigenvalues* . J. Reine Angew. Math. **368** (1986), 184-200.

[23] Lazzeri,F., and Micheletti,A.M., *An application of singularity theory to nonlinear differentiable mappings between Banach spaces* . Nonlinear Anal. **11** (1987), 795-808.

[24] Manes, A., and Micheletti, A.M., *Un' estensione della teoria variazionale classica degli autovalori per operatori ellittici del secondo ordine* . Boll. Un. Mat. Ital. **7** (1973), 285-301.

[25] Micheletti, A.M., *On number of solutions of some nonlinear elliptic equations* . Boll. Un. Mat..Ital. **2** -B (1988), 509-527.

[26] Michor, P., *The division theorem on Banach spaces* . Österr. Ak. Wiss. **II-189-1-3** (1980), 1-18.

[27] Päivärinta, L., and Runst, T., *Multiplicity results for semilinear elliptic boundary value problems in Besov and Triebel– Lizorkin spaces* . Proc. Edinb. Math. Soc. **34** (1991), 393-410.

[28] Podolak, E., *On the range of operator equations with an asymptotically nonlinear term.* Indiana Univ. Math. J. **25** (1976), 1127-1137.

[29] Rolewicz, S., *On a certain class of linear metric spaces* . Bull. Acad. Polon. Sci., Cl. III **5** (1957), 471-473.

[30] Ruf, B., *On nonlinear elliptic boundary value problems with jumping nonlinearities* . Ann. Math. Pura Appl.**128** (1980), 131-151.

[31] Ruf, B., *Multiplicity results for nonlinear elliptic equations* . in 'Nonlinear Analysis, Function Spaces and Applications, Vol. 3', Proceedings of the Spring School held in Litomyšl, 1986 ', Teubner-Texte Math. **93**. Teubner-Verlag, Leipzig, 1986, 109-138.

[32] Ruf, B., *Singularity theory and the geometry of a nonlinear elliptic equation* . Ann. Sc. Norm. Sup. Pisa C. Sc. **17** (1990), 1-33.

[33] Ruf, B., *Singularity theory and forced secondary bifurcation* . preprint Dip. Mat. "F. Enriques", Milano, Quarderno n. 48/1990, 1-32.

[34] Runst, T., *Mapping properties of non-linear operators in spaces of Triebel–Lizorkin and Besov type* . Analysis Math. **12** (1986), 313-346.

[35] Runst, T., *Solvability of semilinear elliptic boundary value problems in Besov–Triebel–Lizorkin spaces* in 'Surveys on Analysis, Geometry and Mathematical Physics ', Teubner-Texte Math. **117**. Teubner-Verlag, Leipzig, 1990, 198-291.

[36] Runst, T., *Singularity theory in quasi–Banach spaces with application to semilinear elliptic problems* , to appear.

[37] Solomini, S., *Some remarks on the number of solutions of some nonlinear elliptic problems* . Ann. Inst. H. Poincaré **2** (1985), 143-156.

[38] Thom, R., *Les singularités des applications différentiables* . Ann. Inst. Fourier **6** (1955-56), 43-87.

[39] Triebel, H., *Theory of function spaces.* Birkhäuser Verl., Boston, Stuttgart; and Akad. Verlagsges. Geest & Portig, Leipzig , 1983.

[40] Whitney, H., *On singularities of mappings of Euclidean spaces I, Mappings of the plane into the plane* . Ann. Math. **62** (1955), 374-410.

Thomas Runst
Mathematisches Institut, FSU Jena
UHH, 17.OG
D-O-6900 Jena, Germany
E-Mail:
RUNST@MATHEMATIK.UNI-JENA.DBP.DE

Function Spaces Related to Singular Integral Operators

Abdellah Youssfi

1 Introduction

Let T be a continuous linear operator from $\mathcal{D}(\mathbb{R}^n)$ to $\mathcal{D}'(\mathbb{R}^n)$. By the Schwartz kernel theorem there is a distribution $K \in \mathcal{D}'(\mathbb{R}^n \times \mathbb{R}^n)$ such that

$$< Tf, g >=< K, f \otimes g > .$$

We say that T is a singular integral operator with order m ($T \in SIO(m)$) if the restriction of K to the set

$$\Omega = \{(x, y) \in \mathbb{R}^n \times \mathbb{R}^n \; x \neq y\}$$

is a function with continuous partial derivatives in the variables x up to order m which satisfy

$$|\partial_x^\alpha K(x, y)| \leq C_\alpha |x - y|^{-n-|\alpha|} \quad \text{for} \quad |\alpha| \leq m \text{ and} \qquad x \neq y. \tag{1.1}$$

Hence if f and g are in $\mathcal{D}(\mathbb{R}^n)$ and $\mathrm{Supp}(f) \cap \mathrm{Supp}(g) = \phi$ then

$$< Tf, g >= \int\int K(x, y) f(y) g(x) dx dy. \tag{1.2}$$

In [5], Coifman–Meyer introduced these operators, which extend the classical Calderon–Zygmund operators. They showed that if $T \in SIO(1)$, $T^t \in SIO(1)$, and T is bounded on L^2 then T is also bounded on L^p for $1 < p < +\infty$.

The problem of characterizing which of these operators are bounded on L^2 was solved by David-Journé [7]. They found two necessary and sufficient conditions that T must satisfy. The principal condition is that $T(1)$ and $T^t(1)$ must be in BMO.

When T vanishes on polynomials whose degrees are smaller than a certain order, the boundedness of T on Sobolev spaces $\dot{H}^s$ was obtained by several authors (see [8], [10],[11] and [13]). Our purpose herein is to obtain the analogue of David-Journé's theorem for the case of Sobolev spaces. To do so, we are led to define BMO^s space apparented to the BMO space. In the case $0 < s < 1$ the problem was studied by [12], and for $s \geq 1$ was begun by the author in [17].

This paper is organized as follows. In Section 2 we recall some properties of Besov and Sobolev spaces. In Section 3 we establish some results related to the BMO^s spaces. In particular we are led to study the boundedness of some paradifferential operators on Sobolev spaces. In section 4 we give a version of David–Journé's theorem in the case of Sobolev spaces. Our criterion involves the BMO^s spaces and some paraproduct operators.

Throughout the paper, C denotes constants which may differ at each appearance, possibly depending on the dimension or other parameters. We denote by $\hat{f}$ the Fourier transform of f and $\check{f}$ it's inverse Fourier transform. We also use the notations:

$\mathcal{D}(\mathbb{R}^n)$ = space of C^∞-functions with compact support, $\mathcal{D}'(\mathbb{R}^n)$ it's dual.

$\mathcal{S}(\mathbb{R}^n)$= the space of Schwartz test functions.

$\mathcal{S}'(\mathbb{R}^n)$= tempered distributions.

$[s]$= the greatest integer smaller than or equal to s.

T^t= transpose operator of T.

2 Preliminaries

In this section we recall the definition of Besov and Sobolev spaces. Let $\varphi \in \mathcal{D}(\mathbb{R}^n)$ be supported in the ball $|\xi| \le 1$ and satisfy $\varphi(\xi) = 1$ for $|\xi| \le \frac{1}{2}$. The function

$$\psi(\xi) = \varphi(\xi/2) - \varphi(\xi)$$

is C^∞ and satisfies the identity

$$\sum_{j\in\mathbb{Z}} \psi(2^{-j}\xi) = 1$$

for $\xi \neq 0$.

We denote by Δ_j and S_j the convolution operators whose symbols are $\psi(2^{-j}\xi)$ and $\varphi(2^{-j}\xi)$ respectively.

For $s \in \mathbb{R}$ and $1 \le p < \infty$, the homogeneous Sobolev space $\dot{H}_p^s$ is defined by

$$||f||_{\dot{H}_p^s} = ||[\sum_{j\in\mathbb{Z}} 4^{sj}|\Delta_j f|^2]^{1/2}||_p \tag{2.1}$$

when $p = 2$ we write $\dot{H}_2^s = \dot{H}^s$. In the particular case $s = 0$ and $p = 1$, the homogeneous Sobolev space $\dot{H}_1^0$ is (modulo polynomials) the Hardy space $\mathcal{H}^1$.

In the case $p = +\infty$, the space $\dot{H}_\infty^s$ is the BMO–Sobolev space (see [9]). We shall say that $b \in \dot{H}_\infty^s$ if there exists a contstant $C > 0$ such that

$$\sum_{j\ge k} \int_B 4^{sj}|\Delta_j(b)|^2 dx \le C|B| \tag{2.2}$$

for all $k \in \mathbb{Z}$ and any ball B of radius 2^{-k}.

Throughout the paper we consider the homogeneous realization of $\dot{H}^s$ and write $\dot{H}^s$ this realization. For $s \in \mathbb{R}$, it was shown in [2] that $\dot{H}^s$ can be realized as distribution spaces modulo polynomials of degre less than $[s - \frac{n}{2}]$. In particular, for $0 < s < \frac{n}{2}$, we have $\dot{H}^s = I_s(L^2(\mathbb{R}^n))$ where

$$I_s(f)(x) = \int \frac{f(x)}{|x-y|^{n-s}} dy$$

is the Riesz potential.

The following Besov norm will be used repeatedly, for $s \in \mathbb{R}$ and $1 \le p, q \le +\infty$

$$||f||_{\dot{B}_p^{s,q}} = [\sum_{j\in\mathbb{Z}} 2^{sjq} ||\Delta_j f||_p^q]^{1/q} \tag{2.3}$$

with modification if $q = +\infty$.

We shall say that a distribution $b \in \dot{B}_\infty^{0,\infty}$ is in the space BMO^s, $s \ge 0$, if the operator

$$\pi(b, f) = \pi_b(f) = \sum_{j\in\mathbb{Z}} \Delta_j(b) S_{j-3}(f) \tag{2.4}$$

is bounded on $\dot{H}^s$. The distribution $\pi_b(f)$ is called the Bony paraproduct of b and f. It is well known fact that $BMO^s = BMO = \dot{H}_\infty^0$ for $s = 0$.

The norm in BMO^s is

$$|||b|||^s = \sup\left(||b||_{\dot{B}_\infty^{0,\infty}}, ||\pi_b||_{\mathcal{L}(\dot{H}^s, \dot{H}^s)}\right) \tag{2.5}$$

More generally, assume that $b \in \dot{B}_\infty^{N,\infty}$ and $\alpha \in \mathbb{N}^n$ with $|\alpha| = N$, π_b^α is defined by

$$\pi_b^\alpha(f) = \sum_{j\in\mathbb{Z}} \Delta_j(b) S_{j-3}(\partial^\alpha f). \tag{2.6}$$

Both Besov and Sobolev spaces enjoy the time–frequency localization property, which is given by the following almost–orthogonality lemma [4] and [15].

Lemma 1 *For $\gamma > 1$ and $1 \le p, q < +\infty$, there exists a positive constant $C = C(n, \gamma, s, p, q)$ such that*

$$||\sum_j f_j||_{\dot{H}_p^s} \le C||[\sum_j 4^{sj}|f_j|^2]^{1/2}||_p \tag{2.7}$$

and

$$||\sum_j f_j||_{\dot{B}_p^{s,q}} \le C[\sum_j 2^{sjq}||f_j||_p^q]^{1/q} \tag{2.8}$$

for all sequences $(f_j)_j$ of functions satisfying

$$\text{Supp}(\hat{f}_j) \subset \{\gamma^{-1}2^j \le |\xi| \le \gamma 2^j\},\ j \in \mathbb{Z}.$$

The following lemma is due to [14] and gives an equivalent norm to $||.||_{\dot{B}_p^{s,q}}$.

Lemma 2 *Let $s \in \mathbb{R}$, $h \in \dot{B}_1^{-s,1}$ such that $\hat{h}(\xi) \ne 0$ for $\frac{1}{2} \le |\xi| \le 2$. For $j \in \mathbb{Z}$, let $h_j(x) = 2^{nj}h(2^j x)$. There exist $C_1, C_2 > 0$ such that*

$$C_1||f||_{\dot{B}_p^{s,q}} \le [\sum_j 2^{sjq}||h_j \star f||_p^q]^{1/q} \le C_2||f||_{\dot{B}_p^{s,q}} \tag{2.9}$$

In particular, if $h \in \mathcal{S}(\mathbb{R}^n)$ satisfies

$$\hat{h}(\xi) \neq 0 \text{ for } \quad \frac{1}{2} \leq |\xi| \leq 2$$

$$\int x^\alpha h(x)dx = 0 \text{ for } \quad |\alpha| \leq [s]+1 \qquad \textit{and}$$

then $h \in \dot{B}_1^{-s,1}$, so that (2.9) holds.

An inequality of the Hardy–Littlewood type is given by the following lemma [16].

Lemma 3 *For $0 \leq s < \frac{n}{p}$ and $1 < p < +\infty$, there exists a positive constant $C > 0$ such that*

$$[\int |f(x)|^p |x|^{-ps} dx]^{1/p} \leq C||f||_{\dot{H}^s_p} \tag{2.10}$$

for all $f \in \mathcal{S}(\mathbb{R}^n)$.

3 Generalized paraproduct and BMO^s spaces.

3.1 The paraproduct operators.

Theorem 1 *Let $b \in \dot{B}_\infty^{0,\infty}$. Then $b \in BMO^s$ if and only if there exists a positive constant $C > 0$ such that*

$$\sum_{j \in \mathbb{Z}} 4^{sj} ||\Delta_j(b) S_{j-3}(f)||_2^2 \leq C||f||^2_{\dot{H}^s} \tag{3.1}$$

for all $f \in \mathcal{S}(\mathbb{R}^n)$.

To prove Theorem 1, it is easy to show, by virtue of Lemma 1, that (3.1) implies that $b \in BMO^s$. The converse is due to the author [18]. More generally, we have the extension of Theorem 1.

Theorem 2 *Suppose that $m \leq s < m+1$ and $h \in \mathcal{D}(\mathbb{R}^n)$ is supported by $\{\gamma^{-1} \leq |\xi| \leq \gamma\}$ with $\gamma > 1$, and let $b_\alpha \in \dot{B}_\infty^{|\alpha|,\infty}$ for $|\alpha| \leq m$ such that the operator*

$$T = \sum_{|\alpha| \leq m} \pi^\alpha_{b_\alpha}$$

is bounded on $\dot{H}^s$. For each $j \in \mathbb{Z}$, let R_j be the convolution operator with symbol $h(2^{-j}\xi)$. Then there exists a positive constant $C > 0$ such that

$$\sum_{j \in \mathbb{Z}} 4^{sj} || \sum_{|\alpha| \leq m} R_j(b_\alpha) S_{j-3}(\partial^\alpha f)||_2^2 \leq C||f||^2_{\dot{H}^s}, \tag{3.2}$$

for all $f \in \mathcal{S}(\mathbb{R}^n)$.

Remark. We can replace in (3.2) $S_{j-3}(\partial^\alpha f)$ by $\partial^\alpha f$, especially we obtain that the operator T is bounded on $\dot{H}^s$ if and only if there exists a constant $C > 0$ such that

$$\sum_{j\in\mathbb{Z}} 4^{sj} \|\sum_{|\alpha|\le m} \Delta_j(b_\alpha)\partial^\alpha f\|_2^2 \le C\|f\|_{\dot{H}^s}^2 \quad \text{if} \quad s \ne m \tag{3.3}$$

and for $s = m$

$$\sum_{j\in\mathbb{Z}} 4^{sj} \|\sum_{|\alpha|\le m-1} \Delta_j(b_\alpha)(\partial^\alpha f) + \sum_{|\alpha|=m} \Delta_j(b_\alpha) S_{j-3}(\partial^\alpha f)\|_2^2 \le C\|f\|_{\dot{H}^s}^2, \tag{3.4}$$

for all $f \in \mathcal{S}(\mathbb{R}^n)$.

To do so, we write

$$\partial^\alpha f = S_{j-3}(\partial^\alpha f) + \sum_{k\ge j-2} \Delta_k(\partial^\alpha f).$$

Since

$$\sum_j 4^{(s-|\alpha|)j} \|\sum_{k\ge j-2} \Delta_k(\partial^\alpha f)\|_2^2 \le C\|f\|_{\dot{H}^s}^2.$$

The hypthesis $b \in \dot{B}_\infty^{|\alpha|,\infty}$ show that

$$\sum_j 4^{sj} \|\Delta_j(b_\alpha)[\sum_{k\ge j-2} \Delta_k(\partial^\alpha f)]\|_2^2 \le C\|f\|_{\dot{H}^s}^2.$$

We now recall the capacity characterization of BMO^s.

Theorem 3 *Let $b \in \dot{B}_\infty^{0,\infty}$, $0 < s < \frac{n}{2}$. Then $b \in BMO^s$ if and only if there exists a constant $C > 0$ such that for any bounded open Ω*

$$\sum_j 4^{sj} \int_\Omega |\Delta_j(b)(x)|^2 dx \le C\text{Cap}_s(\Omega) \tag{3.5}$$

where $\text{Cap}_s(\Omega) = \inf\{\|f\|_{\dot{H}^s}^2, \ f = 1 \ on \ \Omega\}$ *is the Riesz capacity of* Ω.

Proof. It is not hard to show that $b \in BMO^s$ implies (3.5). Conversely, we use the Dahlberg inequality [6], there exists a positive constant $C > 0$ such that

$$\int_0^{+\infty} t\text{Cap}_s(\{x \in \mathbb{R}^n / I_s(g)(x) > t\})dt \le C\|g\|_2^2, \tag{3.6}$$

for all $g \in L^2(\mathbb{R}^n), g \ge 0$. Now we write

$$\begin{aligned}\sum_{j\in\mathbb{Z}} 4^{sj} \int |\Delta_j(b)|^2 |S_{j-3}(f)(x)|^2 dx &\le C\int_0^{+\infty} t \int_{\Omega_t} [\sum_{j\in\mathbb{Z}} 4^{sj} |\Delta_j(b)|^2 dx] dt \\ &\le C\int_0^{+\infty} t \,\text{Cap}_s(\Omega_t) dt\end{aligned}$$

where $\Omega_t = \{x \in \mathbb{R}^n / f^*(x) > t\}$ and f^* is the Hardy–Littlewood maximal function. In fact, a classical result [1] shows that

$$|S_j(f)(x)| \le Cf^*(x) \quad \text{for all} \quad x \in \mathbb{R}^n, j \in \mathbb{Z}.$$

Moreover $(I_s(g))^*(x) \le CI_s(g^*)(x)$ and $\dot{H}^s = I_s(L^2(\mathbb{R}^n))$, then we obtain by (3.6)

$$\int_0^{+\infty} t\mathrm{Cap}_s(\Omega_t)dt \le C||f||^2_{\dot{H}^s},$$

from which the theorem follows.

Theorem 4 *a) For $s > 0$ and $b \in BMO^s$ there exists a positive constant C such that*

$$\sum_{j\in\mathbb{Z}} 4^{sj} \int_Q |\Delta_j(b)(x)|^2 dx \le C|Q|^{1-(2s)/n} \tag{3.7}$$

for all cubes Q of $\mathbb{R}^n$.

b) Let $0 < s < \frac{n}{2}$ and $\omega_s(Q) = \int_Q |x|^{-2s} dx$. Assume that $b \in \dot{B}^{0,\infty}_{\infty}$ and for each cube Q of $\mathbb{R}^n$

$$\sum_{j\in\mathbb{Z}} 4^{sj} \int_Q |\Delta_j(b)(x)|^2 dx \le C\omega_s(Q). \tag{3.8}$$

Then $b \in BMO^s$.

Proof . a) Let $f \in \mathcal{D}(\mathbb{R}^n)$ such that $f = 1$ on the unity cube $Q(0,1)$. For $Q = Q(x_0, r)$ we set

$$f_Q(x) = f(\frac{x - x_0}{r}) \ .$$

Then

$$||f_Q||_{\dot{H}^s} \simeq r^{n/2-s}||f||_{\dot{H}^s}.$$

From (3.3) we get

$$\begin{aligned} \sum_{j\in\mathbb{Z}} 4^{sj} \int_Q |\Delta_j(b)(x)|^2 dx &\le C||f_Q||^2_{\dot{H}^s} \\ &\le C|Q|^{1-(2s)/n}. \end{aligned}$$

b) It is enough to show that

$$\omega_s(\Omega) \le C\mathrm{Cap}_s(\Omega)$$

for all Ω bounded open subset of $\mathbb{R}^n$.

For $f = 1$ on Ω, we have

$$\omega_s(\Omega) \le \int |f(x)|^2 |x|^{-2s} dx.$$

Lemma 3 implies that

$$\omega_s(\Omega) \le ||f||^2_{\dot{H}^s}.$$

This completes the proof.

Corollary 1 *For $s > \frac{n}{2}$ the space BMO^s is trivial ($BMO^s = \{0\}$).*

Proof. Let $Q_k = Q(0, 2^k)$. Then from (3.7) we have

$$\sum_{j\in\mathbb{Z}} 4^{sj} \int_{Q_k} |\Delta_j(b)(x)|^2 dx \le C 2^{k(1-(2s)/n)}.$$

Therefore

$$\sum_{j\in\mathbb{Z}} 4^{sj} \int_{Q_k} |\Delta_j(b)(x)|^2 dx \to 0 \text{ as } \quad k \to +\infty.$$

Remark. Corollary 1 also holds for $s = \frac{n}{2}$ (see Appendix).

3.2 Proof of Theorem 2.

To prove Theorem 2, we shall need two preparatory Lemmas, the second of which is due to the author [18].

Lemma 4 *Let $\alpha \in \mathbb{N}^n$ and $b \in \dot{B}_\infty^{|\alpha|,\infty}$, then π_b^α is bounded on $\dot{H}^t$ for all $t < |\alpha|$.*

Proof. We write

$$\begin{aligned}\sum_{j\in\mathbb{Z}} 4^{tj} ||\Delta_j(b) S_{j-3}(\partial^\alpha f)||_2^2 &\le C \sum_{j\in\mathbb{Z}} 4^{(t-|\alpha|)j} ||S_{j-3}(\partial^\alpha(f)||_2^2 \\ &\le C \sum_{j\in\mathbb{Z}} 4^{(t-|\alpha|)j} [\sum_{k\le j} 4^{|\alpha|k} ||\Delta_k(f)||_2^2].\end{aligned}$$

Since $t < |\alpha|$ implies that

$$\sum_{j\in\mathbb{Z}} 4^{tj} ||\Delta_j(b) S_{j-3}(\partial^\alpha f)||_2^2 \le C ||f||_{\dot{H}^t}^2,$$

from which the lemma follows.

Lemma 5 *Let $h \in \mathcal{S}(\mathbb{R}^n)$, $s \ge 0$ and $m = [s]$. Set*

$$F_m(f)(x, y) = f(y) - \sum_{|\alpha|\le m} \frac{1}{\alpha!} (y - x)^\alpha (\partial^\alpha f)(x) \tag{3.9}$$

for $f \in \mathcal{S}(\mathbb{R}^n)$. There exists a positive constant $C > 0$ such that

$$\sum_{j\in\mathbb{Z}} 4^{sj} \int\int 2^{nj} |h(2^j(x - y)||F_m(S_j f)(x, y)|^2 dx dy \le C ||f||_{\dot{H}^s}^2 \tag{3.10}$$

for all $f \in \mathcal{S}(\mathbb{R}^n)$.

Proof of Theorem 2. Let

$$X_j(f) = \sum_{|\alpha|\le m} R_j(b_\alpha) S_{j-3}(\partial^\alpha f),$$
$$Y_j(f) = \sum_{\nu=-N}^{N} \Delta_j(X_{j+\nu}(f))$$

where $N = \frac{log(\gamma)}{log(2)}$. We shall show that

$$\sum_{j\in\mathbb{Z}} 4^{sj} ||X_j(f) - Y_j(f)||_2^2 \le C||f||_{\dot{H}^s}^2.$$

To do so, we will proceed by induction on m. When $m = 0$, that is $0 \le s < 1$, we write

$$Y_j(f) - X_j(f) = A_j(f) + B_j(f)$$

where

$$A_j(f) = \sum_{\nu=-N}^{N} [\Delta_j(R_{j+\nu}(b_0) S_{j+\nu-3}(f)) - \Delta_j(R_{j+\nu}(b_0)) S_{j+\nu-3}(f)]$$

$$\text{and} \quad B_j(f) = \sum_{\nu=-N}^{N} \Delta_j(R_{j+\nu}(b_0))[S_{j+\nu-3}(f) - S_{j-3}(f)].$$

Since

$$\sum_{\nu=-N}^{N} ||S_{j+\nu-3}(f) - S_{j-3}(f)||_2 \le C \sum_{\nu=-N}^{N} ||\Delta_{j+\nu}(f)||_2,$$

the hypothesis $b_0 \in \dot{B}_\infty^{0,\infty}$ implies that

$$\sum_{j\in\mathbb{Z}} ||B_j(f)||_2^2 \le C||f||_{\dot{H}^s}^2.$$

To establish a similar estimate for $A_j(f)$, let

$$b_j = R_{j+\nu}(b_0) \text{ and } \quad f_j = S_{j+\nu-3}(f).$$

Then

$$A_j(f)(x) = 2^{nj} \int \check{h}(2^j(x-y))(f_j(y) - f_j(x)))b_j(y)dy$$

showing that

$$|A_j(f)(x)| \le C||b_j||_\infty [2^{nj} \int |\check{h}(2^j(x-y))||F_0(f_j)(x,y)|^2 dy]^{1/2}.$$

Since $||b_j||_\infty \le ||b||_{\dot{B}_\infty^{0,\infty}}$ we obtain from (3.10) that

$$\sum_{j\in\mathbb{Z}} 4^{sj} ||A_j(f)||_2^2 \le C||f||_{\dot{H}^s}^2.$$

Therefore Theorem 2 holds for $m = 0$. Now suppose that the theorem holds for any integer $m' < m$, where m is a positive integer. Let $A_j(f) = C_j(f) + D_j(f)$ where

$$\begin{aligned} C_j(f)(x) &= \sum_{|\alpha|\le m}\sum_{\nu=-N}^{N} 2^{nj}\int \check{h}(2^j(x-y))R_{j+\nu}(b_\alpha)(y)F_m(\partial^\alpha f_j)(x,y)dy \\ D_j(f)(x) &= \sum_{\nu=-N}^{N}\sum_{0<|\beta|\le m,|\alpha|\le m} \frac{2^{nj}}{\beta!}(\partial^\alpha f_j)(x)\int \check{h}(2^j(x-y))R_{j+\nu}(b_\alpha)(y)(y-x)^\alpha dy. \end{aligned}$$

By virtue of Lemma 5 we see that

$$\sum_{j\in\mathbb{Z}} 4^{sj}||C_j(f)||_2^2 \le C||f||^2_{\dot{H}^s}.$$

Indeed, we have $b_\alpha \in \dot{B}_\infty^{|\alpha|,\infty}$ and

$$|C_j(f)(x)| \le C\sum_{|\alpha|\le m} ||b_\alpha||_{\dot{B}_\infty^{|\alpha|,\infty}}[2^{(n-|\alpha|)j}\int |\check{h}(2^j(x-y))||F_m(\partial^\alpha f_j)(x,y)|^2dy]^{1/2}.$$

Then

$$\sum_{j\in\mathbb{Z}} 4^{sj}||C_j(f)||_2^2 \le C\sum_{|\alpha|\le m} ||\partial^\alpha f||^2_{\dot{H}^{s-|\alpha|}}.$$

On the other hand, by overlooking ν and setting $b_{j,\alpha} = \Delta_j(b_\alpha)$, $f_j = S_{j-3}(f)$, the expression $D_j(f)$ takes the form $D_j(f) = E_j(f) + E_j'(f)$ where

$$E_j(f)(x) = \sum_{|\alpha|\le m} 2^{nj}\int \check{h}(2^j(x-y))b_{j,\alpha}(y)F_{m-|\alpha|}(\partial^\alpha f_j)(x,y)dy$$

and

$$E_j'(f)(x) = \sum_{|\alpha|\le m,0<|\beta|\le m-|\alpha|} \frac{1}{\beta!}[(\partial^{\alpha+\beta}f_j)(x)][2^{nj}\int \check{h}(2^j(x-y))b_{j,\alpha}(y)(y-x)^\beta dy].$$

But $b_\alpha \in \dot{B}_\infty^{|\alpha|,\infty}$ implies

$$|E_j(f)(x)| \le C\sum_{|\alpha|\le m} 2^{-|\alpha|j}[2^{nj}\int |\check{h}(2^j(x-y))||F_{m-|\alpha|}(\partial^\alpha f_j)(x,y)|^2dy]^{1/2},$$

so that

$$||E_j(f)||_2 \le C\sum_{|\alpha|\le m} 2^{-|\alpha|j}[\int\int 2^{nj}|\check{h}(2^j(x-y))||F_{m-|\alpha|}(\partial^\alpha f_j)(x,y)|^2dydx]^{1/2}.$$

Since $s - |\alpha| < m - |\alpha| + 1$, we obtain by (3.10)

$$\begin{aligned} [\sum_{j\in\mathbb{Z}} 4^{sj}||E_j(f)||_2^2]^{1/2} &\le C\sum_{|\alpha|\le m} ||\partial^\alpha f||_{\dot{H}^{s-|\alpha|}} \\ &\le C||f||_{\dot{H}^s}. \end{aligned}$$

To establish a similar estimate for $E'_j(f)$, we set

$$R_{j,\beta}(b)(x) = 2^{j(n+|\beta|)} \int \check{h}(2^j(x-y))(y-x)^\beta \Delta_j(b)(y)dy$$

and

$$T_\beta(f) = \sum_{|\alpha| \leq m-|\beta|} \pi^\alpha_{b_\alpha}(f).$$

The Lemma 4 and the hypothesis $b_\alpha \in \dot{B}^{|\alpha|,\infty}_\infty$ imply that T is bounded on $\dot{H}^{-s}$ $(s>0)$. Since T is bounded on $\dot{H}^s$ and $L^2(\mathbb{R}^n)$. By interpolation T is bounded on $\dot{H}^{s-|\beta|}$ for $|\beta| \leq m$. Moreover, $\pi^\alpha_{b_\alpha}$ is bounded on $\dot{H}^{s-|\beta|}$ for $m-|\beta| < |\alpha| \leq m$. Therefore,

$$T_\beta = T - \sum_{m-|\beta|<|\alpha|\leq m} \pi^\alpha_{b_\alpha}$$

is bounded on $\dot{H}^{s-|\beta|}$.

The induction hypothesis now implies that for $|\beta| \neq 0$ we have

$$\sum_{j\in\mathbb{Z}} 4^{j(s-|\alpha|)} \|\sum_{|\alpha|\leq m-|\beta|} R_{j,\beta}(b_\alpha)S_{j-3}(\partial^\alpha g)\|_2^2 \leq C\|g\|^2_{\dot{H}^{s-|\beta|}},$$

for all $g \in \mathcal{S}(\mathbb{R}^n)$. But

$$E'_j(f) = \sum_{0<|\beta|\leq m} \sum_{0<|\beta|\leq m} \frac{1}{\beta!} 2^{-j|\beta|} R_{j,\alpha}(b_\alpha)(x) S_{j-3}(\partial^{\alpha+\beta} f)(x).$$

Therefore,

$$\sum_{j\in\mathbb{Z}} 4^{sj} \|E'_j(f)\|_2^2 \leq C \sum_{0<|\beta|\leq m} \|\partial^\beta f\|^2_{\dot{H}^{s-|\beta|}} \leq C\|f\|^2_{\dot{H}^s}.$$

This completes the proof.

3.3 Duality of type $\mathcal{H}^1 - BMO$.

For $0 \leq s < \frac{n}{2}$, whe shall say that a function h is an s–block, if there exist $f, g \in \mathcal{D}(\mathbb{R}^n)$ such that

$$h = f_1 + f_2$$

where $f_1 \in \mathcal{H}^1$, $f_2 = \pi(g, f)$ and

$$\|f_1\|_{\mathcal{H}^1} + \|f\|_{\dot{H}^s}\|g\|_{\dot{H}^{-s}} \leq 1.$$

We denote by $\mathcal{E}_s$ the function space consisting of all functions of the form

$$h = \sum_k \lambda_k m_k$$

where $(m_k)_k$ is a sequence of s–blocks and $(\lambda_k)_k$ is a sequence of complex numbers such that

$$\sum_k |\lambda_k| < +\infty.$$

The norm in $\mathcal{E}_s$ is given by

$$|||h|||_s = \inf\{\sum_k |\lambda_k|\},$$

the infimum is taken over all the representations of h. **Remark.** For $s = 0$, we have $\mathcal{E}_s = \mathcal{H}^1$ is the Hardy space. The duality $\mathcal{E}_s - BMO^s$ and some properties of $\mathcal{E}_s$ are given by the two following theorems [18].

Theorem 5 *For $s \in [0, \frac{n}{2}[$ the dual of $\mathcal{E}_s$ is BMO^s.*

Theorem 6 *a) For $s, t \in [0, \frac{n}{2}[$ with $t \leq s$, there exists a constant $C = C(s,t)$ such that*

$$|||f|||_s \leq C|||f|||_t$$

for all $f \in \mathcal{E}_t$.

b) For $0 \leq s < \frac{n}{2}$ and $p = \frac{n}{n-s}$, the space $\mathcal{E}_S$ is contained in the Sobolev space $\dot{H}_p^{-s}$.

By duality, we obtain

Theorem 7 *a) For $s, t \in [0, \frac{n}{2}[$ with $t \leq s$, there exists a constant $C = C(s,t)$ such that*

$$|||b|||^t \leq C|||b|||^s$$

for all $b \in BMO^s$.

b) For $0 \leq s < \frac{n}{2}$ and $p' = \frac{n}{s}$, then the Sobolev space $\dot{H}_{p'}^s$ is contained in BMO^s.

4 Singular integral and Sobolev spaces.

4.1 Weak boundedness property.

We denote by $\mathcal{G}$ the group of translations and dilations $\lambda(x) = u + tx$ where $t > 0$ and $u \in \mathbb{R}^n$, the action of λ on $\mathcal{D}(\mathbb{R}^n)$ is defined by

$$f_\lambda(x) = f(\lambda^{-1}(x)) = f(\frac{x-u}{t}).$$

Let T be a continuous linear operator from $\mathcal{D}(\mathbb{R}^n)$ to $\mathcal{D}'(\mathbb{R}^n)$. The operator T_λ is defined by

$$< T_\lambda f, g >= t^{-n} < T f_\lambda, g_\lambda >,$$

the kernel of T_λ is $K_\lambda(x,y) = t^n K(tx+u, ty+u)$

We say that T satisfies the weak boundedness property and write $T \in WBP$ if for each bounded subset $\mathcal{B}$ of $\mathcal{D}(\mathbb{R}^n)$ there exists a positive constant $C_{\mathcal{B}} > 0$ such that

$$| < T_\lambda f, g > | \leq C_{\mathcal{B}} \tag{4.1}$$

for all $f, g \in \mathcal{B}$ and for any $\lambda \in \mathcal{G}$.

Lemma 6 *Let $\alpha \in \mathbb{N}^n$ and $b \in \dot{B}_\infty^{|\alpha|,\infty}$. Then*

$$\pi_b^\alpha \in SIO(m) \cap WBP$$

for all $m \in \mathbb{N}$.

Proof. By Lemma 4 we obtain

$$| < \pi_b^\alpha(f), g > | \leq C||f||_{\dot{H}^t} ||g||_{\dot{H}^{-t}}$$

for $-\frac{n}{2} < t < 0$. This shows that $T \in WBP$. On the other hand, the kernel of π_b^α can be written in the form

$$K(x,y) = \sum_{j \in \mathbb{Z}} 2^{(n+|\alpha|)j} \Delta_j(b)(x) \tilde{\phi}(2^j(x-y))$$

where $\tilde{\phi} \in \mathcal{S}(\mathbb{R}^n)$. To prove that $\pi_b^\alpha \in SIO(m)$ we note that

$$\sum_{j \in \mathbb{Z}} 2^{(n+N)j} |h(2^j(x-y))| \leq C(h,N) \; |x-y|^{-n-N} \tag{4.2}$$

for $h \in \mathcal{S}(\mathbb{R}^n)$, $N \in \mathbb{N}$ and $x \neq y$. The hypothesis $b \in \dot{B}_\infty^{|\alpha|,\infty}$ implies that

$$||\partial^\gamma(\Delta_j(b))||_\infty \leq C 2^{-j(|\gamma|+|\alpha|)}.$$

Using (4.2) we obtain

$$|\partial_x^\gamma K(x,y)| \leq C|x-y|^{-n-|\gamma|}.$$

This completes the proof

Remarks

1) Let $\lambda \in \mathcal{G}$. If T is bounded on $\dot{H}^s$ then T_λ is bounded on $\dot{H}^s$ and the norm is independent of λ.

2) If $T \in SIO(m)$, then for any $\lambda \in \mathcal{G}$ we have $T_\lambda \in SIO(m)$, and if $T \in WBP$ then $T_\lambda \in WBP$. The constants in (1.1) and (4.1) are independent of λ.

Lemma 7 *Let $T \in SIO(m)$ $(m \geq 0)$ and suppose that T is bounded on $\dot{H}^s$ $(s \geq 0)$. Then $T \in WBP$.*

Proof: Case $s < \frac{n}{2}$. By duality we have

$$| < T_\lambda(f), g > | \leq C||f||_{\dot{H}^s} ||g||_{\dot{H}^{-s}}$$

for all $f, g \in \mathcal{D}(\mathbb{R}^n)$. Therefore $T \in WBP$.

Case $s \geq \frac{n}{2}$. A necessary condition when $g \in \dot{H}^{-s} \cap \mathcal{D}(\mathbb{R}^n)$ is that

$$\int g(x)dx = 0.$$

Therefore it is not possible to obtain a similar estimate as in the first case. To avoid this difficulty we will proceed as follows.

Let $\mathcal{B}$ a bounded subset of $\mathcal{D}(\mathbb{R}^n)$, there exists $r > 0$ such that

$$\text{Supp}(f) \subset B(0,r) \text{ for all } \quad f \in \mathcal{B}.$$

Fix $a \in \mathbb{R}^n$ such that $|a| = 3r$ and set

$$\Delta_a(g)(x) = g(x) - g(x-a),$$

and by induction

$$\Delta_a^k(g) = \Delta_a(\Delta_a^{k-1}(g)).$$

We put $\nu = [s - \frac{n}{2}]$ and $g_\nu = \Delta_a^{\nu+1}(g)$. It is easy to show that

$$\int x^\alpha g_\nu(x)dx = 0 \text{ for } \quad |\alpha| \leq \nu.$$

Thus $g_\nu \in \dot{H}^{-s}$. Now for $f, g \in \mathcal{B}$, we have

$$| < T_\lambda(f), g_\nu > | \leq C||f||_{\dot{H}^s}||g_\nu||_{\dot{H}^{-s}}.$$

On the other hand, $\text{Supp}(f) \cap \text{Supp}(g_\nu - g) = \phi$, then

$$< T_\lambda(f), g_\nu - g >= \int\int K_\lambda(x,y)f(y)(g_\nu - g)(x)dxdy.$$

Moreover, for $x \in \text{Supp}(g_\nu - g)$ and $y \in \text{Supp}(f)$, we have

$$|x - y| \geq r.$$

Since $T \in SIO(m)$ implies that

$$| < T_\lambda(f), g_\nu - g > | \leq Cr^{-n}||f||_1||g||_1,$$

the lemma follows.

4.2 Action on polynomials.

We denote by $\mathcal{D}_p$ the function space consisting of all $f \in \mathcal{D}(\mathbb{R}^n)$ such that $\int f(x)x^\alpha dx = 0$ for $|\alpha| \leq p$.

Lemma 8 *Assume that $T \in SIO(m)$ and $g \in \mathcal{D}_p$ with $p \leq m-1$. Then*

$$|T^t(g)(x)| \leq C|x|^{-n-p-1} \quad \text{as} \quad |x| \to +\infty \tag{4.3}$$

We now define the natural action on polynomials. Let

$$\mathcal{O}^p = \{f \in C^\infty(\mathbb{R}^n) / f(x) = o(|x|^p) \text{ as } \quad |x| \to +\infty\}.$$

If $f \in \mathcal{O}^p$ and $g \in \mathcal{D}_p$, $< Tf, g >$ can be defined as follows : let $a + b = 1$ be a partition of unity, $a \in \mathcal{D}(\mathbb{R}^n)$ and $a = 1$ on a neighbourhood of $\mathrm{Supp}(g)$. We write

$$< f, T^t(g) >=< T(af), g > + < bf, T^t(g) > .$$

We have $< T(af), g >$ is defined and, by virtue of Lemma 8, the integral

$$< bf, T(g) >= \int b(x)f(x)T^t(g)(x)dx$$

is absolutely convergent. It is easy to show that $< f, T^t(g) >$ is independent of the choice of $a(x)$ and $b(x)$. Then we put $< Tf, g >=< f, T^t(g) >$.

Proof of Lemma 8: For $x_0 \in \mathrm{Supp}(g)$ fixed, let

$$K_p(x,y) = K(x,y) - \sum_{|\alpha| \leq p} \frac{(x-x_0)^\alpha}{\alpha!}(\partial_x^\alpha K)(x_0, y).$$

Let now $y \in \mathbb{R}^n$ such that $|x_0| \leq \frac{1}{2}|y|$ and

$$|x - x_0| \leq \frac{1}{2}|x - y| \quad \text{for} \qquad x \in \mathrm{Supp}(g).$$

The hypothesis $T \in SIO(m)$ implies that for $x \in \mathrm{Supp}(f)$ we have

$$\begin{aligned} |K_p(x,y)| &\leq C|x - x_0|^{p+1}|x_0 - y|^{-n-p-1} \\ &\leq C|x - x_0|^{p+1}|y|^{-n-p-1}. \end{aligned}$$

Indeed, we have for $y \notin \mathrm{Supp}(g)$

$$T^t(g)(y) = G(y) = \int K(x,y)g(x)dx.$$

Using the fact that $g \in \mathcal{D}_p$, we obtain

$$G(y) = \int K_p(x,y)g(x)dx,$$

so that

$$|G(y)| \leq C|y|^{-n-p-1} \int |g(x)||x - x_0|^{p+1}dx \quad \text{as} \qquad |y| \to +\infty.$$

From this the lemma follows.

Remark. In general, $T(x^\alpha)$ is not regular. For example, by considering the multiplier operator

$$T(f)(x) = \chi_{[0,1]^n}(x)f(x)$$

we have $K(x,y) = 0$ for $x \neq y$ and $T \in WBP$ (because $\chi_{[0,1]^n} \in L^\infty$). But $T(x^\alpha) = x^\alpha \chi_{[0,1]^n}$ is not continuous.

Now we are led to define another action of T on plynomials. We denote by R_α the multiplier operator

$$R_\alpha f(x) = x^\alpha f(x).$$

In particular $R_{e_j} f(x) = x^{e_j} f(x)$ for $1 \leq j \leq n$. If $\Gamma_j = [T, R_{e_j}]$, we have

$$\Gamma_j \circ \Gamma_k = \Gamma_k \circ \Gamma_j$$

and by induction we define

$$\Gamma^{\alpha+e_j} = \Gamma_j \circ \Gamma^\alpha \text{ for } \quad \alpha \in \mathbb{N}^n.$$

The kernel $K_\alpha(x,y)$ of Γ^α satisfies

$$|\partial_x^\gamma K_\alpha(x,y)| \leq C_\alpha |x-y|^{-n-|\gamma|+|\alpha|} \quad \text{for} \quad |\gamma| \leq m \text{ and} \quad x \neq y. \tag{4.4}$$

Using (4.4), $\mathcal{A}(T)(x^\alpha) = \Gamma^\alpha(T)(1)$ is well defined (to do so, we use the same arguments as in Lemma 8, see [3]).

To obtain the regularity of $\mathcal{A}(T)(x^\alpha)$ we shall need the following theorem.

Theorem 8 *Let $T \in SIO(m) \cap WBP$ such that $T(x^\gamma) = 0$ for $|\gamma| \leq N-1$ ($N \leq m-1$). Then $T(x^\alpha) \in \dot{B}_\infty^{N,\infty}$ for $|\alpha| = N$.*

Proof : Let $h \in \mathcal{D}(\mathbb{R}^n)$ be supported by the unit ball such that it satisfies $\int h(x)x^\gamma dx = 0$ for $|\gamma| \leq N+1$ and $\hat{h}(\xi) \neq 0$ for $\frac{1}{2} \leq |\xi| \leq 2$. By virtue of Lemma 2 we have

$$||f||_{\dot{B}_\infty^{N,\infty}} \simeq sup_j(2^{Nj}||R_j(f)||_\infty)$$

where $h_j(x) = 2^{nj}h(2^j x)$ and $R_j(f) = h_j \star f$. By replacing $h(x)$ by $h(-x)$, we shall show that

$$|< T(x^\alpha), \mathcal{U}_z h_j >| \leq C2^{Nj} \quad \text{for} \quad j \in \mathbb{Z} \text{ and } \quad z \in \mathbb{R}^n$$

where $\mathcal{U}_z(f)(x) = f(x-z)$. Since $T(x^\gamma) = 0$ for $|\gamma| \leq N-1$, it follows that

$$< T(x^\alpha), \mathcal{U}_z h_j > = < T((x-z)^\alpha), \mathcal{U}_z h_j > .$$

We now return to the definition of $T(x^\alpha)$. Let $a \in \mathcal{D}(\mathbb{R}^\backslash)$ such that $a(x) = 1$ for $|x| \leq 2$ and $a(x) = 0$ for $|x| \geq 3$. We put $a'(x) = x^\alpha a(x)$ and $b' = x^\alpha(1-a(x))$. For $\lambda(x) = z + 2^{-j}x$, we have

$$< T((x-z)^\alpha), \mathcal{U}_z h_j > = 2^{j(n-|\alpha|)}[< T(a'_\lambda), h_\lambda > + < b'_\lambda, T^t(h_\lambda) >].$$

Since $T \in WBP$ it implies that

$$|< T(a'_\lambda), h_\lambda >| \leq C2^{-nj}.$$

By virtue of Lemma 8 we obtain

$$|< b'_\lambda, T^t(h_\lambda) >| \leq C2^{-nj}.$$

From this the lemma follows.

Theorem 9 *Let $T \in SIO(m) \cap WBP$ $(m \geq 1)$. Then $\mathcal{A}(T)(x^\alpha) \in \dot{B}_\infty^{|\alpha|,\infty}$ for $|\alpha| \leq m-1$.*

Proof : It is enough to show that $\mathcal{A}(T)(x^\alpha) = T_\alpha(x^\alpha)$ where T_α is an operator that satisfies the hypothesis of Theorem 8. To do so, by induction and Theorem 8, it is not hard to show that

$$T_\alpha = T - \sum_{|\gamma| \leq |\alpha|-1} \pi_{b_\gamma}^\gamma$$

where $b_\gamma = \frac{1}{\gamma!}\mathcal{A}(T)(x^\gamma)$.

4.3 Boundedness on Sobolev spaces.

In the particular case where T vanishes on polynomials, we have the following theorem. (See [8], [13]).

Theorem 10 *If $T \in SIO(m+1) \cap WBP$ and $T(x^\alpha) = 0$ for $|\alpha| \leq m$, then T is bounded on $\dot{H}^s$ for $0 < s < m+1$.*

A corollary of Lemma 7 and Theorems 9 and 10 is a version of David–Journé's theorem.

Theorem 11 *Let $s > 0$, $m = [s]$ and $T \in SIO(m+1)$. Then T is bounded on $\dot{H}^s$ if only if $T \in WBP$ and the operator*

$$T_0 = \sum_{|\alpha| \leq m} \pi_{b_\alpha}^\alpha$$

is bounded on $\dot{H}^s$ where $b_\alpha = \frac{1}{\alpha!}\mathcal{A}(T)(x^\alpha)$.

In the particular case where $0 < s < 1$ we obtain the Lemarié–Meyer's theorem [12].

Corollary 2 *Let $0 < s < 1$ and $T \in SIO(1)$. Then T is bounded on $\dot{H}^s$ if only if $T \in WBP$ and $T(1) \in BMO^s$.*

We consider now the multiplier operator $T(f) = gf$. We have $K(x,y) = 0$ for $x \neq y$ and we observe that $T(1) = g$, $\Gamma^\alpha(T)(1) = 0$ for $\alpha \neq 0$. On the other hand, we have

$$T \in WBP \text{ if and only if } \quad g \in L^\infty.$$

If we apply Theorem 11 we obtain.

Corollary 3 *We denote by $M(\dot{H}^s)$ the pointwise multipliers of $\dot{H}^s$. Then*

$$M(\dot{H}^s) = BMO^s \cap L^\infty(\mathbb{R}^n).$$

In particular, $M(\dot{H}^s) = \{0\}$ for $s \geq \frac{n}{2}$.

Theorem 12 *Let $T \in SIO(m+1)$, $0 < s < m+1$ and suppose that T is bounded on $\dot{H}^s$. Then*

$$T(1) \in BMO^s.$$

Moreover, if $s = m$ and $|\alpha| = m$ we have

$$b_\alpha = \frac{1}{\alpha!}\mathcal{A}(T)(x^\alpha) \in \dot{H}_\infty^m.$$

Proof. a) **Case** $s \notin \mathbb{N}$. From Theorem 11 and (3.3) it follows that

$$\sum_{j\in\mathbb{Z}} 4^{sj} \| \sum_{|\alpha|\le m} \Delta_j(b_\alpha)\partial^\alpha f\|_2^2 \le C\|f\|_{\dot H^s}^2.$$

Let Ω be a bounded open subset of $\mathbb{R}^n$ and let $f \in \mathcal{D}(\mathbb{R}^n)$ such that $f = 1$ on Ω. Then $\partial^\alpha f = 0$ on Ω for $\alpha \neq 0$, so that

$$\sum_{j\in\mathbb{Z}} 4^{sj} \int_\Omega |\Delta_j(b_0)(x)|^2 dx \le \sum_{j\in\mathbb{Z}} 4^{sj} \| \sum_{|\alpha|\le m} \Delta_j(b_\alpha)\partial^\alpha f\|_2^2.$$

This implies

$$\sum_{j\in\mathbb{Z}} 4^{sj} \int_\Omega |\Delta_j(b_0)(x)|^2 dx \le C\mathrm{Cap}_s(\Omega).$$

b) **Case** $s \in \mathbb{N}$ $(s = m)$. We are led to prove the Carleson conditions

$$\sum_{j\ge k} 4^{sj} \int_B |\Delta_j(b_\alpha)(x)|^2 dx \le C|B|^{1-(2(s-|\alpha|))/n} \tag{4.5}$$

for all $k \in \mathbb{Z}$ and any ball B of radius 2^{-k}. Indeed, if (4.5) holds we have in particular $b_\alpha \in \dot H_\infty^m$ for $|\alpha| = m$. Thus

$$b'_\alpha = \sum_{j\in\mathbb{Z}} 2^{sj}\Delta_j(b_\alpha)$$

is in BMO. A classical result [1] implies that $\pi^0_{b'_\alpha}$ is bounded on L^2. By virtue of Theorem 1, we obtain

$$\sum_j 4^{sj} \|\Delta_j(b_\alpha)S_{j-3}(g)\|_2^2 \le C\|g\|_2^2, \text{ for } \quad g \in L^2(\mathbb{R}^n)$$

Since $|\alpha| = s$ implies that

$$\sum_j 4^{sj} \|\Delta_j(b_\alpha)S_{j-3}(\partial^\alpha f)\|_2^2 \le C\|f\|_{\dot H^s}^2,$$

it follows that $\pi^\alpha_{b_\alpha}$ is bounded on $\dot H^s$. Therefore, the operator

$$T' = \sum_{|\alpha|\le m-1} \pi^\alpha_{b_\alpha}$$

is bounded on $\dot H^s$. Applying (3.4) we see that

$$\sum_{j\in\mathbb{Z}} 4^{sj} \| \sum_{|\alpha|\le m-1} \Delta_j(b_\alpha)\partial^\alpha f\|_2^2 \le C\|f\|_{\dot H^s}^2.$$

Let Ω a bounded open and $f \in \mathcal{D}(\mathbb{R}^n)$ such that $f = 1$ on Ω. Then $\partial^\alpha f = 0$ on Ω for $\alpha \neq 0$ and we have.

$$\sum_{j\in\mathbb{Z}} 4^{sj} \int_\Omega |\Delta_j(b_0)(x)|^2 dx \le C\|f\|_{\dot H^s}^2.$$

This implies

$$\sum_{j\in\mathbb{Z}} 4^{sj}\int_\Omega |\Delta_j(b_0)(x)|^2dx \le C\mathrm{Cap}_s(\Omega).$$

Proof of (4.5). We will proceed by induction on $|\alpha|$. Let $k \in \mathbb{Z}$ and $x_0 \in \mathbb{R}^n$, we set $B = B(x_0, 2^{-k})$.

a) **Case** $\alpha = 0$ **:** Let $f \in \mathcal{D}(\mathbb{R}^n)$ such that $f(x) = 1$ for $|x| \le 2$. We set

$$f_B(x) = f(2^k(x_0 - x)).$$

It is easy to show that

$$|S_{j-3}(\partial^\alpha f_B)(x)| \le C2^{k|\alpha|}2^{(-j+k)\ell}$$

for $\alpha \neq 0$ and $\ell = s - |\alpha| + 1$. Then

$$\sum_{j\ge k} 4^{sj}\int_B |\sum_{0<|\alpha|\le m} \Delta_j(b_\alpha)(x)S_{j-3}(\partial^\alpha f)(x)|^2 \le C4^{ks}|B|,$$

it follows that

$$\begin{aligned}
\sum_{j\ge k} 4^{sj}\int_B |\Delta_j(b_0)|^2dx &\le C[\sum_{j\in\mathbb{Z}} 4^{sj}\int_B |\Delta_j(b_0)f_B(x)\\
&+ \sum_{0<|\alpha|\le m} \Delta_j(b_\alpha)(x)S_{j-3}(\partial^\alpha f_B)(x)|^2 + 4^{ks}|B|]\\
&\le C[||f_B||^2_{\dot H^s} + 4^{ks}|B|]\\
&\le C|B|^{1-(2s)/n}.
\end{aligned}$$

b) Therefore (4.5) holds for $\alpha = 0$. Now suppose that (4.5) holds for any γ such that $|\gamma| < |\alpha|$. Let

$$h_B(x) = 2^{k|\alpha|}(x - x_0)^\alpha f(2^k(x - x_0)).$$

We use the inequality which is similar to (3.4)

$$\sum_{j\in\mathbb{Z}} 4^{sj}||\sum_{|\gamma|\le|\alpha|-1} \Delta_j(b_\gamma)(\partial^\gamma h_B) + \sum_{|\gamma|\ge|\alpha|} \Delta_j(b_\gamma)S_{j-3}(\partial^\gamma h_B)||_2^2 \le C||h_B||^2_{\dot H^s}.$$

Using the same arguments as in the proof of a), we obtain

$$\sum_{j\ge k} 4^{sj}||\sum_{|\gamma|\ge|\alpha|,\gamma\neq\alpha} \Delta_j(b_\gamma)S_{j-3}(\partial^\gamma h_B)||_2^2 \le C|B|^{1-(2s)/n}.$$

The induction hypothesis now implies that

$$\sum_{j\ge k} 4^{sj}||\sum_{|\gamma|\le|\alpha|-1} \Delta_j(b_\gamma)(\partial^\gamma h_B)||_2^2 \le C|B|^{1-(2s)/n},$$

so that we obtain

$$\sum_{j\geq k} 4^{sj} \int_B |\Delta_j(b_\alpha)(x)|^2 dx \leq C2^{-k|\alpha|} \sum_{j\geq k} 4^{sj} ||\Delta_j(b_\alpha)\partial^\alpha h_B||_2^2$$
$$\leq C|B|^{1-(2(s-|\alpha|))/n}.$$

This completes the proof of (4.5).

A corollary of (4.5) is the following.

Corollary 4 *Let $T \in SIO(m+1)$, $0 < s < m+1$ and suppose that T is bounded on $\dot{H}^s$. Then $T(x^\alpha) = 0$ for $|\alpha| < s - \frac{n}{2}$.*

For $s - \frac{n}{2} \in \mathbb{N}$ and $|\alpha| = s - \frac{n}{2}$ Corollary 4 also holds see Appendix). In the one dimensional case, we have the following theorem.

Theorem 13 *Let $s > 0$, $m = [s]$ and $T \in SIO(m+1)$. For $n = 1$, the following are equivalent.*

a) T is bounded on $\dot{H}^s$;

b) $T \in WBP$ and the operator $\pi^m_{b_m}$ is bounded on $\dot{H}^s$ where $b_m = T(x^m)$;

c) $\frac{d^m b_m}{dx} \in BMO^{s-m}$.

From Theorems 11, 12 we obtain the two following corollaries.

Corollary 5 *For $s \in \mathbb{N}$ and $b_\alpha \in \dot{B}^{s,\infty}_\infty$, $|\alpha| = s$, the operator*

$$T = \sum_{|\alpha|=s} \pi^\alpha_{b_\alpha}$$

is bounded on $\dot{H}^s$ if and only if $b_\alpha \in \dot{H}^s_\infty$ for all α.

Corollary 6 *Let $T \in SIO(2)$ .Then T is bounded on $\dot{H}^1$ if and only if $T \in WBP$, $T(1) \in BMO^1$ and*

$$\mathcal{A}(T)(x^{e_j}) \in \dot{H}^1_\infty$$

for $1 \leq j \leq n$.

5 Appendix

To extend Corollary 4 for $s - \frac{n}{2} \in \mathbb{N}$ and $|\alpha| = s - \frac{n}{2}$, we have the following Lemma.

Lemma 9 *Let $\alpha \in \mathbb{N}^n$ and $b \in \dot{B}^{|\alpha|,\infty}_\infty$ such that π^α_b is bounded on $\dot{H}^{|\alpha|+n/2}$. Then $b = 0$ (modulo polynomials).*

Proof. We put

$$b' = \sum_{j\in\mathbb{Z}} 2^{j|\alpha|}\Delta_j(b).$$

Then $b' \in \dot{B}^{0,\infty}_{\infty}$ and the operator $T = \pi_{b'}$ is bounded $\dot{H}^{n/2}$ (this is an easy application of theorem 2). Lemma 4 implies that T is bounded on $\dot{H}^{-n/2}$. So T^t is bounded on L^2 (by interpolation). A classical result [13] shows that T^t is bounded from $\mathcal{H}^1$ to L^1 (because $T \in SIO(1)$ and $T^t \in SIO(1)$). Therefore $T^t(f) \in L^1$ for all $f \in \mathcal{D}_1$.

We now return to the definition of $T(1)$, we have formally

$$< T(1), f >= \int T^t(f)(x)dx$$

for $f \in \mathcal{D}_1$.

On the other hand T^t is bounded on $\dot{H}^{-n/2}$. Thus

$$T^t(f) \in L^1 \cap \dot{H}^{-n/2}$$

for all $f \in \mathcal{D}_1$, so that we obtain

$$\int T^t(f)(x)dx = 0,$$

then $b' = T(1) = 0$ (modulo polynomials).

References

[1] G. Bourdaud: Analyse fonctionnelle dans l'espace euclidien. Pub. Math. Paris VII, 23 (1987).

[2] G. Bourdaud: Realisation des espaces de Besov homogènes. Arkiv for math. 26 (1988), 41–54.

[3] G. Bourdaud: Une algèbre maximale d'opérateurs pseudo-différentiels de type 1, 1. Séminaire E.D.P. Ecole polytechnique (France), Exp. 7 (1987-1988).

[4] G. Bourdaud: Localisation et Multiplicateurs des espaces de Sobolev homogènes. Manuscripta Math. 60, p. 93-103 (1988).

[5] R. Coifman and Y. Meyer : Au delà des opérateurs pseudo-différentiels . Astérisque 57 (1978).

[6] B. E. J. Dahlberg: Regularity properties of Riesz potentials. Ind. Univ. Math. J. 28(1979) p. 257-268.

[7] G. David and J. L. Journé: A boundedness criterion for generalized Calderon–Zygmnd operators. Ann. Math. 120(1984) pp. 371-393.

[8] M. Elhodiabi : Thesis Orsay (1988).

[9] M. Frazier and B. Jawerth: A discrete transform and applications to distribution spaces. J. Funct. Anal. 93 (1990), 34–170.

[10] M. Frazier, R. Torres and G. weiss: The boundedness of Calderon-Zygmund operators on $\dot{F}_p^{s,q}$ spaces. Revista Math Ibero-Americana, 4 (1988).

[11] P.G. Lemarié: Continuité sur les espaces de Besov des opérateurs définis par des intégrales singulières. Ann. Inst. Fourier, Grenoble 35, 4 (1985) 175–187.

[12] M. Meyer : Une classe d'espace fonctionnels de type BMO. Applications aux intégrales singulières. Ark. Math. 27 (1989), 305–318.

[13] Y. Meyer: Opérateurs de Calderon–Zygmund. Tome II. Herman (1990).

[14] J. Peetre: New thoughts on Besov spaces. Duke Univ. Math. Series I Durham, North Carolina (1976).

[15] H. Triebel: Theory of function spaces II (Birkhser, Basel 1992).

[16] A. Youssfi: Localisation des espaces de Lizorkin–Triebel homogènes. Math. Nachr. 147 (1990) 107–121.

[17] A. Youssfi: Continuité–Besov des opérateurs définis par des intégrales singulières. Manuscripta Math. 65 (1989), 289–310.

[18] A. Youssfi: Duality of type $\mathcal{H}^1 - BMO$ and bilinear operators. Preprint

Université Marne-La-Vallée and
Université Paris VII, C.N.R.S. U.R.A. 212
Tour 45–55–5 ème étage
2, place Jussieu
75251 Paris cédex 05
France

Conditional Hypoellipticity and Fourier Multipliers for Weighted L_p-Spaces with an Exponential Weight

Victor I. Burenkov

1 Conditional hypoellipticity. Preliminaries

Our intention is to discuss some results, connected with a notion of conditional hypoellipticity, i.e. hypoellipticity in dependence on behaviour of solutions at infinity.

Let

$$P(D) = \sum_{\alpha \in K} a_\alpha D^\alpha$$

be a linear partial differential operator with constant coefficients a_α ($\alpha = (\alpha_1, \ldots, \alpha_n) \in \mathbb{N}_0^n$, $D^\alpha = D_1^{\alpha_1} \cdots D_n^{\alpha_n}$, $D_j = \frac{1}{i}\frac{\partial}{\partial x_j}$) and

$$\Omega = \Omega_m \times \mathbb{R}^{n-m},$$

where $0 \le m \le n$, Ω_m - open set in $\mathbb{R}^m$ (for $m = 0$ $\Omega = \mathbb{R}^n$, for $m = n$ Ω is an arbitrary open set in $\mathbb{R}^n$).

By Q_m we denote the set of all rectangulars $G = G_m \times \mathbb{R}^{n-m}$ with $G_m = \{x \in \mathbb{R}^m : -\infty < a_k < x_k < b_k < \infty, k = \overline{1,m}\}$, such that $\overline{G_m} \subset \Omega_m$.

Let ϱ denote a positive continuous function on Ω and let $1 \le p \le \infty$. We say that $u \in [L_{p,\varrho}]_m(\Omega)$ iff $\forall G \in Q_m \quad u \in L_{p,\varrho}(G)$, i.e. u is measurable and $\|u\|_{L_{p,\varrho}(G)} = \|\varrho u\|_{L_p(G)} < \infty$. Further on for $l \in \mathbb{N}$, $u \in [W^l_{p,\varrho}]_m(\Omega)$ iff $\forall \alpha \in \mathbb{N}_0^n : |\alpha| \le l$ there exist weak derivatives $D^\alpha u$ and $\forall G \in Q_m$

$$\|u\|_{W^l_{p,\varrho}(\Omega)} = \sum_{|\alpha| \le l} \|D^\alpha u\|_{L_{p,\varrho}(G)} < \infty.$$

Finally

$$[W^\infty_{p,\varrho}]_m(\Omega) = \bigcap_{l \in \mathbb{N}} [W^l_{p,\varrho}]_m(\Omega).$$

All the defined spaces are "local" with respect to variables $x_1, \ldots, x_m$.

For sake of briefness in the sequel we shall omit index m and write $L_{p,\varrho}(\Omega)$, $W^l_{p,\varrho}(\Omega)$ and $W^\infty_{p,\varrho}(\Omega)$ instead of $[L_{p,\varrho}]_m(\Omega)$, $[W^l_{p,\varrho}]_m(\Omega)$ and $[W^\infty_{p,\varrho}]_m(\Omega)$ respectively.

Definition 1 *We say that operator P of order l is conditionally hypoelliptic in $L_{p,\varrho}(\Omega)$ iff*

$$u \in W^l_{p,\varrho}(\Omega) \ \ \& \ \ Pu \in W^\infty_{p,\varrho}(\Omega) \quad \Longrightarrow \quad u \in W^\infty_{p,\varrho}(\Omega). \tag{1}$$

It is a definition of hypoellipticity under additional assumptions, connected with the space $L_{p,\varrho}(\Omega)$: first of all condition $u \in W^l_{p,\varrho}(\Omega)$ describes admissible behaviour of function u at infinity; secondly the property of Pu, that is inherited by u is also expressed in $L_{p,\varrho}$-terms: we assume, that Pu belongs to $W^\infty_{p,\varrho}(\Omega)$, and demand u to belong to the same space.

If we take $m = n$, then definition 1 takes the form of one of the variants of classical definition of hypoellipticity:

$$u \in W^l_p(\Omega) \;\;\&\;\; Pu \in C^\infty(\Omega) \implies u \in C^\infty(\Omega). \tag{2}$$

The more traditional variants contain (instead of $u \in W^l_p(\Omega)$) assumption, that $u \in C^l(\Omega)$ or $u \in L_2(\Omega)$ or $u \in D'(\Omega)$ (Pu is understood in appropriate manner). Necessary and sufficient condition on P to posess property (2) (or any of its variants mentioned above) is as follows: for every multiindex $\alpha \neq 0$

$$\lim_{\xi\to\infty} \frac{P^\alpha(\xi)}{P(\xi)} = 0 \tag{3}$$

(Hörmander's theorem [1,2]).

In non-weighted case ($\varrho \equiv 1$) the complete description of conditionally hypoelliptic operators in $L_2(\Omega)$ was given in [3] (see also [4,5]). Set

$$E_m = \{\alpha \in \mathbb{N}_0^n : \alpha = (\alpha_1, \dots, \alpha_m, 0, \dots, 0)\}, \quad E'_m = \{\alpha \in E_m : \alpha \neq 0\}.$$

The mentioned description has the following form:

1) $P(\xi) \neq 0$ for $\xi \in \mathbb{R}^n$ large enough, and

2) for every $\alpha \in E'_m$ holds (3). (For $m = n$ this condition coincides with Hörmander's one; for $m = 0$ it means that $P(\xi) \neq 0$ for $\xi \in \mathbb{R}^n$ large enough.)

It was natural to investigate conditional hypoellipticity in $L_{2,\varrho}(\Omega)$ with ϱ tending to 0 at infinity, i.e. for a wider class of admissible functions u.

In [6] the case of power weight $\varrho(x) = (1 + \sum_{j=m+1}^n x_j^2)^s$ with $-\infty < s < \infty$ was investigated and necessary and sufficient condition on P was found. It appeared that it has the same form as for the non-weighted case.

For this reason it is of special interest to study the case of an exponential weight

$$\varrho(x) = \exp(A_{m+1}|x_{m+1}|^{\epsilon_{m+1}} + \cdots + A_n|x_n|^{\epsilon_n}) \tag{4}$$

with $A_j \neq 0$, $\epsilon_j > 0$, when as preliminary analysis shows the corresponding condition on P must depend on ϱ.

Our main aim is to present results on conditional hypoellipticity in $L_{p,\varrho}(\Omega)$ with ϱ of form (4).

2 Theorems on conditional hypoellipticity in terms of Fourier multipliers in general weighted $L_{p,\varrho}$-spaces

The results obtained in [3,6] were proved on the base of the method of fractional differentiation of a priori inequalities, worked out in [3]. The same method is applicable for the case of general weights. One of the essential steps in this method is the theorem on repeated norms for Nikol'skij-Besov spaces with fractional order of smoothness [3,7] and its weighted analogue [8]. But the crucial moment for this approach is application of theorems on Fourier multipliers for weighted $L_{p,\varrho}$-spaces.

We remind the definition of Fourier multiplier.

Definition 2 *Let $1 \le p \le \infty$, ϱ - positive measurable function. Then $\mu \in M_{p,\varrho}(\mathbb{R}^n)$ (i.e. μ is an $L_{p,\varrho}$-Fourier multiplier) iff $\mu \in L_\infty(\mathbb{R}^n)$ and for some $c_1 > 0$ for any $\varphi \in C_0^\infty(\mathbb{R}^n)$*

$$||F^{-1}(\mu F\varphi)||_{L_{p,\varrho}(\mathbb{R}^n)} \le c_1 ||\varphi||_{L_{p,\varrho}(\mathbb{R}^n)},$$

where F denotes Fourier transform.

In order to show how necessity in such multipliers appears in general scheme we introduce the following definitions of comparison of power of differential operators with constant coefficients T and P in $L_{p,\varrho}(\Omega)$ for some positive measurable weight function ϱ defined on Ω.

In these definitions we shall use for $G \in Q_m$ the following weighted norms of Nikol'skij-Besov type:

$$||u||_{B^l_{p,\varrho}(G)} = ||u||_{L_{p,\varrho}(G)} + \sum_{j=1}^{n} (\int_0^1 (h^{-l} ||\Delta^\sigma_{j,h} u||_{L_{p,\varrho}(G_{\sigma h})})^p \frac{dh}{h})^{\frac{1}{p}},$$

where $l > 0, 1 \le p \le \infty, \sigma = [l] + 1, \Delta^\sigma_{j,h} u$ is a difference of order σ of a function u with a step h with respect to x_j and $G_\gamma = \{x \in G : \text{ dist } (x, \partial G) > \gamma\}$.

Let us also introduce for $G \in Q_m$ and $|| \cdot ||_{Z(G)}$ with $Z(G) = L_{p,\varrho}(G)$ or $Z(G) = B^l_{p,\varrho}(G)$ the following norm

$$|||Pu|||_{Z(G)} = \sum_{\alpha \in E_m} ||Pu||_{Z(G)}.$$

Definition 3 *Operator T is weaker than operator P in $L_{p,\varrho}(\Omega)$, briefly*

$$T \prec P \quad (L_{p,\varrho}(\Omega)),$$

iff for some $l > 0$ $\forall G \in Q_m$ and for any $u \in D_P(\Omega) \bigcap L_{p,\varrho}(\Omega)$ such that $\Pr_{\mathbb{R}^m}$ supp $u \subset G_m$

$$|||Tu|||_{B^l_{p,\varrho}(G)} \le c_2 |||Pu|||_{L_{p,\varrho}(G)},$$

where c_2 doesn't depend on u.

Definition 4 *Operator T is absorbed by operator P in $L_{p,\varrho}(\Omega)$, briefly*

$$T \propto P \quad (L_{p,\varrho}(\Omega)),$$

iff for every $l > 0$ there exists such $r > 0$ that $\forall G \in Q_m$ and for any $u \in D_P(\Omega) \bigcap L_{p,\varrho}(\Omega)$ such that $\Pr_{\mathbb{R}^m}$ suppu $\subset G_m$

$$|||Tu|||_{B^l_{p,\varrho}(G)} \le c_3(|||Pu|||_{B^r_{p,\varrho}(G)} + |||Tu|||_{L_{p,\varrho}(G)}),$$

where c_3 doesn't depend on u.

Let now $\varrho(x) = \eta(x_{m+1}, \dots, x_n)$, where η is such positive continuous function on $\mathbb{R}^{n-m}$ that

$$\forall t > 0 \qquad \sup_{h \in \mathbb{R}^{n-m} : |h| \le t} \ \sup_{\xi \in \mathbb{R}^{n-m}} \frac{\eta(\xi + h)}{\eta(\xi)} < \infty. \tag{5}$$

From (5) it follows that for some $c_4 > 0,\ b > 0$

$$\sup_{y \in \mathbb{R}^n} \frac{\varrho(x+y)}{\varrho(y)} \le c_4 \exp(b(|x_{m+1}| + \cdots + |x_n|)). \tag{6}$$

The above mentioned method of fractional differentiation of a priori inequalities leads to the following theorem [3,9].

Theorem 1 *Let* $1 \le p \le \infty, 0 \le m \le n$ *and* ϱ *satisfy condition* (5). *If*

$$I \propto P \quad (L_{p,\varrho}(\Omega)) \tag{7}$$

and

$$\forall \alpha \in E'_m \quad P^{(\alpha)} \prec P \quad (L_{p,\varrho}(\Omega)), \tag{8}$$

then P *is conditionally hypoelliptic in* $L_{p,\varrho}(\Omega)$.

Remark 1 *If* $p = 2$ *and* $\varrho \equiv 1$, *then* P *is conditionally hypoelliptic in* $L_2(\Omega)$ *if and only if conditions* (7) *and* (8) *are satisfied* [3].

Conditions providing properties (7) and (8) can be given in terms of space $M_{p,\varrho}(\mathbb{R}^n)$. Set for $\xi \in \mathbb{R}^n$

$$\tilde{P}_m(\xi) = (\sum_{\alpha \in E_m} |P^{(\alpha)}(\xi)|^2)^{\frac{1}{2}}.$$

Theorem 2 *Let* $1 \le p \le \infty$ *and* ϱ *satisfy condition* (5). *Moreover let for some* $\mu \ge 0$ *and some function* $\psi \in C^\infty(\Omega)$ *such that* $(1+|\xi|^2)^\nu(1-\psi(\xi)) \in M_{p,\varrho}(\mathbb{R}^n)$ *for every* $\nu > 0$ *the following condition is fulfilled:*

$$\forall \alpha, \beta \in E_m \quad \psi(\xi) T^{(\alpha)}(\xi) \overline{P^{(\beta)}(\xi)} \tilde{P}_m^{-2}(\xi)(1+|\xi|^2)^{-\mu} \in M_{p,\varrho}(\mathbb{R}^n).$$

Then

$$T \propto P \quad (L_{p,\varrho}(\Omega)).$$

Theorem 3 *Let* $1 \le p \le \infty$, $0 \le m \le n$, ϱ *satisfy condition* (5) *and inequality* (6) *is valid with* $b < 1$. *Let moreover for some* $\mu > 0$

$$\forall \alpha, \beta \in E_m \quad T^{(\alpha)}(\xi) \overline{P^{(\beta)}(\xi)} \tilde{P}_m^{-2}(\xi)(1+|\xi|^2)^{\mu} \in M_{p,\varrho}(\mathbb{R}^n).$$

Then

$$T \prec P \quad (L_{p,\varrho}(\Omega)).$$

Remark 2 *For* $p = 2$ *and* $\varrho \equiv 1$ *condition* $T \propto P \ (L_2(\Omega))$ *is equivalent to condition: for some* $\mu \ge 0$

$$\forall \xi \in \mathbb{R}^n \quad \tilde{T}_m(\xi) \le c_5 \tilde{P}_m(\xi)(1+|\xi|^2)^{\mu},$$

where c_5 *doesn't depend on* ξ. *Moreover condition* $I \propto P$ *is equivalent to condition:*

$$\tilde{P}_m(\xi) \ne 0 \ \textit{for large enough} \ \xi \in \mathbb{R}^n.$$

In its turn condition $T \prec P \ (L_2(\Omega))$ *is equivalent to condition: for some* $\mu > 0$

$$\forall \xi \in \mathbb{R}^n \quad \tilde{T}_m(\xi) \le c_6 \tilde{P}_m(\xi)(1+|\xi|^2)^{-\mu},$$

where c_6 *doesn't depend on* ξ [3].

In order to obtain on the base of theorems 1-3 concrete theorem on conditional hypoellipticity, expressed in terms of characteristic polynomial $P(\xi)$, one needs to know conditions on function, providing that it is an $L_{p,\varrho}$-Fourier multiplier.

3 Fourier multipliers in weighted $L_{p,\varrho}$-spaces with an exponential weight

If $\varrho \equiv 1$ we write $M_{p,1}(\mathbb{R}^n) \equiv M_p(\mathbb{R}^n)$. Let us remind that $\|\mu\|_{M_2(\mathbb{R}^n)} = \|\mu\|_{L_\infty(\mathbb{R}^n)}$, $\|\mu\|_{M_1(\mathbb{R}^n)} = \|\mu\|_{M_\infty(\mathbb{R}^n)} = \int_{\mathbb{R}^n} |dF\mu|$ (Fourier transform of μ in distributional sense is a Borel measure integrable on $\mathbb{R}^n$) and according to famous Marcinkiewicz theorem for $1 < p < \infty$

$$\|\mu\|_{M_p(\mathbb{R}^n)} \le c_7 \sum_{0 \le \alpha \le 1} \|x^\alpha D^\alpha \mu(x)\|_{L_\infty(\mathbb{R}^n)},$$

where c_7 doesn't depend on μ, $x^\alpha = x_1^{\alpha_1} \cdots x_n^{\alpha_n}$, $0 \le \alpha \le 1$ means that for every $j = \overline{1,n}$ $0 \le \alpha_j \le 1$, i.e. α_j is equal to 0 or 1 (it is assumed that μ has continuous derivatives $D^\alpha \mu$ on the set $x \in \mathbb{R}^n : x_1 \cdots x_n \neq 0$). This theorem was generalized in many directions. In particular, there was a number of investigations of space $M_{p,\varrho}(\mathbb{R}^n)$ for a weight of power type and for a weight satisfying A_p-condition. Roughly speaking all these results are in some sense close to initial result of Marcienkiewicz (the recent papers in this direction are [10, 11, 12]).

Let now $0 \le m \le n$ and weight function ϱ has the form (4). This case differs essentially from the cases mentioned above. It appears that it is natural to describe multipliers in terms of some classes of Gevrey type. We start with one of the traditional variants of definition of Gevrey classes. Set for $1 \le k \le n$

$$E^k = \{\alpha \in \mathbb{N}_0^n : \ \alpha = (0, \ldots, 0, \alpha_{n-k+1}, \ldots, \alpha_n)\}.$$

Definition 5 *Let $0 \le m \le n-1$, $\gamma = (\gamma_{m+1}, \ldots, \gamma_n)$ with $\gamma_j > 0$ and $B = (B_{m+1}, \ldots, B_n)$ with $B_j > 0$. Then $\mu \in J_{\gamma,B}(\mathbb{R}^n)$ iff $\mu \in C^\infty(\mathbb{R}^n)$ and for some $c_8 = c_8(\mu) > 0$ for every $\alpha \in E^{n-m}$*

$$\|D^\alpha \mu\|_{L_\infty(\mathbb{R}^n)} \le c_8 B^\alpha \alpha^{\gamma\alpha} = c_8 B_{m+1}^{\alpha_{m+1}} \cdots B_n^{\alpha_n} \alpha_{m+1}^{\gamma_{m+1}\alpha_{m+1}} \cdots \alpha_n^{\gamma_n \alpha_n}.$$

In the next definition the non-weighted spaces of multipliers $M_p(\mathbb{R}^n)$ are used instead of $L_\infty(\mathbb{R}^n)$.

Definition 6 *Let in addition to assumptions of* Definition 4 $1 \le p \le \infty$. *Then $\mu \in J_{\gamma,B;p}(\mathbb{R}^n)$ iff $\mu \in C^\infty(\mathbb{R}^n)$ and for some $c_9 = c_9(\mu) > 0$ for every $\alpha \in E^{n-m}$*

$$\|D^\alpha \mu\|_{M_p(\mathbb{R}^n)} \le c_9 B^\alpha \alpha^{\gamma\alpha}.$$

Obviously $J_{\gamma,B;2}(\mathbb{R}^n) = J_{\gamma,B}(\mathbb{R}^n)$. Let $\dot{J}_{\gamma,B}(\mathbb{R}^n)$ and $\dot{J}_{\gamma,B;p}(\mathbb{R}^n)$ denote subspaces of functions with compact support from $J_{\gamma,B}(\mathbb{R}^n)$ and $J_{\gamma,B;p}(\mathbb{R}^n)$ respectively.

Lemma 1 *For $\gamma = (\gamma_{m+1}, \ldots, \gamma_n) > 0$, $B = (B_{m+1}, \ldots, B_n) > 0$ and $1 \le p \le \infty$*

$$\dot{J}_{\gamma,B;p}(\mathbb{R}^n) = \dot{J}_{\gamma,B}(\mathbb{R}^n).$$

The main result for space $M_{p,\varrho}(\mathbb{R}^n)$ with an exponential weight is as follows.

Theorem 4 *Let $0 \le m < n$, ϱ be an exponential weight, given by formula* (4) *with all negative A_j or with all positive A_j, $\gamma = (\gamma_{m+1}, \ldots, \gamma_n) > 0$, $B = (B_{m+1}, \ldots, B_n) > 0$, $1 \le p \le \infty$ and*

$$0 < \epsilon_j < 1 \quad (j = \overline{m+1, n}). \tag{9}$$

Then in order that inclusion

$$J_{\gamma,B;p}(\mathbb{R}^n) \subset M_{p,\varrho}(\mathbb{R}^n) \tag{10}$$

is valid it is necessary and sufficient, that for any $j = \overline{m+1,n}$

$$0 < \gamma_j < \frac{1}{\epsilon_j}, \quad B_j > 0$$

or

$$\gamma_j = \frac{1}{\epsilon_j}, \quad 0 < B_j < (|A_j|\epsilon_j e)^{-\frac{1}{\epsilon_j}}.$$

(Sufficiency holds also for $\epsilon_j = 1$.*)*

The sharpness of inclusion (10) in confirmed by the following result.

Theorem 5 *Under assumpsions of* Theorem 4

$$\dot{J}_{1/\epsilon,B}(\mathbb{R}^n) \subset \dot{M}_{p,\varrho}(\mathbb{R}^n) \subset \dot{J}_{1/\epsilon,\tilde{B}}(\mathbb{R}^n)$$

if

$$0 < B_j < (|A_j|\epsilon_j e)^{-1/\epsilon_j} < \tilde{B}_j \quad (j = \overline{m+1,n}).$$

In the case $\epsilon_j > 1$ the space $M_{p,\varrho}(\mathbb{R}^n)$ is in some sense trivial: the exact formulation is as follows.

Theorem 6 *Let in assumptions of* Theorem 4

$$\epsilon_j > 1 \quad (j = \overline{m+1,n})$$

instead of (9). *Then for* $m > 0$ $\mu \in M_{p,\varrho}(\mathbb{R}^n)$ *iff* $\forall x \in \mathbb{R}^n$ $\mu(x) = \eta(x_1, \ldots, x_m)$, *where* $\eta \in M_p(\mathbb{R}^m)$, *and for* $m = 0$ $\mu \in M_{p,\varrho}(\mathbb{R}^n)$ *iff* μ *is equivalent to a constant.*

Theorems 4 and 5 are obtained by the author in cooperation with M. Sh. Tujakbaev. Theorems 4 and 6 for $m = 0$ are formulated in [13], for detailed proofs of theorems 4 and 5 see [14] for $m = 0$ and [9] for $m > 0$. Formulations, discussion and sketches of the proofs of theorems 4-6 are given also in [15].

4 Conditional hypoellipticity in weighted $L_{p,\varrho}$-spaces with an exponential weight

Combining theorems 1-4 we can obtain the main result, concerning conditional hypoellipticity in $Lp, \varrho(\Omega)$ with weight (4). Set for $\xi \in \mathbb{R}^n$ and for $0 < m \leq n$

$$\tilde{P}_{n,m}(\xi) = \Big(\sum_{\alpha \in \mathbb{N}_0^n : (\alpha_1, \ldots, \alpha_m) \neq 0} |P^{(\alpha)}(\xi)|^2 \Big)^{\frac{1}{2}}$$

and for $m = 0$ $\tilde{P}_{n,0}(\xi) = 0$; set moreover $\sigma_p = 0$ for $p = 2$, $\sigma_p = 1$ for $1 < p < \infty, p \neq 2$ and $\sigma_p = 2$ for $p = 1$ and $p = \infty$.

Theorem 7 *Let $0 \leq m < n, 1 \leq p \leq \infty$; function ϱ is defined by formula (4), where $0 < \epsilon_j < 1$ $(j = \overline{m+1,n})$, all A_j are positive or all A_j are negative; $\gamma = (\gamma_{m+1}, \ldots, \gamma_n) > 0, B = (B_{m+1}, \ldots, B_n) > 0$, where for every $j = \overline{m+1,n}$*

$$0 < \gamma_j < \frac{1}{\epsilon_j}, \quad B_j > 0$$

or

$$\gamma_j = \frac{1}{\epsilon_j}, \quad 0 < B_j < (|A_j|\epsilon_j e)^{-1/\epsilon_j}.$$

Let moreover $s_1 \geq 0, s_2 > 0, r > 0, c_{10} > 0, c_{11} > 0$ and $\forall \xi \in \mathbb{R}^n : \; |\xi| \geq r$
1) $\tilde{P}_m(\xi) \neq 0$ and for every $\alpha \in E^{n-m}$

$$|D^\alpha(\tilde{P}_m^{-2})(\xi)| \leq c_{10} B^\alpha \alpha^{\gamma\alpha}(1 + |\xi|^2)^{s_1/2}$$

and
2) for every $\alpha \in E^{n-m}$

$$|\tilde{P}_{n,m}(\xi)\tilde{P}_n(\xi) D^\alpha(\tilde{P}_m^{-2})(\xi)| \leq c_{11} B^\alpha \alpha^{\gamma\alpha}(1 + |\xi|^2)^{-(s_2+\sigma_p)/2}.$$

Then operator P is conditionally hypoelliptic in $L_{p,\varrho}(\Omega)$.

Remark 3 *In case $m = 0$ the formulation of this theorem takes the simpliest form - conditions 1) and 2) reduce to condition: $\forall \xi \in \mathbb{R}^n : \; |\xi| \geq r \;\; P(\xi) \neq 0$ and for every $\alpha \in \mathbb{N}_0^n$*

$$|D^\alpha(P^{-2})(\xi)| \leq c_{10} B^\alpha \alpha^{\gamma\alpha}(1 + |\xi|^2)^{s_1/2}.$$

If we want to put $m = n$, we must take $\sigma_p = 0$ for $1 \leq p \leq \infty$ and then conditions 1) and 2) reduce to condition:

$$\frac{\tilde{P}_{n,n}(\xi)}{\tilde{P}_n(\xi)} \leq c_{11}(1 + |\xi|^2)^{-s_2/2},$$

that is one of equivalent conditions of hypoellipticity.

References

[1] L. Hörmander: On the theory of general partial differential operators, Acta Math. Vol. 94 (1955)

[2] L. Hörmander: The analysis of linear partial differential operators.II. Differential operators with constant coefficients, Berlin-Heidelberg-New York-Tokyo: Springer-Verlag (1983)

[3] V. I. Burenkov: Investigation of spaces of differentiable functions with non-regular domain (D. Sc. thesis), Steklov Inst. Math., Moscow (1982)

[4] V. I. Burenkov: Analogue of Hörmander's theorem on hypoellipticity for functions, tending to zero at infinity (in "Applications of methods of theory of functions and functional analysis to problems of mathematical physics"), Proc. of 7-th Soviet-Chekoslovak seminar, Yerevan (1982)

[5] V. I. Burenkov: Criterion of infinite differentiability of solutions of partial differential equations, decreasing at infinity, Intenational Congress of Mathematicians. Abstracts of short communications. Section 11, Warsaw (1983)

[6] M. I. Adnan: Investigation of infinite differentiability of several classes of slowly growing at infinity solutions of partial differential equations (Ph. D. thesis), Friendship of Nations University, Moscow (1984)

[7] V. I. Burenkov: Theorem on repeated norms for Nikol'skij-Besov spaces and its applications, Proc. Steklov Inst. Math. Vol. 181 (1988)

[8] M. I. Adnan, V. I. Burenkov: On inequalities for weighted repeated fractional norms and on weighted spaces of infinitely differentiable functions (in "Differential equations and functional analysis"), Friendship of Nations University, Moscow (1984)

[9] M. Sh. Tujakbaev: On infinite differentiability of classes of solutions of partial differential equations, admitting quick growth at infinity (Ph.D. thesis), Friendship of Nations University, Moscow (1992)

[10] B. Muckenhoupt, R. Wheeden, W.-S. Young: Sufficiency conditions for L^p multipliers with power weights, Trans. Amer. Math. Soc. Vol. 300, No 2 (1987)

[11] B. Muckenhoupt, R. Wheeden, W.-S. Young: Sufficiency conditions for L^p multipliers with general weights, Trans. Amer. Math. Soc. Vol. 300, No 2 (1987)

[12] B. Muckenhoupt: Necessity conditions for L^p multipliers with power weights, Trans. Amer. Math. Soc. Vol. 300, No 2 (1987)

[13] V. I. Burenkov, M. Sh. Tujakbaev: Fourier multipliers in weighted L_p- spaces with exponential weights, Dokl. Ac. sci. USSR. Vol. 320, No 1 (1991)

[14] V. I. Burenkov, M. Sh. Tujakbaev: Fourier multipliers in weighted L_p- spaces with exponential weights.II, Proc. Steklov Inst. Math. Vol. 204 (1993)

[15] V. I. Burenkov: Fourier multipliers in weighted L_p-spaces with exponential weights, Proc. 4-th Polish-Finnish conference on complex analysis, Jyväskylä (1993)

[16] V.I. Burenkov, M. Sh. Tujakbaev: Conditions of infinite differentiability of solutions of partial differential equations, belonging to weighted L_2-spaces with an exponential weight, Dep. in VINITY Ac. sci. USSR. No 2695 (1990)

Victor I. Burenkov
Dept. of Differential Equations and
Functional Analysis
Friendship of Nations University
of Russia
Mikluho-Maklaya st. 6
117 198 Moscow, Russia
Fax: (095)-433-15-11
E-Mail: SATEL@LOG.MIAN.SU

Solutions of Some Weight Problems

Ioseb Genebashvili, Amiran Gogatishvili, Vakhtang Kokilashvili

The goal of this paper is to present solutions of some two or many weight problems for integral transforms defined in homogeneous type spaces. The problem in question consists in finding criteria for couples (triples, quadruples) of weights ensuring validity of two (or more) weight inequalities of strong or weak type.

This will be a partial survey of what has been done at the seminar of weight problems at A. Razmadze Mathematical Institute in Tbilisi since the publication of our joint book with Miroslav Krbec [1].

Many results presented in the sequel are new even in the case of Euclidean spaces.

It is well known that the problems of strong type are substantially more difficult than those of weak type. Having a good covering lemma (for example, Besicovith's lemma), then to get estimates of weak type for maximal functions it just remains to handle simple problems which have nothing in common with weights. The cases of singular integrals and potentials are much more complicated (see, for example, [1]-[3]).

We will consider general spaces, i.e. homogeneous type spaces which need not possess the Besicovich covering property. Such an additional assumption would greatly simplify the things and the problems for maximal functions lose their charm. Our methods permit us to work without the above mentioned property. A prominent example of homogeneous type spaces without the Besicovitch covering property is the Heisenberg group.

For a survey of the foregoing results see [4].

0. Notation. The homogeneous type space (X, d, μ) is a metric space with measure μ such that the class of compactly supported continuous functions is dense in the space $L^1(X, \mu)$. Besides, it is also assumed that there is a nonnegative real-valued function $d : X \times X \to \mathbf{R}^1$ satisfying the following conditions:

(i) $d(x, x) = 0$ for all $x \in X$;
(ii) $d(x, y) > 0$ for all $x \neq y$ in X;
(iii) There is a constant a_0 such that $d(x, y) \le a_0\, d(y, x)$ for all x, y in X;
(iv) There is a constant a_1 such that $d(x, y) \le a_1(d(x, z) + d(z, y))$ for all x, y in X;
(v) For each neighbourhood V of x in X there is an $r > 0$ such that the ball $B(x, r) = \{y \in X : d(x, y) < r\}$ is contained in V;
(vi) The balls $B(x, r)$ are measurable for all x and $r > 0$;
(vii) There is a constant b such that $\mu B(x, 2r) \le b\, \mu B(x, r)$ for all $x \in X$ and $r > 0$ (see [1], p. 2).

In the sequel $\widehat{B}(x,r)$ will denote the set $B(x,r)\times[0,2r)$ for $r>0$ and the one-point set $\{x\}$ for $r=0$. The set $B(x,0)$ will be assumed to be empty, β will be a measure defined on the product of σ-algebras generated by balls in X and by intervals from $[0,\infty)$.
Let φ,ψ and η be nonnegative nondecreasing functions on $[0,\infty)$. For our further discussion we will also need the definitions of a quasiconvex functions. A function ψ is called quasiconvex if there exists a Young function ω and a constant $c>1$ such that

$$\omega(t)\le\psi(t)\le\omega(ct),\quad t\ge 0.$$

A quasiconvex function ψ can be put into correspondence with its complementary function $\widetilde{\psi}$ defined by

$$\widetilde{\psi}(t)=\sup_{s\ge 0}(st-\psi(s)).$$

Let (Y,ν) be a space with a positive σ-additive measure ν. For $1\le p\le\infty$, $1\le s\le\infty$, the Lorentz space L^{ps}_{ν} is the space of all ν-measurable functions f for which $\|f\|_{L^{ps}(Y,\nu)}<\infty$, where

$$\|f\|_{L^{ps}(Y,\nu)}=\Big(s\int_0^{\infty}(\nu\{y\in Y:|f(y)|>\tau\})^{s/p}\tau^{s-1}d\tau\Big)^{1/s}\quad\text{if}\quad 1\le p,s<\infty,$$

and

$$\|f\|_{L^{ps}(Y,\nu)}=\sup\ \tau(\{y\in Y:|f(y)|>\tau\})^{1/p}\quad\text{if}\quad 1\le p<\infty,\ s=\infty.$$

If $1<p<\infty$ and $1\le s\le\infty$, or $p=s=1$, or $p=s=\infty$, then $L^{ps}(Y,\nu)$ is a Banach space with a norm equivalent to $\|\cdot\|_{L^{ps}(Y,\nu)}$.

1. Criteria for general weighted inequalities of weak type. First of all we present the characterization of those weight functions and kernels for which we have general weight weak type inequalities and kernels for which general weak type inequalities hold for integral transforms of the form

$$\mathcal{K}(fd\nu)(x,t)=\int_X f(y)\,k(x,y,t)\,d\nu(y),$$

where X is a homogeneous type space, and $k:X\times X\times[0,\infty)\to\mathbf{R}^1$ a nonnegative measurable kernel.

Theorem 1.1 ([6]). *Let $t^{-\alpha}\theta(t)$ decrease for some $\alpha\in(0,1),\psi\in\Delta_2$. Further, let ν be a finite measure on X, $\sigma:X\to\mathbf{R}^1$ be an almost everywhere positive function which is locally summable with respect to the measure ν.*
Assume that there exist positive constants ε and C_1 such that

$$\int_{X\setminus B(a,r)}\widetilde{\psi}\Big(\varepsilon\,\frac{\varphi(s)\eta(s)}{s}\,\frac{\theta(\beta\widehat{B}(a,N_0(2r+t)))}{\sigma(y)}\,k(a,y,t)\Big)\sigma(y)d\nu(y)\le$$

$$\le C_1\,\varphi(s)\,\theta(\beta\widehat{B}(a,N_0(2r+t)))\tag{1.1}$$

for any $s > 0$, $r \geq 0$, $a \in X$ *and* $t \geq 0$, *where* $N_0 = a_1(1 + 2a_0)$.
Then there exists a positive constant C_2 *such that for any* $\lambda > 0$ *and any nonnegative* ν-*measurable function* $f : X \to \mathbf{R}^1$ *the following inequality holds:*

$$\varphi(\lambda)\, \theta(\beta\{(x,t) \in X \times [0,\infty) : \mathcal{K}(fd\nu)(x,t) > \lambda\}) \leq \tag{1.2}$$
$$\leq C_2 \int_X \psi\Big(\frac{f(x)}{\eta(\lambda)}\Big)\sigma(x)d\nu(x).$$

Assume now that the nonnegative measurable kernel k satisfies the following additional condition: there exist numbers $N \geq N_0$, $N_0 = a_1(1 + 2a_0)$ and c' such that

$$k(a, y, t) \leq c' k(x, y, \tau) \tag{1.3}$$

for $y \in X \setminus B(a,r)$, $(x, \tau) \in \widehat{B}(a, N(r+t))$ for any $a \in X$, $r \geq 0$, $t \geq 0$.

Theorem 1.2 ([6]). *Let* $\varphi\eta$ *and* ψ *be quasiconvex functions,* $\psi \in \Delta_2$, *let the function* $t^{-\alpha}\theta(t)$ *decrease for some* $\alpha \in (0,1)$ *and let* k *satisfy the condition* (1.3). *Then the inequality* (1.2) *is equivalent to any of the following conditions:*

(i) *there exist positive constants* ε *and* C_3 *such that*

$$\int_{X\setminus B(a,r)} \widetilde{\psi}\Big(\varepsilon\, \frac{\varphi(s)\eta(s)}{s}\, \frac{\theta(\beta\widehat{B}(a,r+t))}{\sigma(y)}\, k(a,y,t)\Big)\sigma(y)d\nu(y) \leq$$
$$\leq C_3\, \varphi(s)\, \theta(\beta\widehat{B}(a, r+t)) \tag{1.4}$$

for arbitrary $o > 0$, $r \geq 0$, $a \subset X$ *and* $t \geq 0$;

(ii) *there exist positive constants* ε *and* C_4 *such that*

$$\int_X \widetilde{\psi}\Big(\varepsilon\, \frac{\varphi(s)\eta(s)}{s}\, \frac{\theta(\beta\widehat{B}(a,t))}{\sigma(y)}\, k(a,y,t)\Big)\sigma(y)d\nu(y) \leq$$
$$\leq C_4\, \varphi(s)\, \theta(\beta\widehat{B}(a,t)) \tag{1.5}$$

for any $s > 0$, $a \in X$ *and* $t \geq 0$.

The above formulated results contain the solutions of problems of description of a set of weights ensuring the validity of both weak and extra-weak weighted inequalities for the transform $\mathcal{K}$ in Orlicz spaces which are natural analogies of inequalities of the weak type (p, q).
For the Lorentz spaces we have the following:

Theorem 1.3. *Let* $1 \leq s \leq p < q < \infty$ *and suppose that the kernel* k *satisfies condition* (1.3). *Then the following statements are equivalent:*

(i) *There exists a number* $C_1 > 0$ *such that we have*

$$\beta\{(x,t) \in X \times [0,\infty) : \mathcal{K}(fd\nu)(x,t) > \lambda\} \leq C_1 \lambda^{-q}\, \|f\|^q_{L^{ps}(X,\sigma d\nu)},$$

for any measurable nonnegative $f: X \to \mathbf{R}^1$ *and* $\lambda > 0$;

(ii) *There exists a positive constant* C_2 *such that*

$$(\beta\,\widehat{B}(a,(2r+t)))^{\frac{1}{q}} \left\| \mathcal{X}_{X\setminus B(a,r)}\, \frac{k(a,\cdot,t)}{\sigma} \right\|_{L^{p',s'}(X,\sigma d\nu)} \le C_2$$

for any $a \in X$, $r \ge 0$, $t \ge 0$;

(iii) *There exists a positive constant* C_3 *such that*

$$(\beta\,\widehat{B}(a,t))^{\frac{1}{q}} \left\| \frac{k(a,\cdot,t)}{\sigma} \right\|_{L^{p',s'}(X,\sigma d\nu)} \le C_3\,,$$

for all $a \in X$, $t \ge 0$.

Consider the kernel

$$k(x,y,t) = (\mu\, B(x, d(x,y)+t))^{-\delta}, \quad \delta > 0\,.$$

It is easy to verify that k satisfies the condition (1.3). With an appropriate choice of the space X, the measure ν and a quasimetric d we get the solution of the general weak type problem in Orlicz and Lorentz spaces for general potentials, for Poisson integrals in the upper half-space, the Hardy operator etc. For particular cases see [7-10].
Now for a locally integrable real function $f : X \to \mathbf{R}^1$ we define the generalized fractional maximal operator

$$\widetilde{M}_\gamma f(x,t) = \sup\, (\mu B)^{\gamma-1} \int_B |f(y)|d\mu(y), \quad x \in X, \quad t \ge 0, \quad 0 \le \gamma < 1,$$

where the supremum is taken over the balls B in X, containing x and having radius greater than $t/2$.

Theorem 1.4 ([11]). *Let* ψ *and* $\varphi\eta$ *be quasiconvex. Let* θ *be subadditive on* $[0,\infty)$, *and* $0 \le \gamma < 1$. *Suppose that the functions* σ *and* ν *are locally integrable and a.e. positive on* X. *Then the following statements are equivalent:*
(i) *There exists a positive constant* C_1 *such that*

$$\theta\,(\beta\{(x,t) \in X \times [0,\infty) : \widetilde{M}_\gamma f(x,t) > \lambda\})\,\varphi(\lambda) \le$$

$$\le C_1 \int_X \psi\left(C_1 \frac{|f(x)|\nu(x)}{\eta(\lambda)}\right)\sigma(x)d\mu(x), \tag{1.6}$$

for all f *and* $\lambda > 0$;
(ii) *There exist positive constants* C_2 *and* ε *such that*

$$\int_B \widetilde{\psi}\left(\varepsilon\, \frac{\varphi(\lambda)\eta(\lambda)}{\lambda}\, \frac{\theta(\beta(\widehat{B}))}{\mu(B)^{1-\gamma}}\, \frac{1}{\sigma(x)\nu(x)}\right)\sigma(x)d\mu(x) \le C_2\,\varphi(\lambda)\;\theta(\beta(\widehat{B})).$$

For the classical Hardy-Littlewood maximal function see [12], [13].

In the sequel we will use the following notations:
$\widetilde{M}_\gamma f(x,0) = M_\gamma f(x)$, $\widetilde{M}_0 f(x,t) = \widetilde{M} f(x,t)$, $M_0 f(x) = M f(x)$.

2. The reverse weak type inequalities. Such an inequality for the classical maximal function in the unweighted case has been proved in [14] (see also [15], [16]). Weighted inverse weak inequalities were discussed in the papers [17], [18].
For the homogeneous type spaces we have the following results.

Theorem 2.1 ([19]). *The following two conditions are equivalent:*
(i) *There exists a positive constant C_1 such that*

$$\beta\{(x,t)\in X\times[0,\infty) : \widetilde{M} f(x,t) > \lambda\} \geq C_1\,\lambda^{-1}\int_{\{|f|>\lambda\}} |f(x)|d\mu(x)\,; \tag{2.1}$$

(ii) *There exists a positive constant C_2 such that*

$$\beta\,\widehat{B}(x,r) \geq C_2\,\mu\,B(x,r) \tag{2.2}$$

for every $x\in X$ and $r>0$.

Theorem 2.2. *If the measure β satisfies the doubling condition*

$$\beta(2\widehat{B}) \leq C\,\beta\,\widehat{B}, \tag{2.3}$$

where the positive constant C is independent of B, then the following two conditions are equivalent:

(i) *There is a positive constant C_1 such that*

$$\beta\{(x,t)\in X\times[0,\infty) : \widetilde{M} f(x,t) > \lambda\} \geq C_1\,\lambda^{-1}\int_{\{|f|>\lambda\}} |f(x)|d\mu(x) \tag{2.4}$$

for an arbitrary $\lambda>0$ and f;

(ii) *There exists a positive constant C_2 such that*

$$\frac{\beta\,\widehat{B}}{\mu\,B} \geq C_2 \operatorname*{ess\,sup}_{x\in B} w(x) \tag{2.5}$$

for all balls B.

Theorem 2.3. *The following two conditions are equivalent:*

(i) *There exist positive constants C_1 and C_2 such that*

$$C_1\,\lambda^{-1}\int_{\{|f|>\lambda\}} |f(x)|w(x)d\mu(x) \leq \beta\{(x,t)\in X\times[0,\infty) : \widetilde{M} f(x,t) > \lambda\} \leq$$

$$\leq C_2\,\lambda^{-1}\int_{\{|f|>\frac{\lambda}{2}\}}|f(x)|w(x)d\mu(x)\ ;$$

(ii) $\beta\,\widehat{B}\sim\mu\,B$ *and* $0<b_1\leq w(x)\leq b_2<\infty$ *for all* $x\in X$.

The above theorems yield the following description of the Zygmund class defined on X space.

Theorem 2.4. *If* $f\in L(1+log^{+}L)(X,d\mu)$, *then for an arbitrary measure* β *with the condition* (2.2) *we have*

$$\int_{\{\widetilde{M}f(x,t)>1\}}\widetilde{M}f(x,t)d\beta<\infty. \tag{2.6}$$

Conversely, if (2.6) *holds for some* β *satisfying* $\beta\,\widehat{B}\leq C\,\beta\,B$, *then* $f\in L(1+log^{+}L)(X,d\mu)$.

3. Strong type problems. First of all we consider the unweighted case. We shall say that a function $\varphi:\mathbf{R}^1\to\mathbf{R}^1$ belongs to the class Φ if φ is an even nonnegative nondecreasing function on $(0,\infty)$ such that $\varphi(0+)=0$, $\lim_{t\to\infty}\varphi(t)=0$.

Theorem 3.1. *Suppose that the functions* φ *and* ψ *belong to* Φ, E *is an arbitrary* μ-*measurable subset of* X. *Then the following conditions are equivalent:*

(i) *There is a positive constant* C_1 *such that*

$$\int_E\varphi\,(Mf(x))\,d\mu\leq C_1\int_E\psi\,(C_1f(x))\,d\mu \tag{3.1}$$

for every f *such that* $C_1f\in\psi(L)$;

(ii) *There exists a positive constant* C_2 *such that*

$$t\int_0^t\frac{\varphi(s)}{s^2}ds\leq C_2\,\psi\,(\,C_2t) \tag{3.2}$$

for all $t>0$.

In [20] it was shown that in case $\varphi=\psi$ the condition (ii) is valid iff φ^{α} is quasiconvex for some α, $0<\alpha<1$.

Theorem 3.2. *Suppose that* $\mu X<\infty$, φ *and* ψ *belong to* Φ, g *is a positive function on* $(0,\infty)$. *Then for every fixed* $a\in(0,\infty)$ *the following conditions are equivalent:*

(i) *There exists a positive constant* C_1 *such that*

$$\int_X\varphi\,(Mf(x))\,d\mu\leq\varphi(a)\,\mu\,X+C_1\,g(a)\int_X\psi\,(C_1f(x))\,d\mu$$

for every f such that $C_1 f \in \psi(L)$;

(ii) *There is a positive constant C_2 such that*

$$t\int_a^t \frac{\varphi(s)}{s^2}ds \le C_2\, g(a)\, \psi(C_2 t)$$

Furthermore, the question is posed as follows: What is the full description of such weight functions w for which the weighted inequality of strong type in modulars is valid for the maximal functions?

Theorem 3.3. *Let $\varphi \in \Phi$. Then the necessary and sufficient condition for the validity of the inequality*

$$\int_X \varphi(Mf(x))w(x)\,d\mu \le C\int_X \varphi(Cf(x))w(x)\,d\mu$$

with a constant C independent of f is that φ^α is quasiconvex for some α, $0<\alpha<1$, and there exist positive constants C_1 and ε such that

$$\int_B \tilde{\psi}\left(\varepsilon\,\frac{\varphi(s)}{s}\,\frac{wB}{\mu(B)w(x)}\right)w(x)\,d\mu(x) \le C_1\,\varphi(s)(B)$$

for all $s>0$ and all balls B.

For the classical Hardy-Littlewood functions this problem was solved (using other terms) under the assumption that φ, $\tilde{\varphi} \in \Delta_2$, in the pioneering paper [21]. For further results in this direction see [22] (and also [1]).
Let us note that φ^α is quasiconvex for some α, $0<\alpha<1$, iff $\tilde{\varphi}$ satisfies the Δ_2 condition at infinity.
Now we shall discuss the strong type one-weighted problems for M_γ and for the generalized fractional integrals

$$\mathbf{T}_\gamma f(x) = \int_X \frac{f(y)}{(\mu B(x,d(x,y)))^{1-\gamma}}d\mu, \quad 0<\gamma<1.$$

Definition. An almost everywhere positive function $w : X \to \mathbf{R}^1$ is said to belong to the $A_p(X)$ class (the Muckenhoupt class), $1<p<\infty$, if

$$\sup\left(\frac{1}{\mu B}\int_B w(x)d\mu\right)\left(\frac{1}{\mu B}\int_B w^{-\frac{1}{p-1}}(x)d\mu\right) < \infty\,,$$

where the supremum is taken over all balls $B \subset X$.

In the sequel we put $\|f\|_{L^{p_s}_w} = \|f\|_{L^{p_s}(X,w\,d\nu)}$.

Theorem 3.4 ([23]). *Let $1 < p < \gamma^{-1}$, $0 \le \gamma < 1$, $1 < \nu, r, s < \infty$, $\frac{1}{q} = \frac{1}{p} - \gamma$. Then the conditions below are equivalent:*

(i) *There is a constant $C_1 > 0$ such that*

$$\|M_\gamma(fw^\gamma)\|_{L_w^{q\nu}} \le C_1 \|f\|_{L_w^{p\nu}}$$

for any $f \in L_w^{p\nu}$;

(ii) *There exists a constant $C_2 > 0$ such that*

$$\|M_\gamma(fw^\gamma)\|_{L_w^{q\infty}} \le C_2 \|f\|_{L_w^{pr}}$$

for arbitrary $f \in L_w^{pr}$;

(iii) *There exists a positive constant C_3 such that*

$$(wB)^{1/q} \|\mathcal{X}_B w^{\gamma-1}\|_{L_w^{p's'}} \le C_3 |B|^{1-\gamma} ;$$

(iv) *$w \in A_\beta(X)$, where $\beta = 1 + \frac{q}{p'}$.*

For the case $\gamma = 0$ and $p = s$ this theorem is known [24].

Theorem 3.5 ([23]). *Let $1 < p < \gamma^{-1}$, $0 < \gamma < 1$, $1 < \nu, r, s < \infty$. Then the conditions below are equivalent:*

a) There is a constant $C_1 > 0$ such that

$$\|\mathbf{T}_\gamma(fw^\gamma)\|_{L_w^{q\nu}} \le C_1 \|f\|_{L_w^{p\nu}}$$

for any $f \in L_w^{p\nu}$;

b) There exists a positive constant C_2 such that

$$(wB(x,r))^{1/q} \|\mathcal{X}_{X \setminus B(x,r)} w^{\gamma-1} (\mu B(x, d(x,\cdot)))^{\gamma-1}\|_{L_w^{p's'}} \le C_2 |B|^{1-\gamma};$$

c) $w \in A_\beta(X)$, where $\beta = 1 + \frac{q}{p'}$.

Theorem 3.6 ([25]). *Let $1 < p < \infty$, $1 \le s < \infty$. Then the conditions below are equivalent:*

(i) *There is a constant $C > 0$ such that*

$$\|Mf\|_{L_w^{ps}} \le C \|f\|_{L_w^{p1}}$$

for all $f \in L_w^{p1}(X,\mu)$*;*

(ii) $w \in A_p(X)$.

References

[1] V. Kokilashvili and M. Krbec, *Weighted inequalities in Lorentz and Orlicz spaces.*, World Scientific, Singapore, New Jersey, London, Hong Kong, 1991.

[2] E.T. Sawyer, *A two-weight weak-type inequality for fractional integrals*, Trans. Amer. Math. Soc. 281 (1984), 339-345.

[3] M. Gabidzashvili, *Weighted inequalities for anisotropic potentials*, Trudy Tbilissk. Mat. Inst. im. A. Razmadze Akad. Nauk Gruz. SSR 89 (1986), 25-36.(Russian)

[4] V. Kokilashvili, *Weighted estimates for classical integral operators*, Proceedings of the International Spring School "Nonlinear Analysis, Function Spaces and applications IV ", Roudnice nad Labem (Czechoslovakia), May 21-25, 1990, Teubner-Texte zur Mathematik, Teubner-Verlag, Leipzig, 1990, pp.86-103.

[5] J.O. Strömberg and A. Torchinsky, *Weighted Hardy spaces*, Lecture Notes in Math., vol. 1381, Springer-Verlag, New York, 1989.

[6] I. Genebashvili, A. Gogatishvili, V. Kokilashvili, *Criteria of general weak type inequalities for integral transforms with positive kernels*, Proceedings of Georgian Academy of Sciences. Mathematics (to appear).

[7] I. Genebashvili, M. Gabidzashvili, V. Kokilashvili, *Two-weight inequalities for generalized potentials*, Trudy MIRAN 194 (1991), 89-96.

[8] I. Genebashvili, *Carleson measures and potentials defined on spaces of the homogeneous type*, Soobshch. Akad. Nauk Gruz. SSR 135,3 (1989), 505-508. (Russian)

[9] E. Sawyer, R. Wheeden, *Carleson conditions for the Poisson integral*, Indiana University Math. J. 40,2 (1991), 639-676.

[10] L. Pick, *Weighted modular estimates for Hardy-Littlewood operators* (1991) (preprint)

[11] A. Gogatishvili, *General weak type inequalities for the maximal operators and singular integrals*, Proc. A. Razmadze Math. Inst. Georgian Acad. Sci. (to appear).

[12] L. Pick, *Two weight weak type maximal inequalities in Orlicz classes*, Studia Math. 100,3 (1991), 206-218.

[13] A. Gogatishvili, L. Pick, *Weak and extra-weak type inequalities for the maximal operator and Hilbert transform*, Czechoslovak Math. J. (to appear).

[14] E. Stein, *Note on the class L* log*L*, Studia Math. 31 (1969), 305-310.

[15] O. Tsereteli, *On reversing some theorems of Hardy-Littlewood*, Soobshch. Akad. Nauk Gruz. SSR 56,2 (1959), 269-271. (Russian)

[16] M. de Guzman, *Differentiation of integrals in* R^n, Lecture Notes in Math., vol. 481, Springer-Verlag, Berlin-Heidelberg-New York, 1975.

[17] K.F. Andersen, W.-S. Young, *On the reverse weak type inequality for the maximal function and weighted class* $L(\log L)^k$, Pacific J. Math. 112,2 (1984), 257-264.

[18] B. Muckenhoupt, *Weighted reverse weak type inequalities for the Hardy- Littlewood maximal function*, Pacific J. Math. 117,2 (1985), 371-377.

[19] I. Genebashvili, *Weight reverse inequalities for maximal functions*, Reports of Enlarged Session of the Seminar of I.N. Vekua Inst. Appl. Math., Tbilisi State University, 1992 (to appear).

[20] A. Gogatishvili, V. Kokilashvili, M. Krbec, *Maximal functions in* $\varphi(L)$ *classes*, Dokl. Akad. Nauk SSSR 314,3 (1990), 534-536 (Russian); English transl., Soviet Math. Dokl. 42,3 (1990), 488-490.

[21] R. Kerman and A. Torchinsky, *Integral inequalities with weights for the Hardy maximal function*, Studia Math. 71 (1982), 277-284.

[22] V. Kokilashvili and M. Krbec, *Weighted inequalities for Riesz potentials and fractional maximal functions in Orlicz spaces*, Dokl. Akad. Nauk SSSR 283 (1985), 280-283 (Russian); English Transl., Soviet Math. Dokl. 32 (1985), 70-73.

[23] I. Genebashvili, V. Kokilashvili, *Weighted norm inequalities for fractional maximal functions and integrals defined on the space of homogeneous type*, Bull. Georgian Acad. Sci. 143,1 (1991), 13-15.

[24] R.A. Macias, C. Segovia, *A well quasi-distance for spaces of homogeneous type*, Trab. Mat. Inst. Argent. Mat. 32 (1981), 1-18.

[25] I. Genebashvili, *Weighted norm estimates in Lorentz spaces for maximal functions defined on spaces of homogeneous type*, Bull. Georgian Acad. Sci. 143,3 (1991), 485-487.

Adress of the authors
Mathematical Institute
Georgian Academy of Sciences
Z. Rukhadze Str. 1
38 00 93 Tbilisi, Georgia
E-Mail: KOKIL@IMATH.KHETA.GEORGIA.SU

On Integral Inequalities on the Set of Functions with some Properties of Monotonicity

Michail L. Gol'dman

Hölder, Jensen and Hardy inequalities play a very important role in the mathematical analysis and in the theory of differential equations. These inequalities are exact on the set of all measurable (non-negative) functions (see book [1]: the results on Hardy inequalities, obtained by B. Muckenhoupt, V. Kokilashvili, J. Bredly, V. Maz'ya, A. Rosin and others). But, in the field of mathematical analysis and, first of all, in the theory of function spaces it is often necessary to use such estimates on sets of non-negative functions with additional properties of monotonicity. On these sets such inequalities lose their exactness, that prevent obtaining the exact results on the basis of them. Let us remind that the most important examples of such functions on $R_+ = (0, \infty)$ are distribution functions, non-increasing rearrangements and moduli of continuity in the theory of function spaces:

$$\lambda_f(t) = \text{mes}\,\{x : |f(x)| > t\}\,, \quad f^*(u) = \inf\,\{t \geq 0 : \lambda_f(t) \leq u\}\,;$$

$$f^{**}(u) = \frac{1}{u}\int_0^u f^*(v)\,dv\,; \quad \omega_k^p\,(f, u) = \sup_{|h|\leq u}\,\|\,\Delta_h^k f\,\|_{L_p}\,;$$

and Peetre's $K-$ functional in the interpolation theory

$$K(f, u) = \inf_{f_0+f_1=f}\,(\,\|f_0\|_{F_0} + \,u\,\|f_1\|_{F_1}\,), \quad f \in F_0 + F_1.$$

We consider the cones $\Omega_k^m = \Omega_k \cap \Omega^m$ of functions on R_+, where

$$\Omega_k = \{f \geq 0 : u^{-k} f(u)\; a. \downarrow\}\,, \quad \Omega^m = \{f \geq 0 : u^{-m} f(u)\; a. \uparrow\}$$

($g\ a. \downarrow$ means that g is "almost nonincreasing" on R_+, i.e. there exists such $c \geq 1$ that $g(v) \leq cg(u)$ if $0 < u < v < \infty$; $g\ a. \uparrow$ has the analogous meaning). It is clear, that λ_f and $f^* \in \Omega_0$, $f^{**} \in \Omega_0^{-1}$, $\omega_p^k(f, \cdot) \in \Omega_k^0$, $K(f, \cdot) \in \Omega_1^0$.
This paper is aimed to give the exact estimates for the values

$$G_k^m = \sup_{f\in\Omega_k^m}\left[(\int_0^\infty f^q\,d\gamma)^{1/q}\,(\int_0^\infty f^p\,d\beta)^{-1/p}\right]; \tag{1}$$

$$G_k = \sup_{f\in\Omega_k}\,[\cdots]; \quad G^m = \sup_{f\in\Omega^m}\,[\cdots]; \tag{2}$$

$$H_k^m = \sup_{f\in\Omega_k^m}\{(\int_0^\infty Af^q\,d\gamma)^{1/q}\,(\int_0^\infty f^p\,d\beta)^{-1/p}\}; \tag{3}$$

$$H_k = \sup_{f \in \Omega_k} \{\cdots\}; \quad H^m = \sup_{f \in \Omega^m} \{\cdots\} \tag{4}$$

Here in this work we assume that $p, q \in R_+$, A is the Hardy operator

$$Af(u) = u^{-\delta}\{\int_0^u \{f(v)\, v^{-\alpha}\}^r \frac{dv}{v}\}^{1/r}, \quad r \in R_+; \ \ \alpha, \delta \in R, \tag{5}$$

where β, γ are non-negative σ - finite and σ- additive full measures on R_+ . To obtain the answer we need some notation:

$$d_p(l,\beta) = \{\int_0^1 v^{lp}\, d\beta\}^{1/p},$$

$$e_p(l,\beta) = \{\int_1^\infty v^{lp}\, d\beta\}^{1/p}, \quad l \in R;$$

$$c_p(m,k,\beta) = \{\int_0^\infty (v^{-k} + v^{-m})^{-p} d\beta\}^{1/p} \asymp d_p(k,\beta) + e_p(m,\beta)\ ; \quad m < k.$$

Here $B_1 \asymp B_2$ means $c^{-1} \leq B_1/B_2 \leq c$, where $c \geq 1$ is a constant.

Let us consider the nondegeneration conditions ($m < k$):

$$\beta \in (N) \iff c_p(m,k,\beta) < \infty, \quad d_p(m,\beta) = e_p(k,\beta) = \infty, \tag{6}$$

and define the fundamental function on R_+, generated by the measure β

$$\varphi_\beta(u) = \varphi_\beta(p,m,k;u) = c_p(m,k,\beta) \left\{ \int_0^\infty \frac{d\beta(v)}{[(u/v)^k + (u/v)^m]^p} \right\}^{-1/p}. \tag{7}$$

If $\beta \in (N)$, we have

$$\varphi_\beta \in \Omega_k^m; \quad \lim_{u\to+0}\{u^{-m}\varphi_\beta(u)\} = \lim_{u\to+\infty}\{u^{-k}\varphi_\beta(u)\} = 0; \tag{8$_1$}$$

$$\varphi_\beta = 1; \quad \lim_{u\to+\infty}\{u^{-m}\varphi_\beta(u)\} = \lim_{u\to+0}\{u^{-k}\varphi_\beta(u)\} = \infty. \tag{8$_2$}$$

Let $\{\mu_t\}_{t\in Z}$ be the discretizing sequence for the function φ_β on R_+, that is

(a) $0 < \mu_t \uparrow\uparrow\ ; \quad \mu_0 = 1;$

(b) $\mu_t^{-m}\ \varphi_\beta(\mu_t) \uparrow\uparrow\ ; \quad \mu_t^{-k}\ \varphi_\beta(\mu_t) \downarrow\downarrow\ ;$

(c) $Z = T_1 \cup T_2$, where $T_1 \cap T_2 = \emptyset$, and for $u \in \Delta_t = [\mu_t, \mu_{t+1}]$

$$u^{-m}\varphi_\beta(u) \asymp \mu_t^{-m}\varphi_\beta(\mu_t),\ t \in T_1; \quad u^{-k}\varphi_\beta(u) \asymp \mu_t^{-k}\varphi_\beta(\mu_t),\ t \in T_2.$$

Here

$$a_t \uparrow\uparrow \iff \inf\,[\,a_{t+1}/a_t\,] \ > \ 1, \quad a_t \downarrow\downarrow \iff \sup\,[\,a_{t+1}/a_t\,] \ < \ 1. \tag{9}$$

Such a sequence μ_t exists for every function φ_β, which satisfies the conditions (8). For example, we fix the constant $a > 1$ and put $\mu_0 = 1$,

$$\mu_{t+1} = \max\{\mu'_{t+1}, \mu''_{t+1}\}\ ,\ t \geq 0; \quad \mu_{t-1} = \min\{\mu'_{t-1}, \mu''_{t-1}\}\ ,\ t \leq 0;$$

$$(\mu'_{t\pm1})^{-m}\ \varphi_\beta(\mu'_{t\pm1}) = a^{\pm1}\ \mu_t^{-m}\ \varphi_\beta(\mu_t),\ t\in Z;$$
$$(\mu''_{t\pm1})^{-k}\ \varphi_\beta(\mu''_{t\pm1}) = a^{\mp1}\ \mu_t^{-k}\ \varphi_\beta(\mu_t),\ t\in Z;$$

The conditions $(a), (b), (c)$ will be satisfied, if we put $T_2 = Z - T_1$;

$$T_1 = T_1^+ \cup T_1^- \ ; \quad T_1^+ = \{t\geq 0 : \mu_{t+1} = \mu'_{t+1}\}\ , \quad T_1^- = \{t<0 : \mu_t = \mu'_t\}.$$

Remark 1. First represented in K. Oskolkov's work [2] such sequences were intensively used in the theory of function spaces (G.Kaljabin, V.Koljada, J.Netrusov, M.Gol'dman and others) and in the interpolation theory (J. Brudnij, N.Krugljak, S.Janson, V.Ovchinnikov and others).

Theorem 1. *Let $\beta\in(N)$, $m<k$, $\{\mu_t\}$ be the discretizing sequence for the fundamental function φ_β on R_+. Then*

$$G_k^m \asymp D_k^m = \{\sum_t (\int_{\Delta_t} \varphi_\beta(v)^q\ d\gamma\ )^{x/q}\}^{1/x} \quad , \tag{10}$$

$$\frac{1}{x} = (\frac{1}{q}-\frac{1}{p})_+ \quad (x=\infty \ \ if \ \ p\leq q; \quad x=\frac{pq}{(p-q)}\ if \ \ p>q). \tag{11}$$

Remark 2. We can obtain all the results of our work [3] (part 2) as consequences of this theorem. If (N) - conditions (6) are not satisfied, see the answers in [3].

Theorem 2. *Let $k\in R$; $d_p(k,\beta)<\infty$, $e_p(k,\beta)=\infty$. Let the function*

$$\hat\varphi_\beta(u) = u^k\{\int_0^u v^{kp}\,d\beta\}^{-1/p}$$

be continuous and positive on R_+. Then $G_k \asymp D_k$, where

$$D_k = \sup_{u\in R_+}\left[\{\int_0^u v^{kq}\ d\gamma\}^{1/q}u^{-k}\hat\varphi_\beta(u)\right] \quad if \ \ p\leq q\ ;$$

$$D_k = \{\int_{R_+}[\ \cdots\ ]^\kappa\hat\varphi_\beta(u)^p\ d\beta\}^{1/\kappa} \quad if \ \ p>q; \quad \kappa=\frac{pq}{p-q}\ .$$

Theorem 3. *Let $m\in R$; $d_p(m,\beta)=\infty$, $e_p(m,\beta)<\infty$. Let the function*

$$\check\varphi_\beta(u) = u^m\{\int_u^\infty v^{mp}\,d\beta\}^{-1/p}$$

be continuous and positive on R_+ . Then $G^m \asymp D^m$, where

$$D^m = \sup_{u\in R_+}\left[\{\int_u^\infty v^{kq}\ d\gamma\}^{1/q}u^{-m}\check\varphi_\beta(u)\right] \quad if \ \ p\leq q\ ;$$

$$D^m = \{ \int_{R_+} [\cdots]^\kappa \check{\varphi}_\beta(u)^p \, d\beta \}^{1/\kappa} \quad if \quad p > q; \quad \kappa = \frac{pq}{p-q} .$$

Now we are going to estimate H_k^m. Put

$$\Phi(u) = \{ \int_0^u [\, \varphi_\beta(v) \, v^{-\alpha} \,]^\sigma \, \frac{dv}{v} \, \}^{1/\sigma}; \quad \frac{1}{\sigma} = (\frac{1}{r} - \frac{1}{p})_+ \, , \tag{12}$$

with usual modifications if $\sigma = \infty$ $(p \leq r$, (see(11));

Let

$$g(u) = \{ \int_u^\infty v^{-\delta q} d\gamma \}^{1/q}. \tag{13}$$

Theorem 4. *Let the conditions of* Theorem 1 *be satisfied and assume* $m < \alpha < k$. *Then*

$$H_k^m \asymp E_k^m + F_k^m, \tag{14}$$

where

$$E_k^m = \{ \sum_t (\int_{\Delta_t} [\varphi_\beta(v) v^{-(\alpha+\sigma)}]^q \, d\gamma \,)^{\kappa/q} \, \}^{1/\kappa} \tag{15}$$

$$F_k^m = \sup_t \, [\, g(\mu_t) \, \Phi(\mu_t) \,] \quad if \quad p \leq q \, ; \tag{16}$$

$$F_k^m = \{ \sum_t [\, g(\mu_t)^\kappa - g(\mu_{t+1})^\kappa \,] \, \Phi(\mu_t)^\kappa \, \}^{1/\kappa} \quad if \quad p > q \, . \tag{17}$$

Remark 3. Let $p > q$ in theorem 4. Then

$$F_k^m \asymp \{ \sum_t (\sum_{s \geq t} \omega_s^q \,)^{\kappa/p} \, \omega_t^\kappa \, \Phi(\mu_t)^\kappa \, \}^{1/\kappa} \, ,$$

where

$$\omega_t = \{ \int_{\Delta_t} v^{-\sigma q} d\gamma \, \}^{1/q}, \quad \Delta_t = [\mu_t, \mu_{t+1}] \, .$$

Remark 4. We can replace $\sum_t$ by $\sum_{t \in T_2}$ in (15).

Remark 5. We supposed $m < \alpha < k$. In other cases we have the following:

(i) If $\alpha \geq k$, then $H_k^m = \infty$ $(Af(u) = \infty \quad \forall f \in \Omega_k^m, \, f \not\equiv 0)$.

(ii) If $\alpha < m$, then $Af(u) \asymp f(u) \, u^{-(\alpha+\sigma)}$, $f \in \Omega_k^m$, and theorem 1 gives $H_k^m \asymp E_k^m$.

(iii) In the case $\alpha = m$ we have the following result :

Theorem 5. *Let the conditions of theorem 1 be satisfied and let* $\alpha = m$. *Then*

$$H_k^m \asymp E_{(1)} + E_{(2)} + F.$$

Here

$$\frac{1}{\kappa} = (\frac{1}{q} - \frac{1}{p})_+ , \quad \frac{1}{\sigma} = (\frac{1}{r} - \frac{1}{p})_+ ,$$

$$E_{(1)} = \{ \sum_{t \in T_1} (\int_{\Delta_t} [\varphi_\beta(v)\, v^{-(m+\sigma)}]^q \ln^{q/r} (\frac{2v}{\mu_t})\, d\gamma)^{\kappa/q} \}^{1/\kappa} ;$$

$$E_{(2)} = \{ \sum_{t \in T_2} (\int_{\Delta_t} [\varphi_\beta(v) v^{-(m+\sigma)}]^q\, d\gamma)^{\kappa/q} \}^{1/\kappa} ;$$

and to obtain F *we replace* $\Phi(\mu_t)$ *by* Φ_t *in* (16) *and* (17), *where*

$$\Phi_t = \{ \sum_{s \leq t-1} [\varphi_\beta(\mu_s)\, \mu_s^{-m}]^\sigma \ln^{\sigma/r} (\frac{\mu_{s+1}}{\mu_s}) \}^{1/\sigma}$$

with usual modifications if $\kappa = \infty$ *and/or* $\sigma = \infty$.

Remark 6. We can obtain all the results of our work [3] (part 3) as consequences of theorem 4 and 5. If (N) - conditions (6) are not satisfied, see the answers in [3].

Theorem 6. *Let the conditions of* Theorem 2 *be satisfied and let* $\alpha < k$. *Then*

$$H_k \asymp E_k + F_k ,$$

where

$$E_k = \sup_{u \in R_+} \left[\{ \int_0^u v^{(k-\alpha-\delta)q} d\gamma \}^{1/q} u^{-k}\, \hat{\varphi}_\beta(u) \right] \quad \textit{if} \quad p \leq q ;$$

$$E_k = \{ \int_{R_+} [\cdots]^\kappa\, \hat{\varphi}_\beta(u)^p\, d\beta \}^{1/\kappa} \quad \textit{if} \quad p > q; \quad \kappa = \frac{pq}{p-q} ;$$

$$F_k = \sup_{u \in R_+} \{ g(u)\, \hat{\Phi}(u) \} \quad \textit{if} \quad p \leq q ;$$

$$F_k = \{ \sum_t [g(\nu_t)^\kappa - g(\nu_{t+1})^\kappa]\, \hat{\Phi}(\nu_t)^\kappa \}^{1/\kappa} \quad \textit{if} \quad p > q .$$

To obtain $\hat{\Phi}$ *we replace* φ_β *by* $\hat{\varphi}_\beta$ *in* (12),

$$\{ \int_0^{\nu_t} v^{kp}\, d\beta \}^{1/p} = d_p(k, \beta)\, 2^t , \quad t \in Z .$$

Remark 7. Theorems 2 and 6 contain some results of E.Sawyer [4] (see also V.Stepanov [5]).

Theorem 7. *Let the conditions of* Theorem 3 *be satisfied and assume* $\alpha > m$. *Then*

$$H^m \asymp E^m + F^m ,$$

where

$$E^m = \{ \sum_t 2^{t\kappa} (\int_{\delta_t}^{\delta_{t+1}} [\delta_t^{(m-\alpha)r} - v^{(m-\alpha)r}]^{q/r} v^{-\delta q} d\gamma)^{\kappa/q} \}^{1/\kappa} ,$$

$$F^m = \sup_t \{ g(\delta_t) \check{\Phi}(\delta_t) \} \quad if \quad p \le q ,$$

$$F^m = \{ \sum_t [g(\delta_t)^\kappa - g(\delta_{t+1})^\kappa] \check{\Phi}(\delta_t) \}^{1/\kappa} \quad if \quad p > q .$$

$$\{ \int_{\delta_t}^{\infty} v^{mp} d\beta \}^{1/p} = e_p(m, \beta) 2^{-t}, \quad t \in Z,$$

$$\check{\Phi}(u) = \sup_{0 < v \le u} \left[(v^{(m-\alpha)r} - u^{(m-\alpha)r})^{1/r} v^{-m} \check{\varphi}_\beta(v) \right] \quad if \quad p \le r ;$$

$$\check{\Phi}(u) = \left\{ \int_0^u [\cdots]^\sigma (v^{-m} \check{\varphi}_\beta(v))^{-1} d(v^{-m} \check{\varphi}_\beta(v)) \right\}^{1/\sigma} \quad if \quad p > r, \quad \sigma = \frac{pr}{(p-r)} .$$

References.

[1] V. Mazja. Sobolev spaces (Russian). Izdat. Leningrad. Gos. Univ., Leningrad, 1985.

[2] K. Oskolkov. Approximative properties of integrable functions on the sets of full measures (Russian). Matem. zborn., 103, N4 (1977), 561-589.

[3] M. Gol'dman. On integral inequalities on the cone of functions with properties of monotonicity. (Russian) Dokl. Acad. Nauk SSSR, 320, N5 (1991), 1037-1042.

[4] E. Sawyer. Boundedness of classical operators on classical Lorentz spaces. Studia Math., 96 (1990), 145-158.

[5] V. Stepanov. Integral operators on the cone of monotone functions. Acad. Nauk SSSR, Far-East Branch, Inst. for Applied Mathem. (1991). Preprint.

Michail L. Gol'dman
Krupskaja Str. 8-1-187
117 311 Moscow, Russia

Anisotropic Sobolev Spaces in Probability Theory

Niels Jacob

Abstract. We point out that Dirichlet spaces on $L^2(\mathbb{R}^n)$ can be considered as anisotropic Sobolev spaces, i.e. function spaces. Using properties of these function spaces it is possible to construct Hunt processes and Feller processes associated with certain non-classical pseudo differential operators. Further, one can examine these stochastic processes with the help of these anisotropic Sobolev spaces.

1. Let $s \geq 0$ and $H^s(\mathbb{R}^n)$ the classical Sobolev space. For $0 < s < 1$ we find for the scalar product in $H^s(\mathbb{R}^n)$

$$(u,v)_s = \int_{\mathbb{R}^n} (1+|\xi|^2)^s \hat{u}(\xi)\overline{\hat{v}(\xi)}d\xi = (u,v)_0 + c_{n,s} \int_{\mathbb{R}^n}\int_{\mathbb{R}^n} \frac{(u(x)-u(y))(v(x)-v(y))}{|x-y|^{n+2s}}dxdy, \quad (1.1)$$

where we assumed all functions to be real–valued. Let us denote by E^s the second term on the very right hand side of (1.1). Then E^s is a symmetric, non–negative closed bilinear form on $L^2(\mathbb{R}^n)$ with domain $H^s(\mathbb{R}^n)$. In particular we have $(u,v)_s = E_1^s(u,v) = (u,v)_0 + E^s(u,v)$. Moreover E^s has the following *sub–Markovian* property:

$$u \in D(E^s) \text{ implies } v := (0 \vee u) \wedge 1 \in D(E^s) \text{ and } E^s(v,v) \leq E^s(u,u). \quad (1.2)$$

Definition 1.1. ([2]) *Let G be a locally compact Hausdorff space and μ a Radon measure, $supp\mu = G$. Further let $D(E)$ be a dense subspace of $L^2(G;\mu)$. A non–negative, symmetric, closed bilinear form E defined on $D(E)$ is called a Dirichlet form if it has the sub–Markokvian property. The pair $(E, D(E))$ is called a Dirichlet space on $L^2(G,\mu)$.*

It was M. Fukushima [3] who proved that there is a one–to–one correspondence of all (regular) Dirichlet spaces on $L^2(G,\mu)$ and the set of all Hunt processes $(\Omega, M, (M_t), X_t, P_x)$ with state space G. Note that for this one must take care on certain exceptional sets. Although this is a crucial point in the whole theory, we will not go into these details here.

Let us return to the case $G = \mathbb{R}^n$ and μ the Lebesgue measure. We want to find further concrete examples of Dirichlet spaces on $L^2(\mathbb{R}^n)$.

Definition 1.2. *A function $a^2 : \mathbb{R}^n \to \mathbb{R}$ is called* negative definite *if for all $m \in \mathbb{N}_0$ and all $\xi^1, \ldots, \xi^m \in \mathbb{R}^n$ the matrix $(a^2(\xi^i) + a^2(\xi^j) - a^2(\xi^i - \xi^j))$ is positive Hermitian.*

For real–valued continuous negative definite functions (c.n.d.f. in the following) we have

$$0 \leq a^2(\xi) \leq c(1+|\xi|^2), \quad (1.3)$$

$$|a(\xi) - a(\eta)| \leq a(\xi - \eta) \quad (1.4)$$

and

$$(1+a^2(\xi))(1+a^2(\eta))^{-1} \le 2(1+a^2(\xi-\eta)). \tag{1.5}$$

We will also assume that

$$1+a^2(\xi) \ge c(1+|\xi|^2)^t$$

holds for some $t > 0$ and we refer to [1] where concrete examples are given.

Let a^2 be as above and define for $s \ge 0$

$$H^{s,a^2}(\mathbb{R}^n) = \{u \in L^2(\mathbb{R}^n), ||u||_{a^2,s} < \infty\} \tag{1.6}$$

where

$$||u||^2_{a^2,s} = \int_{\mathbb{R}^n} (1+a^2(\xi))^{2s} |\hat{u}(\xi)|^2 d\xi. \tag{1.7}$$

Then $H^{s,a^2}(\mathbb{R}^n)$ is a Hilbert space with scalar product $(.,.)_{a^2,s}$.

Theorem 1.1. ([2]) *The space $H^{1/2,a^2}(\mathbb{R}^n)$ with its scalar product as well as with the bilinear form*

$$E^{a^2}(u,v) = \int_{\mathbb{R}^n} a^2(\xi)\hat{u}(\xi)\overline{\hat{v}(\xi)}d\xi \tag{1.8}$$

is a Dirichlet space on $L^2(\mathbb{R}^n)$.

Let $(E, D(E))$ be a Dirichlet space on $L^2(\mathbb{R}^n)$ and A its generator, i.e. the uniquely determined non–positive self–adjoint operator with domain $D(A) \subset D(E)$ satisfying $E(u,v) = ((-A)^{1/2}u, (-A)^{1/2}v)_0$ for all $u, v \in D(E)$. In case of Theorem 1.1, i.e. the Dirichlet form E^{a^2}, we find for all $u \in C_0^\infty(\mathbb{R}^n) \subset H^{1/2,a^2}(\mathbb{R}^n)$

$$-Au(x) = -a^2(D)u(x) = -(2\pi)^{-n/2} \int_{\mathbb{R}^n} e^{ix\xi} a^2(\xi)\hat{u}(\xi)d\xi. \tag{1.9}$$

Moreover we have $D(-A) = H^{1,a^2}(\mathbb{R}^n)$ and $-A$ extends to a generator of a Feller semigroup on $C_\infty(\mathbb{R}^n)$.

A natural question is, whether it is possible to prove that certain pseudo differential operators defined on $C_0^\infty(\mathbb{R}^n)$ by

$$-p(x,D)u(x) = -(2\pi)^{-n/2} \int_{\mathbb{R}^n} e^{ix\xi} p(x,\xi)\hat{u}(\xi)d\xi \tag{1.10}$$

do also generate a Dirichlet form or a Feller semigroup provided that $p(x,\xi)$ is for each fixed x a c.n.d.f. with respect to ξ. Further we would like to use properties of $p(x,\xi)$ to get results for the corresponding Hunt process or Feller process, respectively.

2. In our papers [11] - [14] as well as in [9] and in the thesis of W. Hoh [8], it was proved that indeed various classes of pseudo differential operators $p(x,D)$ with a symbol as in (1.10) do generate a Dirichlet space or a Feller semigroup. In proving these results a lot of properties of the spaces $H^{s,a^2}(\mathbb{R}^n)$ are needed. In this section we would like to summarize some of these results and discuss properties of certain pseudo differential operators in these spaces. Using (1.3) and (1.6) we get the embeddings

$$H^{2s}(\mathbb{R}^n) \hookrightarrow H^{s,a^2}(\mathbb{R}^n) \hookrightarrow H^{2st}(\mathbb{R}^n). \tag{2.1}$$

In particular, if $st > \frac{n}{4}$ we find

$$H^{s,a^2}(\mathbb{R}^n) \hookrightarrow C_\infty(\mathbb{R}^n). \tag{2.2}$$

Next we want to find functions which are multipliers on $H^{s,a^2}(\mathbb{R}^n)$. This question can be reduced to the following commutator extimate which is also important for a lot of other considerations:

Theorem 2.1 *Let $s \geq 0$ and $q > \frac{n}{2} + |t+s-\frac{1}{2}| + \frac{1}{2}$. If $b \in L^\infty(\mathbb{R}^n)$ satisfies $|\hat{b}(\xi)| \leq c(1+|\xi|^2)^{-q}$, then $[(1+a^2(D))^s, b] : H^{t+s-\frac{1}{2},a^2}(\mathbb{R}^n) \to H^{t,a^2}(\mathbb{R}^n)$ is for all $t \geq 0$ a continuous operator.*

A first version of this theorem was given in [10], a considerable improvement was given by W. Hoh in [7]. The final form was discussed by W. Hoh in his talk during this conference. Taking $t = 0$ we find

Corollary 2.1 *Let b be as in Theorem 2.1. Then we have $bu \in H^{s,a^2}(\mathbb{R}^n)$ for all $s \geq 0$.*

Using Theorem 2.1 we can handle certain pseudo differential operators $p(x, D)$ with a symbol $p(x,\xi)$ which is with respect to ξ a c.n.d.f. In particular, if $p(x,\xi) = p_1(\xi) + p_2(x,\xi)$, where $0 < c_0 \leq \frac{1+p_1(\xi)}{1+a^2(\xi)} \leq c_1$ and $p_2(x,\xi)$ is in a certain sense small with respect to $c_0(1 + a^2(\xi))$, we could prove in [14], see also [11], [13] , that $-p(x, D)$ generates a Feller semigroup and therefore, if $p(x, D)$ is symmetric, a Dirichlet form E with domain $H^{1/2,a^2}(\mathbb{R}^n)$. In order to get better properties for the Hunt process associated with the Dirichlet form generated by a symmetric operator $-p(x, D)$, one can use the theory of (r,p)-capacities due to M. Fukushima and H. Kaneko, see [6], [16] and [4]. The capacity in $H^{1/2,a^2}(\mathbb{R}^n)$ is defined for an open set by

$$\text{Cap}(\Omega) = \inf\{(u,u)_{a^2,1/2} \quad ; \quad u \in H^{1/2,a^2}(\mathbb{R}^n) \quad , \quad u \geq 1 \ a.e. \text{ on } \Omega\}, \tag{2.3}$$

and for an arbitrary set by $\text{Cap}(A) = \inf\{\text{cap}(B); A \subset B, B \text{ open}\}$. For $r > 0$ and $p \geq 1$ the (r,p)- capacity is defined for an open set by

$$C_{r,p}(\Omega) = \inf\{||u||^p_{r,p}; u \in \mathcal{F}_{r,p}, u \geq 1 \text{ a. e. on } \Omega\}, \tag{2.4}$$

where $||u||_{r,p} = ||f||_{L^p}, u = V_r f = \frac{1}{\Gamma(\frac{r}{2})}\int_0^\infty t^{r/2}e^{-t}T_t f dt$ for $f \in L^p(\mathbb{R}^n)$, and $\mathcal{F}_{r,p} = V_r(L^p(\mathbb{R}^n))$. Here $(T_t)_{t\geq 0}$ is the semigroup generated by $-a^2(D)$ on $L^p(\mathbb{R}^n)$. Note that $\mathcal{F}_{r,p}$ is analogously defined to the Bessel–potential spaces. Again $C_{r,p}$ extends to arbitrary sets and has a lot of nice properties (see [6] or [4]). In particular one has

$$r < r' \text{ implies } C_{r,p}(E) \leq C_{r',p}(E), \quad p < p' \text{ implies } C_{r,p}(E) \leq C_{r,p'}(E). \tag{2.5}$$

This can be used to get refinements for exceptional sets for the Hunt process associated with $-p(x, D)$, see [15], where results due to H. Kaneko [16] could be applied to our concrete example of a pseudo differential operator generating a Dirichlet space. In particular an ergodic theorem could be proved. In proving these results we need besides various estimates the embedding (2.2).

References

[1] Berg, C., and Forst, G.: Potential theory on locally compact Abelian groups. Springer Verlag (1975).

[2] Beurling, A., and Deny, J.: Dirichlet spaces. Proc. Natl. Acad. Sci. U.S.A. 45 (1959) 208–215.

[3] Fukushima, M.: Dirichlet spaces and strong Markov processes. T.A.M.S. 162 (1971) 455–473.

[4] Fukushima, M.: Two topics related to Dirichlet forms: quasi everywhere convergence and additive functionals. In: CIME courses on "'Dirichlet forms"', 1992 (to appear)

[5] Fukushima, M., Jacob, N., and Kaneko, H.: On $(r,2)$–capacities for a class of elliptic pseudo differential operators. Math. Ann. 293 (1992) 343–348.

[6] Fukushima, M., and Kaneko, H.: On (r,p)-capacities for general Markov semi–groups. Research Notes in Math. 124, Pitman, (1985), 41–47.

[7] Hoh, W.: Some commutator estimates for pseudo differential operators with negative definite functions as symbols. Integral Equations and Operator Theory (in press).

[8] Hoh, W.: Zum Martingalproblem für eine Klasse von Pseudodifferentialoperatoren. Dissertation, Universität Erlangen–Nürnberg (1992).

[9] Hoh, W., and Jacob, N.: Some Dirichlet forms generated by pseudo differential operators. Bull. Sc. Math. 116 (1992) 383–398.

[10] Jacob, N.: Commutator estimates for pseudo differential operators with negative definite functions as symbol. Forum Math. 2 (1990) 115–162.

[11] Jacob, N.: Feller semigroups, Dirichlet forms, and pseudo differential operators. Forum Math. 4 (1992) 433–446.

[12] Jacob, N.: A class of elliptic pseudo differential operators generating symmetric Dirichlet forms. Potential Analysis 1 (1992) 221–232.

[13] Jacob, N.: Further pseudo differential operators generating Feller semigroups and Dirichlet forms. Rev. Mat. Iberoamericana (in press).

[14] Jacob, N.: A class of Feller semigroups generated by pseudo differential operators (submitted).

[15] Jacob, N.: An application of $(r, 2)$–capacities to pseudo differential operators (preprint).

[16] Kaneko, H.: On (r, p)–capacities for Markov processes. Osaka J. Math. 23 (1986) 325–336.

Niels Jacob
Mathematisches Institut
Universität Erlangen–Nürnberg
Bismarckstr. 1 1/2
W - 8520 Erlangen

Atomic Decomposition of Besov Spaces on Closed Sets

Alf Jonsson

0. Introduction

We combine ideas from on one hand the papers [2] and [3] by M. Frazier and B. Jawerth dealing with atomic decompositions of distributions in the Besov spaces $B_\alpha^{p,q}(\mathbb{R}^n)$ and the Triebel–Lizorkin spaces $F_\alpha^{p,q}(\mathbb{R}^n)$, and on the other hand the theory of the Besov spaces $B_\alpha^{p,q}(F)$ consisting of functions defined on closed subsets of $\mathbb{R}^n$, developed by the author, H. Wallin and P. Bylund e.g. in [6], [5], and [1]. The aim is to deepen the theory of function spaces on subsets of $\mathbb{R}^n$. The main results are an atomic decomposition of functions in $B_\alpha^{p,q}(F)$, the treatment of these spaces for certain $p < 1$, a new proof of a trace theorem for Besov spaces describing the trace of $B_\alpha^{p,q}(\mathbb{R}^n)$ to closed sets, and a trace theorem for Triebel-Lizorkin spaces (in particular, for Sobolev spaces).

We stress that the considered sets $F \subset \mathbb{R}^n$ are of a very general nature. Sometimes they may be arbitrary closed sets, but in general, some restrictions are imposed, which in particular imply restrictions on the dimension of the set F. In this note, the results are in general given without proofs.

1. Besov spaces $B_\alpha^{p,q}(F)$ and atomic decompositions

We start by recalling the decomposition of the classical Besov spaces $B_\alpha^{p,q}(\mathbb{R}^n)$ given in [2], assuming that $0 < p,q \leq \infty$, $\alpha > 0$, and $\alpha > n(1/p-1)$. (In [2] a decomposition is given for $-\infty < \alpha < \infty$, in which case vanishing moments are required in the definition of the atoms, and one can have vanishing moments for $\alpha > 0$ and $\alpha > n(1/p-1)$ as well.) Let $K \geq [\alpha]+1$, where $[\alpha]$ denotes the integer part of α, and let, for $r > 0$, rQ denote the closed cube obtained by expanding the cube Q with the factor r around its center. A function a is said to be an (α,p)-atom if $a \in C^K(\mathbb{R}^n)$ and there exists a cube Q such that supp a $\subset$ 3Q and

$$|D^j a(x)| \leq |Q|^{(\alpha-|j|)/n-1/p}, \quad x \in \mathbb{R}^n, \quad |j| \leq K. \tag{1}$$

Here $|Q|$ denotes the n-dimensional Lebesgue measure of Q. We will by a_Q denote atoms associated to a cube Q as above, and by π_ν the dyadic net which divides $\mathbb{R}^n$ into cubes called dyadic cubes, with sides of length $2^{-\nu}$, one of them being $\{x;\ 0 \leq x_i \leq 2^{-\nu}, i = 1,2,...,n\}$. In [2, p.795] it is shown that a function f belongs to the inhomogeneous Besov space $B_\alpha^{p,q}(\mathbb{R}^n)$ if and only if it has a representation $f = \sum_{\nu=0}^{\infty} \sum_{Q\in\pi_\nu} s_Q a_Q$, where a_Q are (α,p)-atoms, the convergence is in $\mathcal{S}'$ (the space of tempered distributions), and the numbers s_Q satisfy

$$\left(\sum_{\nu=0}^{\infty} \left(\sum_{Q\in\pi_\nu} |s_Q|^p \right)^{q/p} \right)^{1/q} < \infty. \tag{2}$$

The $B_\alpha^{p,q}(\mathbb{R}^n)$-norm (or quasi-norm if $p < 1$ or $q < 1$) of f is equivalent to the infimum of the sums (2) taken over all possible representations.

We shall now define Besov spaces on closed sets in an analogous way. The definition is somewhat preliminary, cf. Section 3. Throughout the paper, μ will denote a positive Borel measure, finite on finite sets. In the following definition we assume that μ satisfies

$$\mu(B(x,r)) \leq c\, r^d, \quad x \in \mathbb{R}^n, \quad 0 < r \leq 1, \tag{3}$$

where $0 \leq d \leq n$ and $B(x,r)$ denotes the closed ball with center x and radius r.

Definition 1. Let $0 < p,q \leq \infty$, $0 \leq d \leq n$, $\alpha - (n-d)/p > 0$, $K \geq [\alpha]+1$, let F be a closed set, and let μ be a positive Borel measure with supp $\mu =$ F satisfying (3). A function f defined on F belongs to *the Besov space* $B_\alpha^{p,q}(F,\mu)$ if and only if it has a representation

$$f = \sum_{\nu=0}^{\infty} \sum_{Q\in\pi_\nu} s_Q a_Q, \tag{4}$$

where a_Q are (α,p)-atoms, the numbers s_Q satisfy (2), and the convergence (of the sum with respect to ν) is in $L^p(\mu)$. The *norm* (or *quasi-norm* if $p < 1$ or $q < 1$) is defined as the infimum of the sums (2) taken over all possible representations.

The spaces $B_\alpha^{p,q}(F,\mu)$ depend on K, but it follows from the trace theorem (Theorem 1) that different K give equivalent spaces, since this is the case for the spaces $B_\alpha^{p,q}(\mathbb{R}^n)$. Sometimes one can use a K which is not greater than α. If $F = \mathbb{R}^n$, $d = n$, and μ is the n-dimensional Lebesgue measure, the space $B_\alpha^{p,q}(F,\mu)$ coincides with the usual Besov space under the additional assumption $\alpha - n/p + n > 0$ if $p < 1$.

The spaces $B_\alpha^{p,q}(F,\mu)$ do not depend on μ in an essential way, as it is explained by the following proposition (cf. [5] and [1]). Because of this, we shall from now on drop the letter μ from the notation in general and simply write $B_\alpha^{p,q}(F)$.

Proposition 1. *Let p,q,d,α and F be as in Definition 1, and let μ_1 and μ_2 be as μ in that definition. Assume that $f \in B_\alpha^{p,q}(F,\mu_2)$. Then f can be altered on a set with μ_2-measure zero in such a way that f becomes a function in $B_\alpha^{p,q}(F,\mu_1)$ with norm not exceeding the $B_\alpha^{p,q}(F,\mu_2)$-norm of f.*

Let f_N denote the partial sum $\sum_{\nu=0}^{N} \sum_{Q\in\pi_\nu} s_Q a_Q$, where a_Q are (α,p)-atoms and $\{s_Q\}$ satisfies (2). Then it is not hard to show that, under assumptions as in Definition 1, $\{f_N\}$ is a Cauchy sequence in $L^p(\mu)$, and denoting the limit by $\bar{f}$ we have

$$\|\bar{f}\|_{p,\mu} \leq c \left(\sum_{\nu=0}^{\infty} \left(\sum_{Q\in\pi_\nu} |s_Q|^p \right)^{q/p} \right)^{1/q}. \tag{5}$$

This gives one possibility to define the trace to F of a function in $B_\alpha^{p,q}(\mathbb{R}^n)$: We let the trace $R_F(f)$ be the $L^p(\mu)$-limit of the partial sums of an atomic representation of f; it will exist under the assumptions in Definition 1. Of course, one has to show that the trace is independent of the representation, which requires some work, but once this is done the following theorem is almost trivial.

Theorem 1. *Let* $0 < p, q \leq \infty$, $0 \leq d \leq n$, $\alpha - n\,(1/p - 1) > 0$, $\alpha - (n-d)/p > 0$, $F \subset \mathbb{R}^n$, *F closed, and let μ satisfy* supp $\mu =$ F *and* (3). *Then hold*

a) *R_F is a continuous map from $B_\alpha^{p,q}(\mathbb{R}^n)$ onto $B_\alpha^{p,q}(F)$.*

b) *If $f \in B_\alpha^{p,q}(F)$, then there exists a function $Ef \in B_\alpha^{p,q}(\mathbb{R}^n)$ such that $\|Ef\|_{B_\alpha^{p,q}(\mathbb{R}^n)} \leq \|f\|_{B_\alpha^{p,q}(F)}$ and $R_F(Ef) = f$ μ-a.e.*

Proof: Let $f \in B_\alpha^{p,q}(\mathbb{R}^n)$, $f = \sum_{\nu=0}^{\infty} \sum_{Q \in \pi_\nu} s_Q a_Q$ where $\{s_Q\}$ satisfies (2). Then, by the discussion above, $R_F f = \bar{f} = \sum_{\nu=0}^{\infty} \sum_{Q \in \pi_\nu} s_Q a_Q$ with convergence in $L^p(\mu)$, which means that $R_F f \in B_\alpha^{p,q}(F)$. Conversely, if $f \in B_\alpha^{p,q}(F)$, $f = \sum_{\nu=0}^{\infty} \sum_{Q \in \pi_\nu} s_Q a_Q$ in $L^p(\mu)$, then we may look upon this sum as a decomposition in $\mathbb{R}^n$ (setting $a_Q = 0$ if $3Q \cap F = \emptyset$), and by the argument above (let μ be the Lebesgue measure) it has a sum in $L^p(\mathbb{R}^n)$ which we call Ef, and the theorem follows.

2. The trace of Triebel-Lizorkin spaces.

Before discussing Triebel-Lizorkin spaces we give some more definitions.
Let F be a closed subset of $\mathbb{R}^n$, μ a positive Borel measure with support F, $0 < s \leq n$, and $0 \leq d \leq n$. The measure μ satisfies *the condition* (D_s) if, for some constant $c > 0$,

$$\mu(B(x,kr)) \leq ck^s\,\mu(B(x,r)), \quad x \in F,\ r > 0,\ k \geq 1,\ kr \leq 1 \tag{6}$$

and *the condition* (L_d) if, for some constant $c > 0$,

$$\mu(B(x,kr)) \geq ck^d\,\mu(B(x,r)), \quad x \in F,\ r > 0,\ k \geq 1,\ kr \leq 1. \tag{7}$$

We shall use the normalisation

$$c_1 \leq \mu(B(x,1)) \leq c_2, \quad x \in F. \tag{8}$$

The numbers d and s in (6) and (7) are related to the dimension of F; if (6) holds, then the Hausdorff dimension of F is $\leq s$ and if (7) holds, then the Hausdorff dimension of every set $F \cap B(x,r)$, $x \in F$, $r > 0$, is $\geq d$. Any closed set is the support of a measure μ satisfying (D_n) and (8). For these results and related material we refer to [8] (for results connected to the condition (D_s)) and [5].
Using the atomic decomposition of the Triebel-Lizorkin spaces $F_\alpha^{p,q}(\mathbb{R}^n)$ given in [3] as a tool, it is easy to determine the trace of these spaces, once one knows the trace of Besov spaces. This is a consequence of Theorem 13.7 in [3], from which it follows that if F is a closed set without interiour points such that $\complement F \in NST$ (see below), and we know the trace to F of $B_\alpha^{p,p}(\mathbb{R}^n)$, then the trace of $F_\alpha^{p,q}(\mathbb{R}^n)$ is the same. An open set Ω belongs to NST if there exists a positive integer m with the property that for any dyadic cube Q with sides of length $l(Q) \leq 1$ satisfying $\overline{Q} \cap \delta\Omega \neq \emptyset$, there is a cube $P \subset Q$ with $l(P) = 2^{-m} l(Q)$ such

that $P^0 \cap \delta\Omega = \emptyset$ (P° is the interiour of P and $\delta\Omega$ is the boundary of Ω). The condition that $\mathbf{C}F \in NST$ can be expressed in terms of measures μ supported on F in the following way.

Proposition 2. *Let F be a closed subset of $\mathbb{R}^n$ with $F^\circ = \emptyset$. Then $\mathbf{C}F$ belongs to NST if there exists a measure μ with support F satisfying* (8) *and the condition (D_s) for some $s < n$.*

The proof of this proposition is closely related to geometric conditions given in [8] for sets satisfying the refined doubling condition. The trace theorem for Triebel-Lizorkin spaces may now be given as follows.

Theorem 2. *Let $0 < p < \infty$, $0 < q \leq \infty$, $0 \leq d \leq s < n$, $\alpha - n(1/p - 1) > 0$, $s > 0$ and $\alpha - (n-d)/p > 0$. Let F be a closed set, and μ a measure with support F satisfying* (3), (8), *and the condition (D_s). Then hold*

a) *R_F is a continuous map of $F_\alpha^{p,q}(\mathbb{R}^n)$ onto $B_\alpha^{p,p}(F)$.*
b) *If $f \in B_\alpha^{p,p}(F)$, then there exists a function $Ef \in F_\alpha^{p,q}(\mathbb{R}^n)$ such that $||Ef||_{F_\alpha^{p,q}(\mathbb{R}^n)} \leq c||f||_{B_\alpha^{p,p}(F)}$ and $R_F(Ef) = f \quad \mu - a.e.$*

3. A constructive atomic decomposition.

The above results have one serious drawback. The decomposition of $B_\alpha^{p,q}(F)$ is not given in a constructive way, which in particular means that the extension part of the trace theorems is not obtained by means of a linear operator, as is easily seen by the proof of Theorem 1. To cure this we must, in general, impose more restrictions on F. These involve the condition (L_d) given above, and also that F *preserves Markov's inequality* in the following sense:
For all polynomials P in n variables of degree at most k and all balls
$B = B(x,r)$, $x_0 \in F$, $0 < r \leq 1$, holds

$$\max_{F\cap B} |gradP| \leq c\, r^{-1} \max_{F\cap B} |P|,$$

with c depending on f, n, and k, only. The atomic decomposition is obtained by first constructing approximations of f consisting of piecewise polynomials (using orthogonal polynomials), and then from these approximations the atoms. Details can not be given here, a lot of work is needed. We just give the trace theorems one obtains; then the assumptions needed for the construction of the atomic decomposition are also implicitly given.

Theorem 3. *Let $0 < p, q \leq \infty$, $0 \leq d \leq s \leq n$, $s > 0$, $\alpha - n(1/p - 1) > 0$ and $\alpha - (n-d)/p > 0$. Let F be a closed set satisfying Markov's inequality, and let μ be a measure with support F satisfying* (6), (7), *and* (8). *Then $B_\alpha^{p,q}(F)$ is the trace to F of $B_\alpha^{p,q}(\mathbb{R}^n)$ (or $B_\alpha^{p,p}(F)$ the trace of $F_\alpha^{p,q}(\mathbb{R}^n)$ if $s < n$ and $p < \infty$) in the following sense:*

a) *R_F is a continuous linear operator from $B_\alpha^{p,q}(\mathbb{R}^n)$ ($F_\alpha^{p,q}(\mathbb{R}^n)$) onto $B_\alpha^{p,q}(F)$ ($B_\alpha^{p,p}(F)$).*
b) *There is a continuous linear extension operator E from $B_\alpha^{p,q}(F)$ to $B_\alpha^{p,q}(\mathbb{R}^n)$ (from $B_\alpha^{p,p}(F)$ to $F_\alpha^{p,q}(\mathbb{R}^n)$) such that $R_F(Ef) = f$ $\mu - a.e.$*

We conclude by mentioning some particular cases of this theorem. In the case when we can take $d = 0$ and $s = n$ in Theorem 3, F may be a closed set subject only to the condition that F preserves Markov's inequality. This follows since the condition (L_d) vanishes if $d = 0$ and, as already mentioned, it is always possible to construct a measure with support F satisfying (6) and (8). Note that we must have the Besov space case, and that $\alpha - (n-d)/p > 0$

becomes $\alpha p > n$, so the functions in $B_\alpha^{p,q}(\mathbb{R}^n)$ can be considered as continuous functions. On the other hand, if $\gamma = \alpha - (n-s)/p < 1$, then Markov's inequality is not needed (cf. [5]). This can not be seen directly from the statement of the theorem, but the reason is that when constructing atoms we need polynomials of degree zero, only, so Markov's inequality is trivially fulfilled. Together, the above remarks give that if $\alpha p > n$ and $\alpha < 1$, Theorem 3 for Besov spaces holds for an arbitrary closed set.

References

[1] P. Bylund, *The trace of Besov spaces to sets with varying local dimension*, manuscript.

[2] M. Frazier and B. Jawerth, *Decomposition of Besov spaces*, Indiana Univ. Math. J. **34** (1985), 777-799.

[3] M. Frazier and B. Jawerth, *A discrete transform and decompositions of distribution spaces*, J. of Functional Analysis **93** (1990), 34-170.

[4] A. Jonsson, *Markov's inequality and local polynomial approximation*, Lecture Notes in Mathematics 1302, Springer-Verlag, Berlin, 1986, 303-316.

[5] A. Jonsson, *Besov spaces on closed subsets of* $\mathbb{R}^n$, Preprint, to appear in Trans. Am. Math. Soc.

[6] A. Jonsson and H. Wallin, *Function spaces on subsets of* $\mathbb{R}^n$, Mathematical Reports 2, Part 1, Harwood Acad. Publ., 1984.

[7] H. Triebel, *Theory of Function Spaces II*, Birkhäuser Verlag, Basel, 1992.

[8] A.L. Volberg and S.V. Konyagin, *On measures with the doubling condition*, Izv. Acad. Nauk CCCP, Ser. Matem. **51** (1987), No 3, English transl.: Math. USSR Izvestiya **30** (1988), 629-637.

Alf Jonsson
Department of Mathematics
University of Umeå
S-901 87 Umeå, Sweden

Some Remarks Concerning the Hardy Inequality

Alois Kufner

This note deals with the Hardy inequality of order k

$$\left(\int_0^\infty |u(t)|^q w_0(t)dt\right)^{1/q} \leq C\left(\int_0^\infty |u^{(k)}(t)|^p w_k(t)dt\right)^{1/p}, \tag{1}$$

more precisely, with conditions on the parameters p, q and on the weight functions w_0, w_k under which inequality (1) holds for all functions u from a certain class K with a constant $C > 0$ independent of u.

For simplicity, we will assume that

$$1 < p \leq q < \infty$$

and we will consider two classes of functions u, namely

$$K_0 = \{u \in AC^{(k-1)}(0,\infty) : u(0) = u'(0) = u''(0) = \ldots = u^{(k-1)}(0) = 0\}$$

and

$$K_\infty = \{u \in AC^{(k-1)}(0,\infty) : u(\infty) = u'(\infty) = u''(\infty) = \ldots = u^{(k-1)}(\infty) = 0\}.$$

In these cases, necessary and sufficient conditions for the validity of inequality (1) are well-known : If we introduce the functions

$$B_1(x) = \left(\int_x^\infty w_0(t)dt\right)^{1/q}\left(\int_0^x (x-t)^{p'(k-1)} w_k^{1-p'}(t)dt\right)^{1/p'}, \tag{2}$$

$$B_2(x) = \left(\int_x^\infty (t-x)^{q(k-1)} w_0(t)dt\right)^{1/q}\left(\int_0^x w_k^{1-p'}(t)dt\right)^{1/p'}$$

with $p' = \frac{p}{p-1}$, then STEPANOV [1] has shown that inequality (1) holds for every $u \in K_0$ if and only if

$$\sup_{0<x<\infty} B_i(x) = B_i < \infty, \quad i = 1, 2. \tag{3}$$

Analogously, inequality (1) holds for every $u \in K_\infty$ if and only if

$$\sup_{0<x<\infty} \widetilde{B}_i(x) = \widetilde{B}_i < \infty, \quad i = 1, 2, \tag{4}$$

where

$$\widetilde{B}_1(x) = (\int_0^x w_0(t)dt)^{1/q}(\int_x^\infty (t-x)^{p'(k-1)} w_k^{1-p'}(t)dt)^{1/p'}, \tag{5}$$

$$\widetilde{B}_2(x) = (\int_0^x (x-t)^{q(k-1)} w_0(t)dt)^{1/q}(\int_x^\infty w_k^{1-p'}(t)dt)^{1/p'}.$$

If $k = 1$, then $B_1(x) = B_2(x)$ and $\widetilde{B}_1(x) = \widetilde{B}_2(x)$. Consequently, we have only one condition in (3) and/or (4), which coincides with the well-known condition mentioned, e.g., in OPIC and KUFNER [2].

If $k > 1$, then we have a *pair* of conditions in (3) and/or (4); moreover, these conditions are not easy to verify since the evaluation of the integrals in (2) and/or (5) causes problems. Therefore, the author tried to simplify those conditions and in the paper KUFNER and HEINIG [3] the following result was established.

Proposition 1. *Denote*

$$B_3(x) = (\int_x^\infty w_0(t)t^{q(k-1)}dt)^{1/q}(\int_0^x w_k^{1-p'}(t)dt)^{1/p'}. \tag{6}$$

If

$$\sup_{0<x<\infty} B_3(x) = B_3 < \infty, \tag{7}$$

then inequality (1) *holds for every* $u \in K_0$.

An analogous result can be formulated for $u \in K_\infty$.

For $k = 1$, condition (7) is necessary and sufficient for the validity of (1) since in this case $B_3(x) = B_1(x) = B_2(x)$. Unfortunately, for $k \geq 2$, condition (7) is *only sufficient*, as is seen from the following example.

Example 1. Let us take $k = 2$, $p = q = 2$, $w_0(t) = w_2(t) = e^{\alpha t}$. Then inequality (1) holds for every $u \in K_0 = \{u \in AC^{(1)}(0,\infty) : u(0) = u'(0) = 0\}$ if $\alpha < 0$. On the other hand, $B_3^2(x) = \alpha^{-4}[(\alpha x - 1)^2 + 1](1 - e^{\alpha x})$ and consequently, condition (7) is not fulfilled.

On the other hand, condition (7) is also necessary at least for some special weight functions, namely for power functions. Let us give an example.

Example 2. Let us take $p = q > 1$, $w_0(t) = t^\lambda$, $w_k(t) = t^\epsilon$. Then inequality (1), i.e., the inequality

$$\int_0^\infty |u(t)|^p t^\lambda dt \leq C^p \int_0^\infty |u^{(k)}(t)|^p t^\epsilon dt$$

holds for every $u \in K_0$ if and only if

$$\lambda = \epsilon - kp \ \text{ and } \ \epsilon < p - 1. \tag{8}$$

One can check that conditions (8) are equivalent (for our special choice of w_0, w_k) to condition (7) as well as to the pair of conditions (3). Consequently, we can formulate the following problem, which arises quite naturally :

Problem 1. *Determine classes of weight functions w_0, w_k for which the conditions* (3) *and* (7) *are equivalent.*

The solution of this problem would enable us to replace the pair of complicated conditions (3) by the simpler (and single) condition (7).

A similar problem can be formulated with the aim of simplifying conditions (5). Another possibility in this direction is indicated in the following proposition.

Proposition 2. *Denote*

$$W_0(x) = \int_0^x (x-t)^{q(k-1)} w_0(t)dt \tag{9}$$

and suppose that there exists a positive number c_0 such that

$$W_0(2x) \le c_0 W_0(x) \quad \text{for every} \quad x \in (0,\infty). \tag{10}$$

Then inequality (1) *holds for every $u \in K_\infty$ if and only if*

$$\sup_{0<x<\infty} \widetilde{B}_1(x) = \widetilde{B}_1 < \infty, \tag{11}$$

where $\widetilde{B}_1(x)$ is determined by the second formula in (5).

Proof. (i) Necessity. If inequality (1) holds for every $u \in K_\infty$ then conditions (4) are satisfied, and this fact implies (11).

(ii) Sufficiency. We have

$$\int_0^x w_0(t)dt = \int_0^x x^{q(k-1)} w_0(t)dt \cdot x^{-q(k-1)} \ge$$

$$\int_0^x (x-t)^{q(k-1)} w_0(t)dt \cdot x^{-q(k-1)} \ge \frac{1}{c_0} \int_0^{2x} (2x-t)^{q(k-1)} w_0(t)dt \cdot x^{-q(k-1)}$$

due to (10). Further,

$$\int_x^\infty (t-x)^{p'(k-1)} w_k^{1-p'}(t)dt = \int_x^\infty (\frac{t}{x} - 1)^{p'(k-1)} w_k^{1-p'}(t) \cdot x^{p'(k-1)}dt \ge$$

$$\int_{2x}^\infty (\frac{t}{x} - 1)^{p'(k-1)} w_k^{1-p'}(t) \cdot x^{p'(k-1)}dt \ge$$

$$\int_{2x}^\infty w_k^{1-p'}(t)dt \cdot x^{p'(k-1)}dt$$

since in the third integral in the foregoing formula we have $2x < t < \infty$, i.e. $\frac{t}{x} > 2$ and $\frac{t}{x} - 1 > 1$.

Consequently, the function $\widetilde{B}_1(x)$ from (5) satisfies

$$\widetilde{B}_1(x) = (\int_0^x w_0(t)dt)^{1/q}(\int_x^\infty (t-x)^{p'(k-1)} w_k^{1-p'}(t)dt)^{1/p} \geq$$

$$\frac{1}{c_0{}^{1/q}}(\int_0^{2x} (2x-t)^{q(k-1)} w_0(t)dt)^{1/q} \cdot x^{-(k-1)}(\int_{2x}^\infty w_k^{1-p'}(t)dt)^{1/p'} x^{(k-1)}dt =$$

$$\frac{1}{c_0{}^{1/q}}\widetilde{B}_2(2x),$$

and condition (11) implies that

$$\sup_{0<x<\infty} \widetilde{B}_2(x) \leq c_0^p \widetilde{B}_1.$$

Thus, both conditions in (4) are satisfied and consequently, inequality (1) holds for every $u \in K_\infty$.

So we have shown that if the weight function w_0 from the pair w_0, w_k satisfies (10), then the pair of conditions (4) can be reduced to only one of them, namely (11). In a completely analogous way we can reduce the pair (3), if we consider $u \in K_0$:

Proposition 3. *Denote*

$$W_k(x) = \int_0^x (x-t)^{p'(k-1)} w_k^{1-p'}(t)dt \tag{12}$$

and suppose that there exists a positive constant c_k such that

$$W_k(2x) \leq c_k W_k(x). \tag{13}$$

Then inequality (1) holds for every $u \in K_0$ if and only if

$$\sup_{0<x<\infty} B_2(x) = B_2 < \infty,$$

where $B_2(x)$ is determined by the first formula in (2).

Condition (10) [or condition (13)] is the well-known Δ_2-condition which appears frequently in the theory of Orlicz spaces (see, e.g., KUFNER, JOHN and FUČÍK [4]). It remains only to find out which class of functions w_0 and $w_k^{1-p'}$ satisfies this condition. So, we have arrived at the following problem:

Problem 2. *For $g(t) \geq 0$ and $\alpha > 0$, denote*

$$G_\alpha(x) = \int_0^x (x-t)^\alpha g(t)dt. \tag{14}$$

Determine the class of all functions g for which G_α satisfies the Δ_2-condition, i.e. such that there exists a positive constant c^* such that

$$G_\alpha(2x) \leq c^* G_\alpha(x). \tag{15}$$

A partial solution of this problem is given by the following result :

Proposition 4. *If G_α is given by formula* (14) *and condition* (15) *is satisfied for* $\alpha = 0$, *then it is satisfied for every* $\alpha > 0$.

Remark. The Hardy inequality is (for the case $p = q = 2$) closely related to some spectral problems for linear ordinary differential equations. E.g., if we consider on $(0, \infty)$ the eigenvalue problem

$$(-1)^k \frac{d^k}{dx^k}\left(r(x)\frac{d^k y}{dx^k}\right) = \lambda y(x), \tag{16}$$

then a necessary and sufficient condition for the boundedness of the spectrum from below can be expressed in terms of the boundedness of the function

$$B(x) = x^{2k-1}\int_x^\infty \frac{1}{r(t)}dt.$$

It is easy to verify, that $B(x)$ is nothing else than a multiple of the square of the function $\widetilde{B}_2(x)$ from (5), where we take $p = q = 2$, $w_0(t) = 1$, $w_k(t) = r(t)$. Moreover, the boundedness of B (i.e., of $\widetilde{B}_2$) implies also the boundedness of the function $\widetilde{B}_1$, since the function $w_0(t) \equiv 1$ satisfies condition (10). Consequently, the boundedness of the spectrum of the differential operator $(-1)^k(ry^{(k)})^{(k)}$ is connected with the validity of the inequality

$$\int_0^\infty |u(t)|^2 dt \le C^2 \int_0^\infty |u^{(k)}(t)|^2 r(t) dt$$

for $u \in K_\infty$.

References

[1] STEPANOV, V.D.: Two-weighted estimates for Riemann-Liouville integrals. Preprint no. 39, Czech.Acad.Sci. Prague 1988

[2] KUFNER, A.; HEINIG, H.P.: Hardy's inequality for higher order derivatives (Russian). Trudy Mat.Inst. Steklov 192 (1990), 105–113

[3] OPIC, B.: KUFNER, A.: Hardy-type inequalities. Pitman Research Notes in Mathematics Series 219, Longman Scientific and Technical, Harlow 1990

[4] KUFNER, A.; JOHN, O.; FUČÍK, S.: Function spaces. Academia Prague and Noordhoff International Publishing Leyden 1977.

Alois Kufner
Mathematical Institute,
Czech Academy of Sciences
Žitná 25
115 67 Prague 1, Czech Republic
E-Mail: KUFNER@CSEARN.BITNET

Inequalities for Integral and Discrete Norms of Entire Functions of Exponential Type

Tamara V. Tararykova

1 Preliminaries

Let for $\nu > 0$ E_ν denote the space of entire functions of n variables of exponential type ν in each variable. We shall discuss some results connected with estimates of integral norms $||f(t)||_{L_q(\mathbb{R}^n)}$ via discrete norms $||f(k)||_{l_p(\mathbb{Z}^n)}$ and vice versa and interference inequalities.

Estimates of discrete norms via integral norms are more simple. They are valid for arbitrary $\nu > 0$. It was proved by M.Plancherel and G.Polya [1] that for $1 \le p \le \infty$ and every function $f \in E_\nu$ the following inequality is valid

$$||f(k)||_{l_p(\mathbb{Z})} \le A_p(\nu)||f(t)||_{L_p(\mathbb{R})}, \tag{1}$$

where $A_p(\nu) = \frac{8}{\pi}\left(\frac{e^{\nu p}-1}{\nu p}\right)^{1/p}$. S. M. Nikol'skij [2] proved inequality (1) with $A_p(\nu) = 1 + \nu$ and used it in the theory of function spaces [2] and [3].

As for the estimate inverse to (1) it is valid not for every $\nu > 0$. M. Cartwright [4] for $p = \infty$ and M.Plancherel and G.Polya [1] for $1 \le p < \infty$ proved that if $\nu < \pi$ then there exists such $B_p(\nu) > 0$ that for every $f \in E_\nu$

$$||f(t)||_{L_p(\mathbb{R})} \le B_p(\nu)||f(k)||_{l_p(\mathbb{Z})}. \tag{2}$$

A simple example ($f(z) = \sin \pi z$) shows that for $\nu \ge \pi$ for any $1 \le p \le \infty$ inequality (2) is not valid for all functions $f \in E_\nu$. But for $\nu = \pi$ under additional assumption

$$\lim_{t\to\infty} f(t) = 0 \tag{3}$$

for $1 < p < \infty$ it is still valid (see B.Ya.Levin [5]). For $p = 1$ and $p = \infty$ inequality (2) is not valid even under this additional assumption.

For case $p = \infty$ it was found by S. N. Bernstein [6] that under condition (3) the following inequality is valid instead of (2)

$$||f(t-\tfrac{1}{2}) + f(t+\tfrac{1}{2})||_{L_\infty(\mathbb{R})} \le \tfrac{8}{\pi}||f(k)||_{L_\infty(\mathbb{Z})}. \tag{4}$$

It means that possible growth of one summand is cancelled by another summand. This effect was named an interference phenomena and correspondingly inequality (4) is called an interference inequality. Inequality (4) is still valid if we substitute (3) by the following weaker condition

$$\lim_{t\to\infty} \frac{f(t)}{t} = 0. \tag{5}$$

The condition (5) is sharp in the sense that it cannot be changed by condition "the ratio $\frac{f(t)}{t}$ is bounded when $|t| \geq 1$", as example $f(z) = z \sin \pi z$ shows. Moreover the constant $\frac{8}{\pi}$ in (4) is the best possible.

For case $p = 1$ it was proved by G. Z. Ber [7] that for functions $f \in E_\pi$, satisfying (3),

$$\|f(t - \tfrac{1}{2}) + f(t + \tfrac{1}{2})\|_{L_1(\mathbb{R})} \leq C_1 \|f(k)\|_{l_1(\mathbb{Z})}, \tag{6}$$

where

$$C_1 = \left\| \frac{\cos \pi t}{\pi(t^2 - \frac{1}{4})} \right\|_{L_1(\mathbb{R})}$$

is the best possible constant.

Later on an interference phenomena has been studied for convolutions $f * \mu$, where μ is a measure ,satisfying certain conditions, i.e. the validity of inequality

$$\|f(t) * \mu(t)\|_{L_p(\mathbb{R})} \leq C_p \|f(k)\|_{l_p(\mathbb{Z})}, \tag{7}$$

where C_p doesn't depend on f, was investigated. For $\mu = \delta_{-\frac{1}{2}} + \delta_{\frac{1}{2}}$, where δ_x is Dirac's measure, $f(t) * \mu(t) = f(t - \frac{1}{2}) + f(t + \frac{1}{2})$ and for $p = 1$ (7) coincides with (6).

A. F. Timan [8] and [9, chapter 4] proved that for $p = \infty$ inequality (7) is valid for functions $f \in E_\pi$, satisfying (5), iff

$$\hat{\mu}(-\pi) = \hat{\mu}(\pi) = 0, \tag{8}$$

where $\hat{\mu}$ denotes the Fourier transfrom of measure μ.

The analogues result for $p = 1$ was obtained by G. Z. Ber [7]. Moreover the best possible constant in (7) was calculated:

$$C_1 = \|\frac{\sin \pi t}{\pi t} * \mu(t)\|_{L_1(\mathbb{R})} \tag{9}$$

(for $\mu = \delta_{-\frac{1}{2}} + \delta_{\frac{1}{2}}$ it takes the form (7)).

In a number of papers different generalizations of the above results were given(N. I. Ahiezer, N. I. Ahiezer and B. Ya. Levin, R. Boas, B. Ya. Levin and Dinh Than Hoa, Y. N. Logvinenko, MacIntyre, A. F. Timan and others).

Here we present some results, obtained by the author [10]–[13], connected with inequalities

$$\|f(t)\|_{L_q(\mathbb{R}^n)} \leq B_{p,q}^{(n)} \|f(k)\|_{l_p(\mathbb{Z}^n)} \tag{10}$$

and

$$\|f(t) * \mu(t)\|_{L_q(\mathbb{R}^n)} \leq C_{p,q}^{(n)} \|f(k)\|_{l_p(\mathbb{Z}^n)}, \tag{11}$$

where $1 \leq p \leq q \leq \infty$. We are mainly interested in the cases, when it is possible to calculate the best possible constants in these inequalities. Furthermore, we investigate interference phenomena not only for the case $\nu = \pi$, but also for the case $\pi < \nu < 2\pi$ (for case $\nu \geq 2\pi$ this phenomena in some sense becomes trivial: one needs to have such strong conditions on μ that $f * \mu \equiv 0$). In all the variants we consider the multidimensional case.

2 Convolutions of functions with sequences

Let f be a function defined on $\mathbb{R}^n$ and $a = \{a_k\}_{k\in\mathbb{Z}^n}$ be a sequence. We consider a convolution of a function f with a sequence a, defined $\forall t \in \mathbb{R}^n$ by equality

$$(f * a)(t) = \sum_{k\in\mathbb{Z}^n} f(t-k)a_k = (f * A)(t),$$

where $A = \sum_{k\in\mathbb{Z}^n} a_k\delta_k$.

Some well-known interpolation formulas may be written in a more compact form using this definition. For example, if $n = 1$, $f \in E_\nu$ and $\sum_{k=-\infty}^{\infty} \frac{|f(k)|}{1+|k|} < \infty$, then $\forall t \in \mathbb{R}$

$$f(t) = \operatorname{sinc} \pi t * f(k),$$

where $\operatorname{sinc} x = \frac{\sin x}{x}$ (Shannon-Kotelnikov theorem).

Definition *For* $1 \le p, q \le \infty$ (l_p, L_q) *is a set of all functions* $f \in L^{loc}(\mathbb{R}^n)$ *for which for some* $c > 0$ *for every sequence* a *with compact support (i.e. for some* $s \in \mathbb{N}$ $a_k = 0$ *for* $|k| \ge s$*)*

$$\|f * a\|_{L_q(\mathbb{R}^n)} \le c\|a\|_{l_p(\mathbb{Z}^n)}. \tag{12}$$

The minimal c *in this inequality is* $\|f\|_{(l_p,L_q)}$ *- norm in the space* (l_p, L_q).

For $p = 1$ the following direct analogue of Young's inequality is valid:

$$\|f * a\|_{L_q(\mathbb{R}^n)} \le \|f\|_{L_q(\mathbb{R}^n)}\|a\|_{l_1(\mathbb{Z}^n)} \tag{13}$$

for $1 \le q \le \infty$.

Easy examples show that for $p > 1$ direct analogue of Young's inequality is not valid. Instead we have the following inequality in which the norms of "amalgama" type appear.

Theorem 1 *For* $1 \le p \le q \le \infty, \frac{1}{r} = 1 - \frac{1}{p} + \frac{1}{q}$

$$\|f * a\|_{L_q(\mathbb{R}^n)} \le \|\|\|f(k+t)\|_{l_{r,k}(\mathbb{Z}^n)}\|_{L_{q,t}(Q_{\frac{1}{2}})}\|a\|_{l_p}, \tag{14}$$

where for $\delta > 0$

$$Q_\delta = \{t \in \mathbb{R}^n : |t_j| < \delta, j = 1, \ldots, n\}.$$

Corollary 1 *For* $1 \le p \le q \le \infty$

$$\|f\|_{L_q(\mathbb{R}^n)} \le \|f\|_{(l_p,L_q)} \le \|\|\|f(x+t)\|_{l_{r,k}(\mathbb{Z}^n)}\|_{L_{q,t}(Q_{\frac{1}{2}})}$$

and for $1 \le q \le \infty$

$$\|f\|_{(l_1,L_q)} = \|f\|_{L_q(\mathbb{R}^n)}.$$

Remark 1 *For* $1 \le q < p \le \infty$ $\|f\|_{(l_p,L_q)} = 0$.

Theorem 2 *For* $1 < p < \infty$ *for some* $c > 0$

$$\|f\|_{(l_p,L_p)} \le c \max_{0\le m\le 1} \|\|\|f(\tau) * (ik)^m e^{ik\cdot t}\|_{C_t(\bar{Q}_\pi)}\|_{L_{p,\tau}(Q_{\frac{1}{2}})}, \tag{15}$$

where $m \in \mathbb{Z}^n$, *notation* $0 \le m \le 1$ *means that* $0 \le m_j \le 1$ *for* $j = 1, \ldots, n$ *(i.e.* $m_j = 0$ *or* $m_j = 1$*) and* $(ik)^m = (ik_1)^{m_1} \ldots (ik_n)^{m_n}$.

Corollary 2 *For* $n = 1, 1 \le p, q \le \infty$

$$\operatorname{sinc} \pi t \in (l_p, L_p)$$

iff $1 \le p < q \le \infty$ *or* $1 < p = q < \infty$.

3 Estimates of integral norms via discrete ones

Theorem 3 *If* $0 < \nu < \pi, 1 \leq p \leq 2, \frac{1}{q} \leq \frac{1}{p} - \frac{1}{2}$, *then* $\forall f \in E_\nu$

$$\|f(t)\|_{L_q(\mathbb{R}^n)} \leq \left(\frac{\nu}{\pi}\right)^{n(\frac{1}{p}-\frac{1}{q})} \|f(k)\|_{l_p(\mathbb{Z}^n)}, \tag{16}$$

in particular

$$\|f(t)\|_{L_\infty(\mathbb{R}^n)} \leq \left(\frac{\nu}{\pi}\right)^{\frac{n}{2}} \|f(k)\|_{l_1(\mathbb{Z}^n)}, \tag{17}$$

where the constant $(\frac{\nu}{\pi})^{\frac{n}{2}}$ *is the best possible.*

Theorem 4 *Let* $1 \leq p < q \leq \infty$ *or* $1 < p = q < \infty$, *then for all* $f \in E_\nu$ *such that condition* (3) *is fulfilled*

$$\|f(t)\|_{L_q(\mathbb{R}^n)} \leq \|\text{sinc}\,\pi\tau\|^n_{(l_p, L_q)} \|f(k)\|_{l_p(\mathbb{Z}^n)} \tag{18}$$

(here $\tau \in \mathbb{R}$*), in particular for* $1 < q \leq \infty$

$$\|f(t)\|_{L_q(\mathbb{R}^n)} \leq \|\text{sinc}\,\pi\tau\|_{L_q(\mathbb{R})} \|f(k)\|_{l_1(\mathbb{Z}^n)}. \tag{19}$$

In both the inequalities (18) *and* (19) *the constants are the best possible; in* (19) *the equality holds for*

$$f(t) = \text{sinc}\,\pi t \equiv \prod_{j=1}^{n} \text{sinc}\,\pi t_j, \ \forall t \in \mathbb{R}^n.$$

4 Interference inequalities

Let for $t \in \mathbb{R}^n$

$$\eta^*(t) = \sup_{\tau: 1-\epsilon_0 \leq \frac{\tau_k}{t_k} \leq 1+\epsilon_0} |\eta(\tau)|$$

where $\epsilon_0 \in (0,1)$ is fixed (for $t_k = 0$ we set $\tau_k = 0$).

We shall also denote for $\Omega \subset \mathbb{R}^n, l > 0, 1 \leq p, \theta \leq \infty$ by $\mathring{B}^l_{p,\theta}(\Omega)$ the space of functions f defined on Ω, for which their extensions by zero outside of Ω belong to the Nikol'skij-Besov space $B^l_{p,\theta}(\mathbb{R}^n)$ (see [3] for definition and properties of these space).

Theorem 5 *Let* $\pi \leq \nu < 2\pi$, μ *- Borel measure such that* $\int_{\mathbb{R}^n} |d\mu| < \infty$, $\hat{\mu} \in \mathring{B}^{\frac{n}{2}}_{2,1}(Q_\pi)$, $D^m \hat{\mu} \in C(\bar{Q}_\pi)$ *for* $0 \leq m \leq 1$,

$$\hat{\mu}(\lambda) = 0, \qquad \forall \lambda \in \bar{Q}_\nu \setminus Q_{2\pi-\nu} \tag{20}$$

and

$$\sum_{k \in \mathbb{Z}^n} \eta^*(t-k) < \infty, \qquad \forall t \in \mathbb{R}^n, \tag{21}$$

where $\eta(t) = \text{sinc}\,\pi t * \mu(t)$.

Then for every $f \in E_\nu$, *such that*

$$f^*(t) * |\text{sinc}\,\pi t| < \infty, \ \ \forall t \in R^n, \tag{22}$$

for $1 \leq p \leq q \leq \infty$

$$\|f(t) * \mu(t)\|_{L_q(R^n)} \leq \|\operatorname{sinc} \pi t * \mu(t)\|_{(l_p, L_q)} \|f(k)\|_{l_p(Z^n)}, \tag{23}$$

in particular for $1 \leq q \leq \infty$

$$\|f(t) * \mu(t)\|_{L_q(R^n)} \leq \|\operatorname{sinc} \pi t * \mu(t)\|_{L_q(R^n)} \|f(k)\|_{l_1(Z^n)}. \tag{24}$$

In both the inequalities (23) *and* (24) *constants are the best possible (equality holds for* $f(t) = \operatorname{sinc} \pi t$*).*

Corollary 3 *If* $1 \leq p \leq 2, f \in E_\pi$ *and* $\int_{\mathbb{R}^n} |f(t)|(1+|t_1|)^{-1} \ldots (1+|t_n|)^{-1} dt < \infty$, *then*

$$\|f(t) * \operatorname{sinc} \pi t\|_{L_\infty(\mathbb{R}^n)} \leq \|f(k)\|_{l_p(\mathbb{Z}^n)}, \tag{25}$$

where the constant 1 *is the best possible.*

Remark 2 *For* $\nu = \pi$ *condition* (20) *means, that* $\hat{\mu}$ *is equal to zero on* ∂Q_π *and coincides for* $n = 1$ *with* (8). *We also note, that interference phenomena works not only for* $\nu = \pi$ *but also for* $\pi < \nu < 2\pi$ *- in this case one needs to demand, that* $\hat{\mu} = 0$ *not only on* ∂Q_π, *but on the "strip"* $\bar{Q}_\nu \setminus Q_{2\pi-\nu}$ *around* ∂Q_π. *Theorem 5 is valid also for* $\nu \geq 2\pi$, *but in this case condition* (20), *which reads "*$\hat{\mu} = 0$ *on* Q_ν*", becomes too strong: it follows that* $f * \mu \equiv 0$ *on* $\mathbb{R}^n$.

Remark 3 *Some generalizations of above results for weighted spaces where obtained in* [14] *and* [15].

References

[1] M. Plancherel, G. Polya: Fonctions éntieres et integrals de Fourier multiples, Comm.Helv. Vol. 9-10 (1937)

[2] S. M. Nikol'skij: Inequalities for entire functions of exponential type and applications to theory of differentiable functions of several variables, Proc. Steklov. Inst. Math. Vol. 38 (1951)

[3] S. M. Nikol'skij: Approximation of functions of several variables, Moscow: Nauka (1977)

[4] M. Cartwright: On certain integral functions of order 1, Quarterly J. of Math. (Oxford series) Vol. 7 (1936)

[5] B. Ya. Levin: Entire functions (lecture course), Moscow State University (1971)

[6] S. N. Bernstein: Extension of properties of trigonometric plynomials to entire functions of exponential type, Collection of Works, Vol. 2 (1949)

[7] G. Z. Ber: On interference phenomena in integral metrics and approximation by enrire functions of exponential type (in "Theory of functions, functional analysis and applications"), Kharkov University Vol. 34, No 5 (1980)

[8] F. A. Timan: Theory of approximation of functions of real variable, Moscow: Fizmatgiz (1960)

[9] F. A. Timan: On interferece phenomena in behaviour of entire functions of exponential type, Dokl. Ac. sci. USSR Vol. 89, No. 1 (1953)

[10] T. V. Tararykova: On convolution of functions with sequences, Dep. in VINITY Ac. sci. USSR No 5622-82 (1982)

[11] T. V. Tararykova: Abstract interference theorems, Dokl. Ac. sci. Armenia. Vol. 74, No 1 (1983)

[12] T. V. Tararykova: On estimates of integral norms of entire functions of exponential type via discrete norms, (in "Functional Analysis. Linear spaces"), Ul'yanovsk (1983)

[13] T. V. Tararykova: Interpolation of operator valued functions and applications (Ph. D. thesis), Voronež State University (1983)

[14] T. V. Tararykova, B. R. Rafaelov-Karapetyan: Youngs's inequality for convolutions of functions with sequences in weighted spaces, 13-th school on theory of operators in function spaces. Abstracts, Samara (1988)

[15] T. V. Tararykova: Estimates of integral norms via discrete ones in spaces with power weights, 14-th school on theory of operators in function spaces. Abstracts. Part 3, Novgorod (1989)

Tamara V. Tararykova
Dept. of Mathematics
Univ. of Voronež
Universitetskaya sq. 1
394000 Voronež, Russia
E-Mail:
DMA@VUCNIT.VORONEZH.SU

List of Participants

1. **Herbert Amann**
Mathematisches Institut, Universität Zürich
Rämistraße 74, CH - 8001 Zürich
Fax: Swizerland 1-2529525
E-Mail: K610620@CZHRZU1A.BITNET

2. **Günter Berger**
Fachbereich Mathematik/Informatik,
Institut für Mathematik
Universität Leipzig
Augustusplatz 10
D-04109 Leipzig, Germany

3. **Juha Berkovits**
Department of Mathematics, Univ. of Oulu
SF 90570 Oulu, Finland
Fax: 5531730
E-Mail: MATHDEPT@FINOU.OULU.FI

4. **Oleg V. Besov**
Steklov Mathematical Institute RAN
Vavilov 42
117 966 Moscow GSP-1, Russia
Fax: (095) 135-05-55

5. **Bogdan Bojarski**
Mathematical Institute, Academy of Sciences
ul. Sniadeckich 8
00 - 950 Warsaw, Poland
Fax: 48-22-293997
E-Mail: IMPANW@PLWATU21.BITNET

6. **Gérard Bourdaud**
URA 212, U.F.R. de Mathematiques,
Université Paris VII
Tour 45-55, 5éme - étage
2, place Jussieu
75 251 Paris cédex 05, France

7. **Victor I. Burenkov**
Dept. of Differential Equations and Functional Analysis
Friendship of Nations University
Mikluho-Maklaya st. 6
117 198 Moscow, Russia
Fax: (095)-433-15-11
E-Mail: SATEL@LOG.MIAN.SU

8. **Pavel Drábek**
Department of Mathematics
University of West Bohemia
Americká 42
30614 Plzeň, Czechoslovakia
Fax: 0042-19-220019

9. **Yulij A. Dubinskij**
Energeticheskaja, d. 16, korp. 2, kw. 57
111 116 Moscow, Russia
Tel.Nr: 362 52 04

10. **Reinhard Farwig**
Institut für Mathematik, RWTH Aachen
Templergraben 55
D-52062 Aachen, Germany
Fax: 0241-80-3952
E-Mail: FARWIG@RWTH-AACHEN.DBP.DE

11. **Hans G. Feichtinger**
Institut für Mathematik,
Universität Wien
Strudlhofgasse 4
A-1090 Wien, Austria
Fax: Int-43-1-3102683
E-Mail: A8131DAN@AWIUNI11.EDVZ.UNIVIE.AC.AT
or FEI@TYCHE.MAT.UNIVIE.AC.AT

12. **El'fat Galeev**
Dept. of Mechanics and Mathematics,
Moscow State University
119 899 Moscow, Russia

13. **Michael Geisler**
Mathematisches Institut, FSU Jena
UHH, 17.OG
D-07743 Jena, Germany
E-Mail: GEISLER@MATHEMATIK.UNI-JENA.DBP.DE

14. **Michail L. Gol'dman**
Krupskaja Str. 8-1-187
117 311 Moscow, Russia
E-mail: glas!clearh
to Lidia Popova

15. **Vagif Guliev**
Faculty of Mechanics and Mathematics,
Baku State University
L. Khalihor str. 23
370 145 Baku, Azerbeidžan

16. **Dorothee Haroske**
Mathematisches Institut, FSU Jena
UHH, 17.OG
D-07743 Jena, Germany
E-Mail: HAROSKE@MATHEMATIK.UNI-JENA.DBP.DE

17. **Walter Hoh**
Mathematisches Institut,
Universität Erlangen-Nürnberg
Bismarckstr. 1 $\frac{1}{2}$
D-91054 Erlangen, Germany
Fax: 09131-852684

18. **Wladimir A. Iwanov**
Guryevskiy Proyezd 29-1-156
115 597 Moscow, Russia
Fax: (095)-465-96-85

19. **Niels Jacob**
Mathematisches Institut,
Universität Erlangen-Nürnberg
Bismarckstr. 1 $\frac{1}{2}$
D-91054 Erlangen, Germany
Fax: 09131-852684

20. **Alf Jonsson**
Dept. of Mathematics , University of Umeå
S-90187 Umeå, Sweden

21. **Gennady A. Kalyabin**
Aviation Institute
Moscowskoye Chaussee 34
443 086 Samara, Russia
Fax: 7-8462-361836
E-Mail: RENAT@SAICNIT.SAMARA.SU

22. **Arto Kittilä**
Dept. of Mathematics , University of Oulu
SF-90570 Oulu, Finland
Fax: 5531730
E-Mail: KITTILA@PHOENIX.OULU.FI

23. **Vakhtang Kokilashvili**
Mathematical Institute,
Georgian Academy of Sciences
Z. Rukhadze Str. 1
38 00 93 Tbilisi, Georgia
E-Mail: KOKIL@IMATH.KHETA.GEORGIA.SU

24. **Miroslav Krbec**
Mathematical Institute,
Czech. Academy of Sciences
Žitná 25
115 67 Prague 1, Czech Republic
E-Mail: KRBECM@CSEARN.BITNET

25. **Lev D. Kudryavzev**
Steklov Mathematical Institute RAN
Vavilov 42
117 966 Moscow GSP-1, Russia
Fax: 7-095-135-05-55
E-Mail: SATEL@LOG.MIAN.SU

26. **Alois Kufner**
Mathematical Institute,
Czech. Academy of Sciences
Žitná 25
115 67 Prague 1, Czech Republic
E-Mail: KUFNER@CSEARN.BITNET

27. **Jan Lang**
Mathematical Institute,
Czech. Academy of Sciences
Žitná 25
115 67 Prague 1, Czech Republic

28. **Hans-Gerd Leopold**
Mathematisches Institut, FSU Jena
UHH, 17.OG
D-07743 Jena, Germany
E-Mail: LEOPOLD@MATHEMATIK.UNI-JENA.DBP.DE

29. **Tran Duc Long** (current address)
Cite' Daksi, Bat. 22, App. 767
Constantine 25.001, Algerie

30. **Mircea Malarski**
Mathematisches Institut, FSU Jena
UHH, 17.OG
D-07743 Jena, Germany
E-Mail: MALARSKI@MATHEMATIK.UNI-JENA.DBP.DE
or NMM@CNVE.RZ.UNI-JENA.DE

31. **Julian Musielak**
Institute of Mathematics, Adam Mickiewicz University
Matejki 48/49
60 - 769 Poznan, Poland
Fax: (61) 535 535
E-Mail: MUSIELAK@PLUAM11.BITNET

32. **Vesa Mustonen**
Dept. of Mathematics, University of Oulu
SF 90570 Oulu, Finland
Fax: 5531730
E-Mail: MATHDEPT@FINOU.OULU.FI

33. **Jindřich Nečas**
Faculty of Mathematics and Physics,
Charles Univ. Prague
Sokolovska 83
..... Prague 8, Czech Republic
Tel: 2316034

34. **Yurij V. Netrusov**
LOMI
Fontanka 27
191011 St. Petersburg, Russia
E-Mail: NETRUSOV@LOMI.SPB.SU

35. **Sergej M. Nikol'skij**
Steklov Mathematical Institute RAN
Vavilov 42
117 966 Moscow GSP-1, Russia
Fax: (095)-135-05-55

36. **Kaj Nyström**
Matematiska Inst., Umeå Unversitet
S-90 187 Umeå, Sweden

37. **Bohumir Opic**
Mathematical Institute,
Czech. Academy of Sciences
Žitná 25
115 67 Prague 1, Czech Republic
E-Mail: OPIC@CSEARN.BITNET

38. **Peter Oswald**
Institut für angewandte Mathematik, FSU Jena
UHH, 17.OG
D-07743 Jena, Germany
E-Mail: POSWALD@MATHEMATIK.UNI-JENA.DBP.DE

39. **Muhtarbaj Otelbaev**
Cabanbei-Batur 55-40
480100 Alma-Ata, Kazakhstan

40. **Lassi Päivärinta**
Dept. of Mathematics, Univ. of Oulu
SF 905 70 Oulu, Finland
Fax: 5531730

41. **Luboš Pick** (current address)
School of Mathematics, University of Wales
College of Cardiff
Senghennydd Road
Cardiff, CF2 4AG, U.K.
E-Mail: PICKL@TAFF.CARDIFF.AC.UK
or LUBOS@CM.CF.AC.UK

42. **Stanislav I. Pohožaev**
Steklov Mathematical Institute RAN
Vavilov 42
117 966 Moscow GSP-1, Russia
Fax: 0007-095-135-05-55
E-Mail: POHOZAEV@LOG.MIAN.SU

43. **Jiří Rákosník**
Mathematical Institute,
Czech. Academy of Sciences
Žitná 25
115 67 Prague 1, Czech Republic
E-Mail: RAKOSNIK@CSEARN.BITNET

44. **Boris Rubin**
Department of Mathematics,
The Hebrew University of Jerusalem
Givat Ram
91 904 Jerusalem, Israel
Fax: 972-2-630702
E-Mail: BORIS@HUMUS.HUJI.AC.IL

45. **Thomas Runst**
Mathematisches Institut, FSU Jena
UHH, 17.OG
D-07743 Jena, Germany
E-Mail: RUNST@MATHEMATIK.UNI-JENA.DBP.DE

46. **Hans-Jürgen Schmeißer**
Mathematisches Institut, FSU Jena
UHH, 17.OG
D-O7743 Jena, Germany
Fax: (03641) 8222345
E-Mail: SCHMEISZER@MATHEMATIK.UNI-JENA.DBP.DE

47. **Thomas Schott**
Mathematisches Institut, FSU Jena
UHH, 17.OG
D-O7743 Jena, Germany
E-Mail: SCHOTT@MATHEMATIK.UNI-JENA.DBP.DE

48. **Elmar Schrohe**
FB Mathematik, Johannes-Gutenberg-Univ. Mainz
Saarstr. 21
D-55122 Mainz, Germany
Fax: 06131-394389
E-Mail: SCHROHE@MAT.MATHEMATIK.UNI-MAINZ.DE

49. **Winfried Sickel**
Mathematisches Institut, FSU Jena
UHH, 17.OG
D-O7743 Jena, Germany
E-Mail: SICKEL@MATHEMATIK.UNI-JENA.DBP.DE

50. **Leszek Skrzypczak**
Institute of Mathematics,
Adam-Miekiewicz-Univ. Poznan
Matejki 48/49
60-769 Poznan, Poland
E-Mail: LSKRZYP@PLUAM11.BITNET

51. **Tamara V. Tararykova**
Dept. of Mathematics, Univ. of Voronež
Universitetskaya sq. 1
394000 Voronež, Russia
E-Mail: DMA@VUCNIT.VORONEZH.SU

52. **Sergej A. Telyakovskij**
Steklov Mathematical Institute RAN
Vavilov 42
117 966 Moscow GSP-1, Russia
Fax: (095)-135-05-55
E-Mail: SATEL@LOG.MIAN.SU

53. **Matti Tienari**
Dept. of Mathematics, Univ. of Oulu
SF-90 570 Oulu, Finland
Fax: 5531730

54. **Hans Triebel**
Mathematisches Institut, FSU Jena
UHH, 17.OG
D-O7743 Jena, Germany
Fax: (03641) 8222345
E-Mail: TRIEBEL@MATHEMATIK.UNI-JENA.DBP.DE

55. **Abdellah Youssfi**
URA 212, U.F.R. de Mathematiques,
Université Paris VII
Tour 45-55, 5éme - étage
2, place Jussieu
75 251 Paris cédex 05, France
E-Mail: YOUSSFI@MATHP7.JUSSIEU.FR